Android 开发从入门到进阶实战

钱慎一　刘芳华　编著

化学工业出版社

·北京·

内容简介

本书系统讲述了Android程序开发的相关知识，从Android的基本概念讲起，依次对Android应用程序的组成与执行、界面布局、常见资源和控件、事件处理机制与多线程、Android控件、菜单和对话框、Activity组件、Intent机制、Service组件、BroadcastReceiver组件、数据存储、SQLite数据库、数据共享机制、在线音视频、网络编程、定位服务等内容进行了讲解，最后通过一个实例对书中各章节知识点的综合应用进行讲解。

本书结构编排合理，内容丰富实用，由浅入深，通俗易懂，知识点与案例结合紧密，所选案例新颖丰富，紧贴实战。同时配备了极为丰富的学习资源，主要有同步教学视频、所有实例的程序代码、课后习题及答案等。

本书适合从事Android移动编程和应用开发的人员学习使用，还可作为高等院校和培训学校相关专业的教学及参考用书。

图书在版编目（CIP）数据

Android开发从入门到进阶实战／钱慎一，刘芳华编著．—北京：化学工业出版社，2020.9
ISBN 978-7-122-37102-7

Ⅰ.①A… Ⅱ.①钱… ②刘… Ⅲ.①移动终端－应用程序－程序设计 Ⅳ.①TN929.53

中国版本图书馆CIP数据核字（2020）第091969号

责任编辑：耍利娜　　　　　　　　　文字编辑：林　丹　师明远
责任校对：王　静　　　　　　　　　装帧设计：王晓宇

出版发行：化学工业出版社（北京市东城区青年湖南街13号　邮政编码100011）
印　　装：大厂聚鑫印刷有限责任公司
787mm×1092mm　1/16　印张28¼　字数750千字　2021年1月北京第1版第1次印刷

购书咨询：010-64518888　　　　　　　售后服务：010-64518899
网　　址：http://www.cip.com.cn
凡购买本书，如有缺损质量问题，本社销售中心负责调换。

定　价：98.00元　　　　　　　　　　　　　　　　　版权所有　违者必究

前言

关于本书

移动终端的快速发展使得Android系统应用的需求激增，很多在校生和广大程序开发者都加入了Android开发阵营。为了帮助开发者更快地进入Android开发行列，笔者精心编写了本书。本书从读者的实际需求出发，科学安排知识结构，内容由浅入深，循序渐进，逐步展开，反映了当前Android技术的发展和应用水平。通过大量简单易懂的实例，帮助读者快速掌握知识点，每部分既相互连贯又自成体系，读者既可以按照本书编排的章节顺序进行学习，也可以根据自己的需求对某一章节进行有针对性的学习。书中所有的实例都已调试运行通过，读者可以直接参照使用。

本书特点

本书采用通俗的语言、合理的结构对Android程序设计的知识进行了细致的剖析，在讲解理论知识时，辅以典型的案例进行补充说明，以实现既"授人以鱼"又"授人以渔"的目的。本书每个章节都有二维码，手机扫一扫，即可以随时随地看视频，体验感非常好；从配套到拓展，资源库一应俱全。读者跟着案例边学边做，学习可以更高效。具体来说，主要有如下特点。

① 内容精练，全面实用。本着"理论知识够用，举例丰富实用"的原则，指导初学者采取有效的方法和良好的途径进行学习。

② 过程简洁，步骤详细。讲解步骤做到详细但不烦琐，避免直接使用大量代码占用读者的阅读时间，而是对关键代码进行详细的讲解，做到清晰和透彻。

③ 易教易学，通俗易懂。作者均是一线工作人员及教学人员，项目经验丰富，传授知识的能力强；精选案例具有实战性和代表性，能够使读者快速上手。

④ 实例完整，资源丰富。配套资源不仅包含书中的所有实例程序代码，还包含主要操作步骤的教学视频，最大限度地满足读者的阅读需求。

适读人群

- 从事Android程序开发的工作人员
- 培训班中学习Android开发的学员
- 对程序设计有着浓厚兴趣的爱好者
- 零基础想转行到IT行业的社会人员
- 高等院校、职业院校相关专业师生

本书由钱慎一、刘芳华编著，他们在长期的工作中积累了大量经验，在写作过程中精益求精。此外，本书也得到了郑州轻工业大学教务处以及其他老师的大力支持，在此向他们表示感谢。

本书在编写过程中力求严谨细致，但由于时间与精力有限，疏漏之处在所难免，望广大读者批评指正。读者可以联系QQ1908754590获取相关学习资源，并与笔者探讨交流。

编著者

目录

第 1 章 Android入门必备

视频讲解 1段，15分钟
视频页码 001

- 1.1 智能手机操作系统简介 ………………………………………………… 001
- 1.2 Android的基本概念 ……………………………………………………… 002
  - 1.2.1 Android的发展历程 ………………………………………… 003
  - 1.2.2 Android的优点 ……………………………………………… 003
- 1.3 Android系统架构 ………………………………………………………… 004
- 1.4 开发前的准备工作 ……………………………………………………… 005
- 1.5 搭建开发环境 …………………………………………………………… 006
  - 1.5.1 Android Studio的安装 ……………………………………… 007
  - 1.5.2 Android Studio设置 ………………………………………… 009
- 1.6 构建Android应用程序 ………………………………………………… 011
  - 1.6.1 使用Android Studio创建应用程序 ………………………… 011
  - 1.6.2 运行Android应用程序 ……………………………………… 012
- 强化训练 ……………………………………………………………………… 015

第 2 章 Android应用程序精讲

视频讲解 1段，25分钟
视频页码 017

- 2.1 Android应用程序组成 ………………………………………………… 017
  - 2.1.1 R.Java文件详解 ……………………………………………… 020
  - 2.1.2 组件标识符 …………………………………………………… 022
  - 2.1.3 AndroidMainfest.xml详细介绍 ……………………………… 023
- 2.2 Android应用程序的执行 ……………………………………………… 026
- 2.3 Android应用程序的主要组件 ………………………………………… 029
  - 2.3.1 Activity ……………………………………………………… 030
  - 2.3.2 Service ……………………………………………………… 030
  - 2.3.3 BroadcastReceiver …………………………………………… 031
  - 2.3.4 ContentProvider ……………………………………………… 031
  - 2.3.5 Intent和IntentFilcter ………………………………………… 031
- 强化训练 ……………………………………………………………………… 032

第 3 章 界面布局精讲

 2段，58分钟
 036，040

- 3.1 UI概述 ·· 034
- 3.2 线性布局LinearLayout ··· 035
- 3.3 相对布局RelativeLayout ··· 039
- 3.4 绝对布局AbsoluteLayout ·· 043
- 3.5 表格布局TableLayout ·· 045
- 3.6 约束布局ConstraintLayout ·· 047
- 强化训练 ·· 052

第 4 章 常见资源和控件精讲

 3段，43分钟
 063，064

- 4.1 常见资源 ·· 054
  - 4.1.1 字符串资源 ·· 055
  - 4.1.2 颜色资源 ··· 056
  - 4.1.3 尺寸资源 ··· 057
  - 4.1.4 形状 ··· 058
  - 4.1.5 按钮背景及点击状态 ·· 060
- 4.2 TextView和EditText控件 ·· 060
- 4.3 Button和ImageButton控件 ··· 063
- 4.4 ImageView控件 ·· 064
- 4.5 RadioButton和ChekBox控件 ··· 066
- 4.6 AnalogClock和DigitalClock控件 ······································ 067
- 4.7 Toast控件 ·· 067
- 强化训练 ·· 069

第 5 章 事件处理机制与多线程

 4段，43分钟
074，075

- 5.1 事件处理机制 ·· 072
  - 5.1.1 基于监听接口的事件处理 ······································ 073

IV

  5.1.2 基于回调机制的事件处理 ························ 077
  5.1.3 回调方法应用案例 ································ 080
5.2 Android多线程机制 ··· 083
  5.2.1 多线程机制的特点 ································ 083
  5.2.2 多线程的实现 ······································ 084
 强化训练 ··· 095

第 6 章　Android控件进阶

视频讲解 3段，32分钟
视频页码 098，100

6.1 进度条ProgressBar ··· 097
6.2 列表视图ListView ·· 099
6.3 下拉列表Spinner ··· 103
6.4 网格视图GridView ··· 105
6.5 日期和时间选择器DatePicker和TimePicker ······· 108
6.6 控件的综合应用案例 ······································· 111
 强化训练 ··· 113

第 7 章　菜单和对话框的应用

视频讲解 2段，17分钟
视频页码 116，123

7.1 选项菜单和子菜单 ··· 115
  7.1.1 选项菜单Options Menu ························· 116
  7.1.2 监听菜单事件 ······································ 121
  7.1.3 与菜单项关联的Activity的设置 ············· 123
  7.1.4 子菜单Sub Menu ································· 124
7.2 上下文菜单Context Menu ································ 128
7.3 Android中的常用对话框 ·································· 130
  7.3.1 提示对话框 ··· 131
  7.3.2 单选对话框 ··· 132
  7.3.3 多选对话框 ··· 134
  7.3.4 列表对话框 ··· 135
  7.3.5 自定义对话框 ······································ 136
7.4 进度对话框ProgressDialog ······························ 138
7.5 日期对话框和时间对话框 ································ 139
 强化训练 ··· 143

第 8 章 可视化使者之Activity组件

视频讲解 2段，28分钟

视频页码 148, 151

- 8.1 Activity生命周期 ········· 146
- 8.2 Activity管理栈 ········· 150
- 8.3 创建、配置和使用Activity ········· 150
  - 8.3.1 创建Activity ········· 150
  - 8.3.2 配置Activity ········· 151
  - 8.3.3 启动关闭Activity ········· 153
  - 8.3.4 需要传递参数的Activity启动 ········· 156
  - 8.3.5 启动其他Activity并返回结果 ········· 158
- 8.4 启动模式 ········· 163
- 8.5 Fragment的使用 ········· 164
  - 8.5.1 Fragment简介 ········· 164
  - 8.5.2 创建Fragment ········· 166
- 强化训练 ········· 170

第 9 章 信息传递者之Intent机制

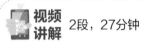
视频讲解 2段，27分钟

视频页码 179, 187

- 9.1 Intent概述 ········· 172
  - 9.1.1 Intent属性 ········· 173
  - 9.1.2 Intent解析 ········· 174
- 9.2 Intent Filter ········· 175
  - 9.2.1 动作检测 ········· 175
  - 9.2.2 种类检测 ········· 176
  - 9.2.3 数据检测 ········· 176
  - 9.2.4 通用情况 ········· 177
  - 9.2.5 使用Intent匹配 ········· 178
- 9.3 Intent的调用 ········· 178
  - 9.3.1 显式调用 ········· 179
  - 9.3.2 隐式调用 ········· 179
  - 9.3.3 在Intent中传递数据 ········· 182
  - 9.3.4 在Intent中传递复杂对象 ········· 184
  - 9.3.5 实现Activity之间的协同 ········· 187

9.4　常用Intent组件的使用 ……………………………………………………… 189
　　强化训练 …………………………………………………………………………… 192

第 10 章　骨干成员之Service组件

视频讲解 2段，36分钟
视频页码 198，203

10.1　Service概述 ……………………………………………………………………… 195
10.2　Service的生命周期 ……………………………………………………………… 196
　　10.2.1　startService启动服务 …………………………………………………… 197
　　10.2.2　bindSerivce启动服务 …………………………………………………… 200
10.3　Service的使用方法 ……………………………………………………………… 202
　　10.3.1　编写不需和Activity交互的本地服务 …………………………………… 202
　　10.3.2　编写本地服务和Activity交互 …………………………………………… 203
　　10.3.3　编写传递基本型数据的远程服务 ………………………………………… 206
　　10.3.4　编写传递复杂数据类型的远程服务 ……………………………………… 210
10.4　IntentService …………………………………………………………………… 215
　　强化训练 …………………………………………………………………………… 220

第 11 章　开发利器之BroadcastReceiver组件

视频讲解 2段，22分钟
视频页码 224，228

11.1　BroadcastReceiver概述 ………………………………………………………… 223
11.2　广播消息 ………………………………………………………………………… 223
　　11.2.1　自定义BroadcastReceiver ……………………………………………… 224
　　11.2.2　普通广播 …………………………………………………………………… 226
　　11.2.3　有序广播 …………………………………………………………………… 228
11.3　处理系统广播消息 ……………………………………………………………… 230
11.4　BroadcastReceiver的生命周期 ………………………………………………… 236
　　强化训练 …………………………………………………………………………… 236

第 12 章　数据存储精讲

视频讲解 3段，42分钟
视频页码 243，252，257

12.1　数据存储概述 …………………………………………………………………… 239

12.2 SharedPreferences ·········· 240
- 12.2.1 使用SharedPreferences ·········· 240
- 12.2.2 PreferenceActivity ·········· 245

12.3 文件 ·········· 251
- 12.3.1 应用程序文件读写 ·········· 252
- 12.3.2 操作资源文件 ·········· 254
- 12.3.3 操作SD卡上的文件 ·········· 255

▶ 强化训练 ·········· 264

第 13 章 SQLite数据库精讲

视频讲解 3段，51分钟
视频页码 281，282

13.1 SQLite概述 ·········· 271
13.2 使用SQLite数据库 ·········· 275
- 13.2.1 SQLiteDatabase ·········· 276
- 13.2.2 SQLiteOpenHelper ·········· 280

▶ 强化训练 ·········· 287

第 14 章 数据共享机制精讲

视频讲解 2段，38分钟
视频页码 293，303

14.1 ContentProvider概述 ·········· 289
14.2 自定义ContentProvider ·········· 292
14.3 监听ContentProvider中数据的变化 ·········· 299
14.4 系统ContentProvider ·········· 301

▶ 强化训练 ·········· 309

第 15 章 在线音视频的应用与管理

视频讲解 2段，15分钟
视频页码 312，316

15.1 录制音频 ·········· 311
- 15.1.1 使用Intent录制音频 ·········· 312
- 15.1.2 使用MediaRecorder录制音频 ·········· 314

15.2 应用音频 ·········· 320

	15.2.1	常见的音频格式	320
	15.2.2	使用Intent播放音频	321
	15.2.3	使用MediaPlayer播放音频	325

15.3 录制视频 328
| | 15.3.1 | 使用Intent录制视频 | 328 |
| | 15.3.2 | 使用MediaRecorder录制视频 | 331 |

15.4 应用视频 339
	15.4.1	常见的视频格式	339
	15.4.2	使用Intent播放视频	339
	15.4.3	使用VideoView播放视频	340
	15.4.4	使用MediaPlayer播放视频	341

▶ 强化训练 346

第 16 章 网络编程精讲

视频讲解 2段，21分钟
视频页码 352，354

16.1 Android网络编程基础 348

16.2 基于HTTP协议的网络编程 350
	16.2.1	HTTP介绍	350
	16.2.2	使用HttpURLConnection访问网络	351
	16.2.3	使用HttpClient访问网络	357

16.3 基于Socket的网络编程 364
| | 16.3.1 | 套接字Socket | 365 |
| | 16.3.2 | Socket编程 | 365 |

16.4 基于WebView的简单浏览器 368

▶ 强化训练 373

第 17 章 定位服务精讲

视频讲解 2段，17分钟
视频页码 380，388

17.1 定位服务相关类 377
17.2 定位实例 380
17.3 Baidu Map使用 384
| | 17.3.1 | 申请Map API KEY | 384 |

17.3.2　开发和测试环境搭建 ·················· 387
17.4　地图定位 ·· 391
　　强化训练 ··· 394

第 18 章
Android应用项目的设计与开发

18.1　系统概述 ·· 396
　　18.1.1　项目总体需求 ······················· 397
　　18.1.2　项目功能分析 ······················· 397
　　18.1.3　运行环境 ······························· 397
18.2　系统框架设计 ······································ 397
18.3　本地歌曲列表 ······································ 399
　　18.3.1　创建Fragment ······················· 400
　　18.3.2　获取本机音乐列表 ················ 400
　　18.3.3　显示歌曲 ······························· 406
　　18.3.4　刷新歌曲 ······························· 413
18.4　网络歌曲列表 ······································ 417
　　18.4.1　音乐接口介绍 ······················· 417
　　18.4.2　JSON解析 ······························ 417
　　18.4.3　封装工具类 ··························· 419
　　18.4.4　获取音乐数据及显示 ············ 422
18.5　音乐播放 ·· 425
　　18.5.1　使用Service播放音乐 ············ 426
　　18.5.2　发送通知 ······························· 430
　　18.5.3　广播接收者 ··························· 434
　　本章小结 ··· 436

附录
配套学习资源

第1章 Android 入门必备

扫一扫 看视频

内容导读

智能手机给人们的生产生活带来了巨大的便利，其中 Android 操作系统由于其巨大的市场占有率受到了各方人士的关注。本章将从智能手机操作系统的发展历程讲起，对其分类及其优点、Android 操作系统的发展和优越性、Android 系统架构等多个角度进行介绍，让读者对智能手机操作系统及 Android 操作系统有总体的了解，然后搭建 Android 开发环境，并创建第一个自己的 Android 应用程序，带领读者进入纷繁复杂而又绚丽多彩的 Android 编程世界。

学习目标

- 了解智能手机操作系统；
- 熟悉 Android 系统的发展、优点；
- 了解 Android 系统的架构；
- 掌握 Android 开发环境的搭建，创建第一个 Android 应用程序并运行。

1.1 智能手机操作系统简介

智能手机是指"像个人电脑一样，具有独立的操作系统，可以由用户自行安装软件、游戏等第三方服务商提供的程序，通过此类程序来不断对手机的功能进行扩充，并可以通过移动通信网络来实现无线网络接入的这样一类手机的总称"。由于智能手机多使用ARM而非X86的CPU体系架构，因此智能手机操作系统和开发环境与普通计算机有很大不同。目前主流的智能手机操作系统有Android、iOS、Symbian、Windows Phone和BlackBerry OS等，它们占据了智能手机市场99%以上的份额。下面对这些手机操作系统逐一进行简介。

（1）Symbian

Symbian系统是塞班公司为手机设计的操作系统。2008年12月2日，塞班公司被诺基亚收购。2011年12月21日，诺基亚官方宣布放弃塞班（Symbian）品牌。由于缺乏新技术支持，塞班的市场份额日益萎缩。截至2012年2月，塞班系统的全球市场占有率仅为3%，中国市场占有率则降至2.4%。2013年1月24日晚间，诺基亚宣布，今后将不再发布塞班系统的手机，意味着塞班这个智能手机操作系统在长达14年的发展历史之后，终于迎来了谢幕。

（2）iOS

苹果iOS是由苹果公司开发的手持设备操作系统，在2007年1月9日的Macworld大会上公布了该系统。iOS与苹果的Mac OS X操作系统一样，也是以Darwin为基础的，因此同样属于类Unix的商业操作系统。原本这个系统名为iPhone OS，直到2010年6月7日WWDC大会上宣布改名为iOS。截至2017年3月，根据Gartner的数据显示，iOS已经占据了全球智能手机系统市场份额的13.1%，在美国的市场占有率为36.5%。

（3）Windows Phone

Windows Phone是微软发布的一款手机操作系统，2010年2月，微软正式向外界展示Windows Phone操作系统。2010年10月，微软公司正式发布Windows Phone智能手机操作系统的第一个版本Windows Phone 7，简称WP7，并于2010年底发布了基于此平台的硬件设备。2012年6月21日，微软在美国旧金山召开发布会，正式发布全新操作系统Windows Phone 8（以下简称WP8）。Windows Phone 8放弃WinCE内核，改用与Windows 8相同的NT内核。Windows Phone 8系统也是第一个支持双核CPU的WP版本，宣布Windows Phone进入双核时代，同时宣告着Windows Phone 7退出历史舞台。

（4）Android

Android是一种基于Linux的自由及开源代码的操作系统，主要使用于移动设备，如智能手机和平板电脑，由Google公司和开放手机联盟领导及开发。Android操作系统最初由Andy Rubin开发，主要支持手机，2005年8月由Google收购注资。2007年11月，Google与84家硬件制造商、软件开发商及电信运营商组建开放手机联盟，共同研发改良Android系统。随后Google以Apache开源许可证的授权方式，发布了Android的源代码。第一部Android智能手机发布于2008年10月。后来Android逐渐扩展到平板电脑及其他领域上，如电视、数码相机、游戏机等。

1.2 Android的基本概念

Android操作系统由于具有良好的性价比、开放的环境、友好的人机交互接口等优点，取得了极高的市场占有率，本节主要介绍Android的发展历程及其优点。

1.2.1 Android的发展历程

Android的诞生还要从Andy Rubin说起。Rubin是硅谷著名的极客（对计算机和网络技术有狂热兴趣并投入大量时间钻研的人），他家的门铃是硅谷最昂贵的玩具——视网膜扫描仪。Rubin很喜欢机器人，这也就是他为创立的新公司取名Android的原因。Rubin最初的目标是想把Android打造成一个可以对任何软件设计人员开放的移动终端平台。很快这个公司就获得了众人的青睐，很多人表示打算买下他的公司。而Rubin唯独向Google抛出了橄榄枝，他发了一封邮件给拉里·佩奇，告诉他有人要跟他合伙。几周后，Google抢先把Rubin的公司买下。Google收购Android的时候没有宣布任何计划，只是向《商业周刊》表示："我们收购Android是因为它拥有天才般的工程师，这些工程师具有非常棒的技术。我们非常兴奋让他们加入Google。"

随着Rubin加入Google，2007年网络上就盛传全球最大的在线搜索服务商Google将进军移动通信市场，并推出自主品牌的移动终端。Google手机的图片更是满天飞，光外形就有翻盖、滑盖、旋屏、触控等多种版本。更有人将其与苹果公司于2007年年初推出的iPhone相提并论。

2007年11月，Google终于揭开谜底。Google宣布与其他33家手机厂商（包括摩托罗拉、华为、宏达国际电子、三星、LG等）、手机芯片供应商、软硬件供应商、移动运营商联合组成开放手机联盟（Open Handset Alliance，OHA），并发布了名为Android的开放手机软件平台。参与开放手机联盟的这些厂商，都会基于Android平台来开发新的手机业务。Android向手机厂商和移动运营商提供一个开放的平台，供他们开发创新性的应用软件。

Android作为Google企业战略的重要组成部分，将进一步推进"随时随地为每个人提供信息"这一企业目标的实现。Google的目标是让移动通信不依赖于设备甚至平台，基于此Android将进一步补充Google长期以来的移动发展战略：通过与全球各地的手机厂商和移动运营商结成合作伙伴，开发既有用又有吸引力的移动服务，并推广这些产品。

1.2.2 Android的优点

与其他智能手机操作系统相比，Android具有以下几个无可比拟的优点。

（1）开放性

Google与开放手机联盟合作开发了Android，Google通过与运营商、设备制造商、开发商和其他有关各方结成深层次的合作伙伴关系，希望通过建立标准化、开放式的移动电话软件平台，在移动产业内形成一个开放式的生态系统。

（2）应用程序无界限

Android上的应用程序可以通过标准API访问核心移动设备功能。通过互联网，应用程序可以产生它们的功能，可供其他应用程序使用。

（3）应用平等

移动设备商的应用程序可以被替换或扩展，即使是拨号程序或主屏幕这样的核心组件。

（4）快速方便的应用开发

Android平台为开发人员提供了大量的使用库和工具，开发人员可以快速地创建自己的应用。

1.3 Android系统架构

通过上面内容的学习，我们对Android的优点有了初步的了解，而这些优越性取决于Android优秀的体系架构，其体系架构图如图1-1所示。

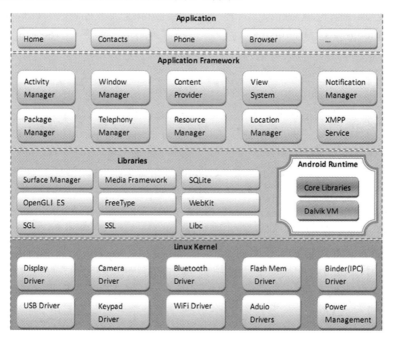

图1-1　Android体系架构图

由上图可以看出，Android体系架构采用了软件叠层的技术，整个架构由应用层、应用程序框架层、Android运行时、系统库以及Linux内核五层构成。下面对这五层进行简要的分析和介绍。

（1）应用层

Android平台缺省包含了一系列的核心应用程序，包括电子邮件、短信、日历、地图、浏览器、联系人管理程序等，这些应用程序都是用Java语言编写并运行在虚拟机上的。当然，程序员也可以用自己写的程序来替换Android提供的应用程序，这需要应用程序框架层来保证。

（2）应用程序框架层

这一层是进行Android开发的基础，开发人员可以使用这些框架来开发自己的应用程序，这样便简化了程序开发的架构设计，但是必须遵守其框架的开发原则。应用程序框架层包含了视图系统、内容提供器、资源管理器、通知管理器、活动管理器、窗口管理器、电话管理器、包管理器位置管理、XMPP服务等几大部分。

（3）Android运行时

Android虽然采用Java语言来开发、编写应用程序，但却不使用J2ME执行Java程序，而是用 Android 自有的 Android 运行时 (Android Runtime) 来执行。Android 运行时包括核心库和 Dalvik 虚拟机两部分。

核心库包含两部分内容：一部分提供了 Java 编程语言核心库的大多数功能；另一部分为 Android 的核心库。与标准的 Java 不一样，Android 不是用一个 Dalvik 虚拟机来同时执行多个 Android 应用程序的，而是每个 Android 应用程序都用一个自有的 Dalvik 虚拟机来执行。

Dalvik 虚拟机（Dalvik Virtual Machine）是一种基于寄存器的 Java 虚拟机，是专为移动设备而设计的，它在编写时就已经设想用最少的内存资源来执行，以及支持同时执行多个虚拟机的特性。在设计方面，Dalvik 虚拟机有许多地方参考了 Java 虚拟机，不过 Dalvik 虚拟机执行的中间码并非是 Java 虚拟机执行的 Java 字节码，同时也不直接执行 Java 的类，而是依靠转换工具 dx 将 Java 字节码转换为 Dalvik 虚拟机执行时特有的 dex（Dalvik Excutable）格式。

（4）系统库

应用程序框架层是贴近于应用程序的软件组件服务，而更底层则是 Android 的库函数（C/C++），这一部分是应用程序框架的支撑。这一层主要包括以下功能。

- Surface Manager：同时执行多个应用程序时，Surface Manager 会负责管理器显示与存取操作间的互动，另外也负责将 2D 绘图与 3D 绘图进行显示上的合成。
- 媒体框架：提供了对各种音、视频的支持。Android 支持多种音频、视频、静态图像格式，如 MPEG-4、H.264、MP3、AAC、ARM、JPG、PNG、GIF 等。
- SQLite：SQLite 是一套轻量级的数据库引擎，可供其他应用程序调用。
- OpenGL| ES：Android 是依据 OpenGL ES 1.0 API 标准来实现的 3D 绘图函数库的，该函数库可以用软件方式执行，也可以用硬件加速方式执行，其中 3D 软件光栅处理方面已进行高度优化。
- FreeType：提供点阵字、向量字的描绘显示。
- WebKit：它是一套网页浏览器的软件引擎，该引擎的功能不仅可供 Android 内建的网页浏览器所调用，也可以提供内嵌性网页显示效果。
- SGL：提供 Android 在 2D 绘图方面的绘图引擎。
- SSL：安全通信协议，保证网络通信的安全。
- Libc：提供了针对移动设备而优化了的 C 库。

（5）Linux 内核层

之前我们提到了 Android 平台的一个主要优点就是开放性，采用 Linux 内核则是 Android 平台开放性的基础。在 Android 平台中操作系统采用了 Linux 2.6 版的内核，它包括了显示驱动、摄像头驱动、蓝牙驱动、Flash 内存驱动、Binder(IPC) 驱动、USB 驱动、键盘驱动、Wi-Fi 驱动、Audio 驱动、电源管理等。Linux 内核层在软件层和硬件层之间建立了一个抽象层，使应用开发人员无须关心硬件细节。不过对手机开发商而言，如果想要 Android 平台运行到自己的硬件平台上就必须对 Linux 内核层进行修改，通常要做的工作就是为自己的硬件编写驱动程序。

1.4 开发前的准备工作

在开发 Android 应用之前，需要搭建 Android 开发环境，本节将主要对开发 Android 应用程序所需要的软件进行详细介绍。

（1）JDK8.0

JDK 的版本采用 JDK8.0，到 Java 的官方网站即可下载。

（2）Android Studio

Android Studio 是 Android 的官方 IDE，是专为 Android 而打造的，可以加快开发速度，帮助用户为每款 Android 设备构建最优应用。它提供专为 Android 开发者量身定制的工具，其中包括丰富的代码编辑、调试、测试和性能分析工具。

由于 Android Studio 官网访问速度慢，推荐"Android Studio 中文社区"，资源热度很高，下载很快，并且更新及时，下载界面如图1-2所示。

平台	Android Studio 软件包	大小	SHA-1 校验和
Windows（64位）	android-studio-bundle-162.4069837-windows.exe 包含 Android SDK（推荐）	1,926 MB (2,020,009,280 bytes)	dc23bc968d381a5ca7fdd12bc7799b95ec0d11f1e4007387cb0de55c78b475ba
	android-studio-ide-162.4069837-windows.exe 无 Android SDK	451 MB (473,299,352 bytes)	f0b72473cb94ba4bcbc80eeb84f4b53364da097efa255f7cab71bcb10a28775a
	android-studio-ide-162.4069837-windows.zip 无 Android SDK，无安装程序	468 MB (490,882,918 bytes)	b61d6f08758b5b2e6dad604d8a8d61acf549f746b07dbb0c2265daad01a7d2b7
Windows（32位）	android-studio-ide-162.4069837-windows32.zip 无 Android SDK，无安装程序	467 MB (490,323,833 bytes)	db7526187d492287b6e2979249d27a67f1dd62d6e095cca7508e05edce74e272
Mac	android-studio-ide-162.4069837-mac.dmg	463 MB (486,148,957 bytes)	da0d39221b8cb7b4b5cbe483dd4174dd1f7dc688e74f7de7cc47cc8ffb6296715
Linux	android-studio-ide-162.4069837-linux.zip	468 MB (490,782,431 bytes)	1383cfd47441e5f820b6257a1bdd683e0e980bc76c7f2027ef84dc2e6ad2f17f

图1-2　Android Studio下载界面

如果是首次安装，建议下载"android-studio-bundle-162.4069837-windows.exe 包含 Android SDK（推荐）"，这样可以一次性安装好 Android Studio 和 Android SDK，节省安装的时间。

（3）Android SDK

Android SDK 是 Android 应用程序开发工具包，类似于 Java 的 JDK，可以到 Android 的官方网站下载。

准备并安装好这些工具，就可以搭建 Android 开发环境了。

注意：以上提供的下载地址会由于官方的更新而产生变动，有时下载到的版本会不同，但下载方式相同，如遇问题可以参考官方的帮助文档。

1.5　搭建开发环境

鉴于读者已经有了一定的 Java 开发基础，所以 JDK 的安装过程在此不再详细说明。安装完 JDK，并配置完环境变量之后，就可以接着下一步的 Android Studio 安装了。

1.5.1 Android Studio 的安装

在指定地址下载可得到一个 android-studio-bundle-162.4069837-windows.exe 文件，双击开始安装，如图1-3所示。单击"Next"按钮，进入图1-4，选择安装组件。

图1-3 欢迎界面

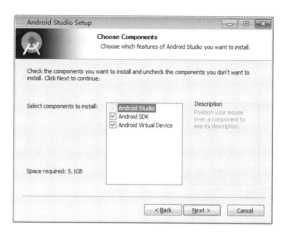

图1-4 选择安装组件

单击"Next"按钮，进入图1-5许可协议。单击"I Agree"按钮，进入图1-6安装设置。

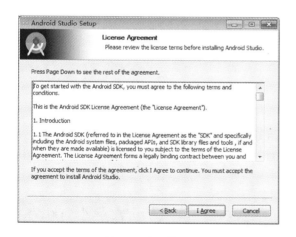

图1-5 许可协议

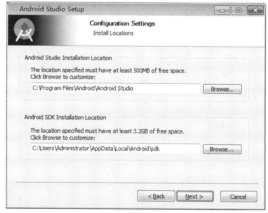

图1-6 安装设置

建议将文件安装在除C盘外的其他分区，这样在重装系统时可以避免安装文件丢失，如将 Android Studio 安装在 D:\Android\Android Studio，Android SDK 安装在 D:\Android\sdk。

单击"Next"按钮，进入图1-7，选择快捷方式位置。

单击"Install"按钮，进入图1-8，开始安装。安装完成后，单击"Next"按钮，进入图1-9，完成Android Studio的安装。单击"Finish"按钮，进入图1-10，完成安装。

图1-10所示是Android Studio的一个人性化设计，将用户的个人设置保存到一个指定的位置，如果是重装Android Studio，不会损坏这些设置文件，可以直接导入个人设置，非常方便。如果是第一次安装，请选择第二个单选按钮，这样将使用默认设置。

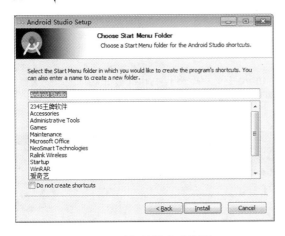

图1-7 选择快捷方式位置

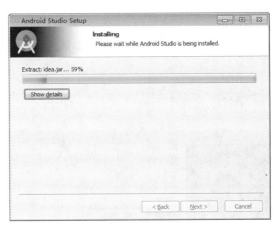

图1-8 安装对话框

图1-9 Android Studio安装完成

图1-10 完成安装

单击"OK"按钮,进入图1-11,打开Android Studio集成开发环境设置的欢迎对话框。单击"Next"按钮,进入图1-12,选择安装类型。

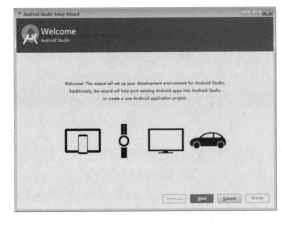

图1-11 集成开发环境设置的欢迎对话框

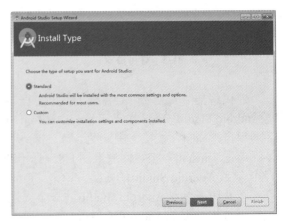

图1-12 选择安装类型

第 1 章　Android入门必备

两个选项：Standard，适用于大多数用户；Custom，自定义设置和安装组件。选择"Custom"，单击"Next"按钮，进入图1-13，选择IDE主题。

单击"Next"按钮，进入图1-14，选择安装的SDK组件。

图1-13　选择IDE主题

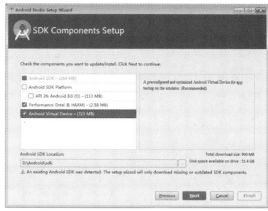

图1-14　选择安装的SDK组件

如果在图1-6中修改了SDK的安装路径，需要选择为安装的路径。单击"Next"按钮，进入图1-15，设置模拟器。

单击"Next"按钮，进入图1-16，确认设置。

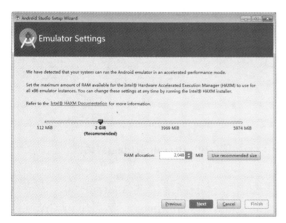

图1-15　模拟器设置

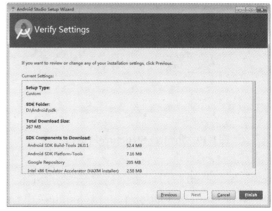

图1-16　确认设置

单击"Finish"按钮，进入图1-17，开始下载并安装选中的组件。

由于从网上下载组件，需要等待一段时间。下载后，单击"Finish"按钮，即可完成Android Studio的安装。

1.5.2　Android Studio设置

安装完成后，会自动启动Android Studio，如图1-18所示。

在Android Studio欢迎界面中，依次单击"Configure|Project Defaults|Project Structure"，打开Project Structure对话框，如图1-19所示。

在图1-19中，设置Android SDK和Java JDK的路径，然后单击"OK"返回欢迎界面。

在Android Studio欢迎界面中，依次单击"Configure|SDK Manager"，打开Default Settings对话框，如图1-20所示。

图1-17　下载组件

图1-18　Android Studio欢迎界面

图1-19　Project Structure对话框

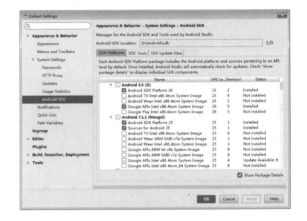

图1-20　Default Settings对话框

在图1-20中可以看到已经安装的SDK组件，如果要体验新推出的Android 8.0，需要选中Android SDK Platform 26和Google APIs Intel x86 Atom System Image，然后单击"Apply"进行下载安装。其中，第一个文件为Android 8.0的软件开发工具包，是必需的；第二个文件为模拟器的镜像文件。镜像文件有多个，需要根据用途选择相应的文件。

- TV：智能电视；
- Wear：可穿戴设备；
- Google Play：提供对Google在线应用程序商店的支持；
- ARM：大多数手机采用ARM架构，早期只提供基于ARM架构的镜像，由于在PC上面运行速度太慢，所以推荐使用Intel x86架构的镜像；
- Intel x86：运行速度快，但是要求PC的CPU必须为Intel x86架构，且支持VT（Intel Virtualization Technology，虚拟技术）。还有32位和64位之分，请自行选择。

第 1 章　Android入门必备

1.6　构建Android应用程序

通过上面的学习，相信读者已经可以完成Android开发环境的基本设置。接下来我们将带领大家一步一步构建第一个Android应用程序"Hello world!"，并编译、运行、体验，开始Android开发之旅。

1.6.1　使用Android Studio创建应用程序

启动Android Studio，在图1-18中单击"Start a new Android Studio project"，打开创建新项目对话框，如图1-21所示。

属性说明：
- Application name：应用程序名称。本书约定，每章创建一个应用，名称为"ChapterXX"，故本应用程序名称为"Chapter01"。
- Company domain：公司域名，用来自动生成包名。本书示例统一使用"androidstudy.cn"。
- Project location：项目存储位置。

设置完成后，单击"Next"按钮，进行目标Android设备选择，如图1-22所示。

图1-21　创建新项目对话框

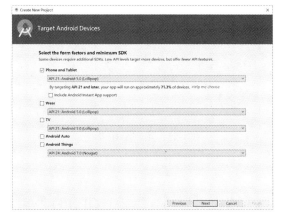
图1-22　目标Android设备选择

目标Android设备分为五种，分别为Phone and Tablet（手机和平板电脑）、Wear（可穿戴设备）、TV（智能电视）、Android Auto（汽车）、Android Things（嵌入式设备）。

本书主要介绍手机和平板电脑系统的开发，选择"Phone and Tablet"，需要选择应用程序支持的SDK最低版本，选择后，会根据Android每个版本的市场占有率提示本应用可以运行的Android设备的比例。

单击"Next"按钮，如是第一次创建，系统将安装必需的组件，如图1-23所示；否则，选择添加Activity，如图1-24所示。

图1-23 安装必需的组件

图1-24 添加Activity

在图1-24中,选择"Empty Activity",单击"Next"按钮,进行Activity定制,如图1-25所示。

单击"Finish"按钮,完成应用的创建。创建完成后,打开Android Studio集成开发环境,如图1-26所示。

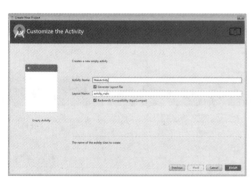
图1-25 定制Activity

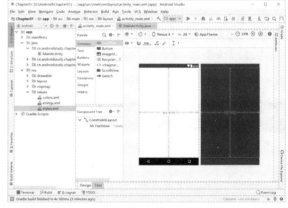
图1-26 Android Studio主界面

1.6.2 运行Android应用程序

运行Android项目是否需要首先启动模拟器呢?答案是可以的,这样运行Android项目之后会自动发布到已经启动的模拟器上。如果运行之前有多于一个模拟器已经启动,那么在运行的时候会有一个界面提示选择要发布的目标模拟器。如果运行之前没有启动任何模拟器,那么运行代码后,会自动启动一个默认的模拟器。

若是第一次运行Android项目,需要首先创建一个虚拟设备,在图1-26的工具栏单击图标,打开Android虚拟设备管理对话框,如图1-27所示。

单击"Create Virtual Device...",选择虚拟设备的硬件,如图1-28所示。

单击"Next",选择虚拟设备的Android版本,如图1-29所示。

单击"Next",确认虚拟设备的设置,如图1-30所示。

第 1 章　Android入门必备

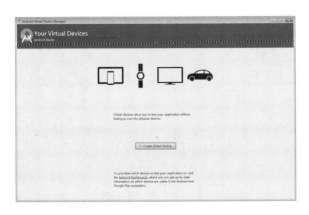

图1-27　Android虚拟设备管理

图1-28　选择硬件

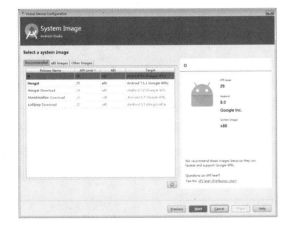

图1-29　选择虚拟设备的Android版本

图1-30　虚拟设备确认

单击"Finish"，即可完成Android虚拟设备的创建，同时打开图1-31，可以对创建的虚拟设备进行启动、编辑、删除等操作。

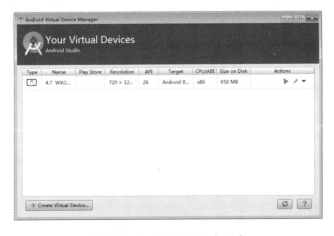

图1-31　Android虚拟设备列表

单击虚拟设备的▶图标，即可启动该模拟器，如图1-32所示。模拟器启动好后，就可以在模拟器上运行Android应用程序。在图1-26的工具栏单击图标▶，运行创建的app项目，选择Android设备，如图1-33所示。选择刚刚打开的虚拟设备，单击"OK"按钮，程序运行效果如图1-34所示。

由于运行Android模拟器对电脑的配置要求较高，建议使用自己的Android手机进行测试，不同品牌不同版本的手机操作略有不同，大概步骤如下。

首先，将手机通过数据线连接到电脑USB口，自动安装驱动。

然后，驱动安装成功后，将会在图1-33中已连接设备下面看到，运行时选择自己的手机。

最后，在手机端依次打开"设置|开发者选项"，启用"USB调试"即可。

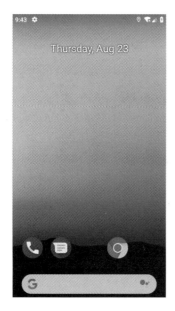

图1-32　Android模拟器

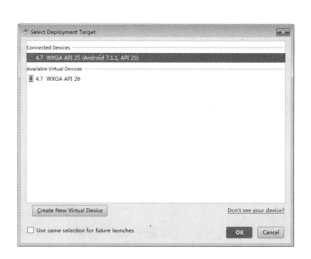

图1-33　选择Android设备

图1-34　app运行效果

第 1 章 Android入门必备

强化训练

本章首先介绍了智能手机操作系统的发展历程、分类及其优缺点，然后对Android操作系统及其发展和优越性进行了介绍，最后讲述了Android如何通过系统架构来体现其优越性，使读者对手机操作系统及Android有了总体的了解。

要开发Android应用程序，首先要进行相关的准备工作，本章首先详细阐述了如何获取JDK、Android Studio等；其次讲解了如何搭建Android开发环境；最后叙述了在搭建好的平台上如何创建一个简单的Android程序，并对这个程序进行了简单的分析，使读者对Android应用程序开发有一个初步的了解。

学完本章知识后，不妨通过以下习题来巩固所学的内容吧。

一、填空题

1. 目前，常见的智能手机操作系统有Symbian、＿＿＿＿＿＿＿、iOS、Windows Phone、BlackBerry OS等。

2. 与其他智能手机操作系统相比，Android操作系统具有以下的优点：＿＿＿＿＿＿＿、＿＿＿＿＿＿＿、＿＿＿＿＿＿＿、＿＿＿＿＿＿＿。

3. Android体系架构采用了软件叠层的技术，整个架构由＿＿＿＿＿＿＿、＿＿＿＿＿＿＿、＿＿＿＿＿＿＿、＿＿＿＿＿＿＿以及＿＿＿＿＿＿＿五层构成。

4. 开发Android应用程序所需要的软件为：＿＿＿＿＿＿＿、＿＿＿＿＿＿＿、＿＿＿＿＿＿＿。

5. 在Android Studio欢迎界面中，依次单击"＿＿＿＿＿＿＿"，打开Default Settings对话框，如果要体验新推出的Android 8.0，需要选中Android SDK Platform 26和Google APIs Intel x86 Atom System Image，然后单击"＿＿＿＿"进行下载安装。其中，第一个文件为Android 8.0的软件开发工具包，是必需的；第二个文件为模拟器的＿＿＿＿＿＿＿。

6. 目前Android设备分为五种，分别为＿＿＿＿＿＿＿、＿＿＿＿＿＿＿、＿＿＿＿＿＿＿、＿＿＿＿＿＿＿、＿＿＿＿＿＿＿。

二、选择题

1. Android最初是由（   ）发明的。
   A. James Gosling
   B. John Gage
   C. Andy Rubin
   D. Eric Schmide

2. 2005年8月Android由（   ）公司收购注资。
   A. 诺基亚
   B. Microsoft
   C. Google
   D. 苹果

3. 在安装Android Studio之前需要安装下列哪个软件并进行配置（   ）。
   A. Netframework

B. JDK
C. Photoshop
D. Office

4. 下面哪个属于Android体系结构中的应用程序。（　　）

A. SQLite
B. OpenGL ES
C. 浏览器
D. LibWebCore

5. 系统库中的（　　）提供了Android在2D绘图方面的绘图引擎。

A. SGL
B. FreeType
C. SQLite
D. Libc

6. Android采用（　　）语言来开发、编写应用程序。

A. python
B. SQL
C. JAVA
D. C#

三、操作题

1. 搭建Android开发环境是开发Android程序前必须完成的内容，请在计算机上搭建Android开发环境（包括JDK和Android Studio的安装与配置）。

2. Android Studio是Android的官方IDE，请使用Android Studio创建第一个Android应用程序"Hello world!"，并创建模拟器，在模拟器上运行Android程序，验证Android开发环境是否搭建成功。

第 2 章

Android 应用程序精讲

扫一扫 看视频

内容导读

在编写复杂的 Android 应用程序之前，需要对相关知识和理论进行系统学习，对 Android 应用程序的组成、主要组件及程序的内部编译执行流程等有一个初步的了解。本书将使用 Android Studio 为编程 IDE，初始接触时会感觉在 Android Studio 中 Android 项目文件比较多、结构比较复杂。本章通过对一个相对简单的 Android 应用程序的深入剖析，使读者对在 Android Studio 中 Android 应用程序项目的构成以及执行流程有个初步的了解。

学习目标

- 熟悉 Android 应用程序的目录结构及相关文件的作用；
- 熟悉 Android 应用程序的执行流程；
- 了解 Android 应用程序的基本组件及相互关系与作用。

2.1 Android 应用程序组成

在 Android Studio 中，一个 Android 应用程序项目包含有相对比较复杂的目录结构和文件，如资源文件以及功能清单文件 AndroidManifest.xml、Activity、Service、BroadcastReceiver、ContentProvider 等，还需要通过 Intent 进行通信。如果刚接触这么复杂的目录结构和文件，可能会产生一些困惑。本节将在 Android Studio 这一个强大的 Android 集成开发工具中为读者实际讲解。这里将从程序的目录结构入手进行分析。如图2-1所示。

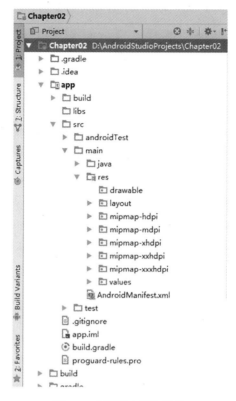

图2-1　应用程序目录结构

　　如果读者以前使用过Eclipse，那需要分清楚两个概念——Project和Module。在Android Studio中，Project的真实含义是工作空间，相当于Eclipse中的Workspace；Module为一个具体的项目，相当于Eclipse中的Project，一个Project可以包含多个Module。

　　在本章中创建了一个Project，名为Chapter02，系统自动创建了一个Module，名为app，运行的也是app。Chapter02中文件夹的功能和作用介绍如下。

- .idea：Android Studio生成的工程配置文件，类似Eclipse的project.properties。
- app：Android Studio创建工程中的一个Module。
- .gradle：构建工具系统的jar和wrapper等，jar告诉了Android Studio如何与系统安装的gradle构建联系。
- External Libraries：不是一个文件夹，依赖lib文件，如SDK等。

下面主要介绍app中每个文件夹的功能和作用。

- build：构建目录，该目录不用开发人员维护。
- libs：存放依赖的包。
- src/androidTest：专门存放测试类。是运行在emulator和device上的测试方式，测试用例中所有的行为都是经过android framework验证的。
- src/test：专门存放测试类。脱离emulator和device独立运行在jvm的测试方式。
- src/main/java：专门存放用户编写的Java源代码的包。
- src/main/res：资源目录。该目录可以存放一些图标、界面文件、应用中用到的文字信息。
- AndroidManifest.xml：该文件是Android项目的系统清单文件，它用于控制Android应用的名称、图标、访问权限等整体属性。

资源被编译到最终的 APK 文件里。Android 创建了一个被称为 R 的类，这样在 Java 代码中可以通过它关联到对应的资源文件。接下来对 res/ 的子目录做详细的说明。

- drawable：存放各种位图文件（PNG、JPG、GIF等），除此之外可能是一些其他的 drawable 类型的 XML 文件。
- mipmap-hdpi：高分辨率，一般把图片放这里。
- mipmap-mdpi：中等分辨率，很少，除非兼容的手机很旧。
- mipmap-xhdpi：超高分辨率，手机屏幕材质越来越好，以后估计会慢慢向这个等级过渡。
- mipmap-xxhdpi：超超高分辨率，这个等级在高端机上有所体现。
- mipmap-xxxhdpi：超超超高分辨率，这个等级在高端机上有所体现。

drawable 和 mipmap 区别其实不大，只是使用 mipmap 会在图片缩放时提供一定的性能优化，分辨率不同系统会根据屏幕分辨率来选择 hdpi、mdpi、xhdpi、xxhdpi、xxxhdpi 下的对应图片，所以解压 APK 可以看到上述目录同一名称的图片在五个文件夹下都有，只是大小和像素不一样而已。

（1）res/values 文件夹下常放的文件

strings.xml 用来定义字符串和数值，在 Activity 中使用 getResources().getString(resourceId) 或 getResources().getText(resourceId) 取得资源。打开 app 项目的 strings.xml，可以看到如下内容：

```xml
<resources>
    <string name="app_name">Chapter02</string>
</resources>
```

每个 string 标签声明了一个字符串，name 属性指定其引用名。为什么需要把应用中出现的文字单独放在 strings.xml 文件中呢？原因有如下两点。

一是为了国际化，Android 建议将在屏幕上显示的文字定义在 strings.xml 中。比如我们开发的应用本来是面向国内用户的，那么要在屏幕上使用中文，如今我们要让应用走向世界，打入日本市场，那么就需要在手机屏幕上显示日语，如果没有把文字信息定义在 strings.xml 中，就需要修改程序内容了。但当我们把所有屏幕上出现的文字信息都集中存放在 strings.xml 文件之中，只需要再提供一个 strings.xml 文件，把里面的汉字信息都修改为日语，再运行程序时，Android 操作系统会根据用户手机的语言环境和国家来自动选择相应的 strings.xml 文件，这时手机界面就会显示出日语。

二是为了减小应用的体积，降低数据冗余。假设在应用中要使用"我们一直在努力"这段文字 10000 次，如果不将"我们一直在努力"定义在 strings.xml 文件中，而是在每次使用时直接写上这几个字，这样下来程序中将有 70000 个字，这 70000 个字占 136KB 的空间。而由于手机的资源有限，其 CPU 的处理能力及内存是非常有限的，136KB 对手机程序来说是个不小的空间，我们在做手机应用时一定要记住"能省内存就省内存"。如果将这几个字定义在 strings.xml 中，在每次使用到的地方通过 Resources 类来引用该文字，只需占用 14B 空间。作为手机应用开发人员，我们一定要养成良好的编程习惯。

styles.xml 用来定义样式。打开本项目的 styles.xml 文件，内容如下：

```xml
<resources>
    <!-- Base application theme. -->
    <style name="AppTheme" parent="Theme.AppCompat.Light.
```

```
DarkActionBar">
        <!-- Customize your theme here. -->
        <item name="colorPrimary">@color/colorPrimary</item>
        <item name="colorPrimaryDark">@color/colorPrimaryDark</item>
        <item name="colorAccent">@color/colorAccent</item>
    </style>
</resources>
```

注意：Android中的资源文件不要以数字作为文件名，这样会导致错误。

（2）res/layout目录下的布局文件

本例中的布局文件是ADT默认自动创建的activity_main.xml文件。可以用两种方式——Graphical Layout或者XML清单显示其中的内容。在Android Studio中，这两种查看方式可以随意切换。双击打开此XML文件，内容如下：

```
<?xml version="1.0" encoding="utf-8"?>
<android.support.constraint.ConstraintLayoutxmlns:android="http://schemas.android.com/apk/res/android"
    xmlns:app="http://schemas.android.com/apk/res-auto"
    xmlns:tools="http://schemas.android.com/tools"
    android:layout_width="match_parent"
    android:layout_height="match_parent"
    tools:context="cn.androidstudy.chapter02.MainActivity">
    <TextView
        android:layout_width="wrap_content"
        android:layout_height="wrap_content"
        android:text="Hello World!"
        app:layout_constraintBottom_toBottomOf="parent"
        app:layout_constraintLeft_toLeftOf="parent"
        app:layout_constraintRight_toRightOf="parent"
        app:layout_constraintTop_toTopOf="parent" />
</android.support.constraint.ConstraintLayout>
```

与在网页布局中使用HTML文件一样，Android在XML文件中使用XML元素来设定屏幕布局。每个文件包含整个屏幕或部分屏幕，被编译进一个视图资源，可以被传递给Activity.setContentView或被其他布局文件引用。文件保存在工程的res/layout目录下，它被Android资源编辑器编译。

2.1.1 R.Java文件详解

Android创建了一个被称为R的类，在Java代码中可以通过它关联到对应的资源文件，R.java文件的位置如图2-2所示。

第2章 Android应用程序精讲

图2-2 R.java位置

R.Java 文件中有一系列静态内部类，每个静态内部类分别对应一种资源，如 layout 静态内部类对应 layout 中的界面文件，其中每个静态内部类中的静态常量分别定义一条资源标识符，如 "public static final int activity_main= 0x7f04001b；" 对应的是 layout 目录下的 activity_main.xml 文件。具体的对应关系如图2-3所示。

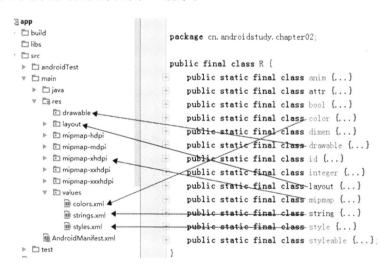

图2-3 R.Java中的资源的对应关系图

现在已经理解了 R.Java 文件中内容的来源，即当开发者在res 目录中任何一个子目录中添加相应类型的文件之后，会在R.Java 文件中相应的匿名内部类当中自动生成一条静态int 类型的常量，对添加的文件进行索引。如果在 layout 目录下再添加一个新的界面，那么在 "public static final class layout" 中也会添加相应的静态 int 常量。相反的，当我们在res 目录下删除任何一个文件，其在 R.Java 中对应的记录会被自动删除。再比如我们在strings.xml 添加一条记录，在 R.Java 的 string 内部类中也会自动增加一条记录。R.Java 文件会给我们开发程序带来很大的方便，比如在程序中使用 " public static final int ic_launcher=0x7f020000；" 就可以找到其对应的ic_launcher这幅图片。R.Java 文件除了有自动标识资源的 "索引" 功能之外，还有另一个主要

021

的功能：当 res 目录中的某个资源在应用中没有被使用到，在该应用被编译的时候系统就不会把对应的资源编译到该应用的 APK 包中，这样可以节省 Android 手机的资源。

2.1.2 组件标识符

通过对 R.Java 文件的介绍，我们已经了解了 R 文件的索引作用，它可以检索到我们应用中需要使用的资源。下面介绍如何通过 R.Java 文件来引用到所需要的资源。

（1）在 Java 程序当中按照 Java 的语法来引用

① R.resource_type.resource_name。比如说前文的 ic_launcher.png 文件的资源标识符可以通过如下方式获取：R.mipmap.ic_launcher

注意：resource_name 不需要文件的后缀名。

② android.R.resource_type.resource_name。Android 系统本身自带了很多的资源，我们也可以进行引用，只是需要在前面加上"android."以声明该资源来自 Android 系统。

（2）在 XML 文件中引用资源的语法

① @[package:]type/name   使用自己包中的资源可以省略 package 。

在 XML 文件中，如 activity_main.xml 以及 AndroidMainfest.xml 文件中通过"@mipmap/ic_launcher"的方式获取。其中"@"代表的是 R.Java 类，"mipmap"代表 R.Java 中的静态内部类"mipmap"，"/ic_launcher"代表静态内部类"mipmap"中的静态属性"ic_launcher"，而该属性可以指向 res 目录下的"mipmap-*dpi"中的 ic_launcher.png 图标。

其他类型的文件也类似。凡是在 R 文件中定义的资源都可以通过"@Static_inner_classes_name/resourse_name"的方式获取。如"@id/button""@string/app_name"。

② 如果访问的是 Android 系统中带的文件，则要添上包名"android：".如 android:textColor="@android:color/red"。

（3）"@+id/string_name"表达式

顺便说一下，在布局文件当中我们需要为一些组件添加 id 属性作为标识，可以使用如下的表达式"@+id/string_name"。其中"+"表示在 R.Java 的名为 id 的内部类中添加一条记录。如"@+id/button"的含义是在 R.Java 文件中的 id 这个静态内部类添加一个常量 button，该常量就是该资源的标识符。如果 id 这个静态内部类不存在，则会先生成它。通过该方式生成的资源标识符仍然可以以"@id/string_name"的方式引用。示例代码片段如下。

```
< RelativeLayout
  android:layout_width = "fill_parent"
  android:layout_height = "wrap_content"
>
< Button
  android:layout_width = "wrap_content"
  android:layout_height = "wrap_content"
  android:text = "@string/cancle_button"
  android:layout_alignParentRight ="true"
  android:id = "@+id/cancle" />
< Button
```

```
    android:layout_width = "wrap_content"
    android:layout_height = "wrap_content"
    android:layout_toLeftOf = "@id/cancle"
    android:layout_alignTop = "@id/cancle"
    android:text = "@string/ok_button" />
</ RelativeLayout >
```

其中，android:id=" @+id/cancle"将其所在的button标识为cancle，在第二个button中通过"@id/cancle"对第一个button进行引用。

2.1.3 AndroidMainfest.xml详细介绍

每个应用程序都有一个功能清单文件AndroidManifest.xml（一定是这个名字）在它的根目录里。这个清单文件给Android系统提供了这个应用程序的基本信息，系统在运行任何程序代码之前必须知道这些信息。今后我们开发Activity、Broadcast、Service之后都要在AndroidManifest.xml中进行定义。另外，如果需要使用到系统自带的服务，如拨号服务、应用安装服务、GPRS服务等，都必须在AndroidManifest.xml中声明权限。

AndroidManifest.xml主要包含以下功能：
- 命名应用程序的Java应用包，这个包名用来唯一标识应用程序；
- 声明应用程序所必须具备的权限，用于访问受保护的部分API，以及和其他应用程序交互；
- 声明应用程序其他的必备权限，用于组件之间的交互；
- 列举测试设备Instrumentation类，用来提供应用程序运行时所需的环境配置及其他信息，这些声明只在程序开发和测试阶段存在，发布前将被删除；
- 声明应用程序所要求的Android API的最低版本；
- 列举application所需要链接的库。

下面以app项目的功能清单文件为例进行讲解。

```xml
<?xml version="1.0" encoding="utf-8"?>
<manifestxmlns:android="http://schemas.android.com/apk/res/android"
    package="cn.androidstudy.chapter02">

    <application
        android:allowBackup="true"
        android:icon="@mipmap/ic_launcher"
        android:label="@string/app_name"
        android:roundIcon="@mipmap/ic_launcher_round"
        android:supportsRtl="true"
        android:theme="@style/AppTheme">
        <activity android:name=".MainActivity">
            <intent-filter>
                <action android:name="android.intent.action.MAIN"/>

                <category android:name="android.intent.category.LAUNCHER"/>
```

```xml
            </intent-filter>
        </activity>
    </application>

</manifest>
```

下面详细讲解各个标签。

(1) <manifest>元素

```xml
<manifest xmlns:android="http://schemas.android.com/apk/res/android"
    package="cn.androidstudy.chapter02">
```

该元素是AndroidManifest.xml文件的根元素，为必选。根据XML文件的语法，"xmlns:android"指定该文件的命名空间。功能清单文件会使用"http://schemas.android.com/apk/res/android"所指向的一个文件。"package"属性指定Android应用所在的包。

(2) <application>元素

```xml
<application
    android:allowBackup="true"
    android:icon="@mipmap/ic_launcher"
    android:label="@string/app_name"
    android:roundIcon="@mipmap/ic_launcher_round"
    android:supportsRtl="true"
    android:theme="@style/AppTheme">
    <activity android:name=".MainActivity">
        <intent-filter>
            <action android:name="android.intent.action.MAIN" />
            <category android:name="android.intent.category.LAUNCHER" />
        </intent-filter>
    </activity>
</application>
```

<application>是非常重要的一个元素，今后我们开发的许多组件都会在该元素下定义，该元素为必选元素。<application>的"icon"属性用来设定应用的图标。<application>的"label"属性用来设定应用的名称。指定其属性值所用的表达式"@string/app_name"的含义与前文的表达式"@mipmap/ic_launcher"一样，同样是指向R.Java文件中的string静态内部类中的app_name属性所指向的资源。在这里它指向的是"strings.xml"文件中的一条记录"app_name"，其值为"Chapter02"，因此，这种表达方式等价于android:label = "Chapter02"。

(3) <activity>元素

<activity>元素的作用是注册一个Activity信息。当我们在创建"Chapter02"这个项目时，指定了"Activity name"属性为"MainActivity"，在生成项目时自动创建了一个Activity名称就是"MainActivity.Java"，Activity在Android中属于组件，它需要在功能清单文件中进行配置。<activity>元素的"name"属性指定的是该Activity的类名。<activity>元素的"label"属性表示Activity所代表的屏幕的标题，其属性值的表达式在前文已经介绍过了，此处不再赘述。该属性值在 AVD 运行程序到该 Activity 所代表的界面时，会在标题上显示该值。

（1）<intent-filter>元素

翻译成中文是"意图过滤器"。应用程序的核心组件（活动、服务和广播接收器）通过意图被激活，意图代表的是用户要做的一件事情，代表用户的目的，Android 寻找一个合适的组件来响应这个意图，如果需要会启动这个组件一个新的实例，并传递给这个意图对象。后面会进行详细的介绍。

组件通过意图过滤器（intent filters）通告它们所具备的功能——能响应的意图类型。由于 Android 系统在启动一个组件前必须知道该组件能够处理哪些意图，那么意图过滤器需要在<manifest>中以<intent-filter>元素指定。一个组件可以拥有多个过滤器，每一个描述该组件所具有的不同能力。一个指定目标组件的显式意图将会激活那个指定的组件，意图过滤器不起作用。但是一个没有指定目标组件的隐式意图只在它能够通过组件过滤器时才能激活该组件。

第一个过滤器

```
<intent-filter>
    <action android:name="android.intent.action.MAIN" />
    <category android:name="android.intent.category.LAUNCHER" />
</intent-filter>
```

是最常见的，它表明这个 Activity 将在应用程序加载器中显示，就是用户在设备上看到的可供加载的应用程序列表。换句话说，这个 Activity 是应用程序的入口，是用户选择运行这个应用程序后所见到的第一个 Activity。

（5）权限（Permissions）

项目的功能清单文件中并没有出现<Permissions>元素，但是 Permission 也是一个非常重要的节点，在后面的学习中会经常用到。Permission 是代码对设备上数据的访问限制，这个限制被引入来保护可能会被误用而曲解或破坏用户体验的关键数据和代码，如拨号服务、短信服务等。每个许可被一个唯一的标签所标识，这个标签常常指出受限的动作。

如申请发送短信服务的权限需要在功能清单文件中添加如下语句：

```
<uses-permission android:name="android.permission.SEND_SMS"/>
```

一个功能（feature）最多只能被一个权限许可保护。如果一个应用程序需要访问一个需要特定权限的功能，它必须在<manifest>元素内使用 <uses-permission> 元素来声明这一点。这样，当应用程序安装到设备上之后，安装器可以通过检查签署应用程序认证的机构来决定是否授予请求的权限，在某些情况下，会询问用户。如果权限已被授予，那应用程序就能够访问受保护的功能特性；如果没有，访问将失败，但不会给用户任何通知。因此在使用一些系统服务，如拨号、短信、访问互联网、访问 SD Card 时，一定要记得添加相应的权限，否则会出现一些难以预料的错误。

应用程序还可以通过权限许可来保护它自己的组件（活动、服务、广播接收器、内容提供者）。它可以利用 Android 已经定义（列在 android.Manifest.permission 里）或其他应用程序已声明的权限许可，或者定义自己的许可。一个新的许可通过<permission>元素声明。比如，一个 Activity 可以用下面的方式保护：

```
<manifest...>
<permission android:name="com.example.project.DEBIT_ACCT".../>
```

```
...
<application...>
<activity android:name="com.example.project.FreneticActivity"...>
android:permission="com.example.project.DEBIT_ACCT"
...>
...
</activity>
</application>
...
<uses-permission android:name="com.example.project.DEBIT_ACCT"/>
...
</manifest>
```

注意：在这个例子里，这个DEBIT_ACCT许可并非仅仅在<permission>元素中声明，如果该应用程序的其他组件要使用该组件，那么它同样要声明在<uses-permission>元素里。

(6) 库（Libraries）

每个应用程序都链接到缺省的Androi库，这个库包含了基础应用程序开发包（实现了基础类，如活动、服务、意图、视图、按钮、应用程序、内容提供者等）。然而，一些包处于它们自己的库中。如果用户的应用程序使用了其他开发包中的代码，它必须显式地请求链接到它们。<manifest>必须包含一个单独的<uses-library>元素来命名每一个库。如在进行单元测试的时候需要引入其所需要的库。

代码片段如下：

```
<application android:icon="@drawable/icon"
           android:label="@string/app_name">
    <uses-library android:name="android.test.runner" />
</application>
```

2.2 Android应用程序的执行

经过环境配置和代码编程之后，程序的执行往往是比较令人期盼和激动的。在熟悉了Android项目组成结构以及相关文件之后，下面我们将进一步对Android程序的具体编译执行过程进行详细分析，带领读者深入Android的编程世界。

首先，要生成APK，Android程序的编译过程如图2-4所示。

① 资源打包。使用IDE中的资源打包工具（Android Asset Packaging Tool，即图中的aapt）会将应用中的资源文件进行编译，这些资源文件包括AndroidManifest.xml文件、为Activity定义的XML文件等。在这个编译过程中会产生一个R.java文件，这样用户就可以在自己的Java代码中引用这些资源了。

② aidl（Android Interface Definition Language）。aidl工具会将用户项目中的所有.aidl接口转换成Java接口。

③ 编译项目中的所有的Java代码。包括R.java和.aidl文件，都会被Java编译器编译，然后输出.class文件。

④ 生成dex文件。使用Dalvik VM Executes（Android Dalvik执行程序）将上一步骤产生的.class文件转成Dalvik字节码，也就是.dex文件。同时项目中包含的所有第三方类库和.class文件也会被转换成.dex文件，这样方便下一步被打包成最终的.apk文件。

⑤ 生成APK。所有的不能编译的资源（比如图片）、编译后的资源文件和.dex文件会被apkbuilder工具打包成一个.apk文件。

⑥ 签名APK。使用jarsigner对APK进行签名。.apk文件构建好之后，如果要把它安装到设备上面去的话，它就必须用一个debug或者发行key来对这个APK文件签名。

⑦ 对齐APK。当应用程序在设备上运行，减少内存使用。对齐的应用在系统中执行时，通过共享内存IPC读取资源就能得到较高的性能。最后，如果应用程序已经被签名成为发行模式的APK，用户还需要使用aipalign工具对.apk进行对齐优化。这样可以减少应用程序在设备上的内存消耗。

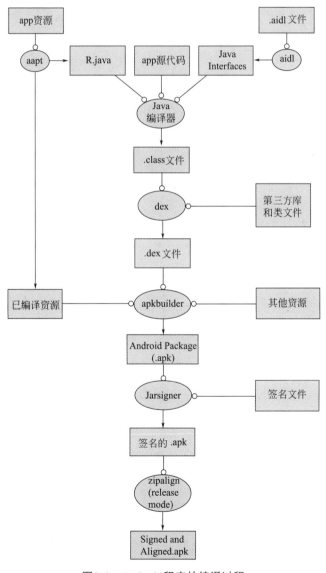

图2-4　Android程序的编译过程

然后，将生成的 APK 安装到手机或者模拟器上。发布程序到手机上之后，当双击该应用的图标时，系统会将这个点击事件包装成一个 Intent，该 Intent 包含两个参数：

```
{ action : "android.intent.action.MAIN" ,
  category : "android.intent.category.LAUNCHER" }
```

这个意图被传递给 app 应用之后，在应用的功能清单文件中寻找与该意图匹配的意图过滤器，如果匹配成功，找到相匹配的意图过滤器所在的<activity>元素，再根据<activity>元素的"name"属性来寻找其对应的 Activity 类。接着，Android 操作系统创建该 Activity 类的实例对象，对象创建完成之后，会执行到该类的 onCreate 方法，此 onCreate 方法是重写其父类 Activity 的 onCreate 方法而实现。onCreate 方法用来初始化 Activity 实例对象。下面是 HelloWorld.Java 类中 onCreate 方法的代码。

```
@Override
    protected void onCreate(Bundle savedInstanceState) {
        super.onCreate(savedInstanceState);
        setContentView(R.layout.activity_main);
    }
```

其中，super.onCreate(savedInstanceState);的作用是调用其父类 Activity 的 onCreate 方法来实现对界面的画图工作。在实现自己定义的 Activity 子类的 onCreate 方法时一定要记得调用该方法，以确保能够绘制界面。

setContentView(R.layout. main)的作用是加载一个界面。该方法中传入的参数是"R.layout. activity_main"，其含义为 R.Java 类中静态内部类 layout 的静态常量 activity_main 的值，而该值是一个指向 res 目录下的 layout 子目录下 activity_main 文件的标识符，因此代表着显示 activity_main 所定义的画面。

关于 Activity 类的执行流程及其生命周期会在后面的部分详细讲解。

Android 程序执行的整个序列图如图 2-5 所示。

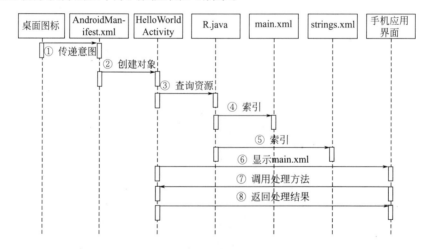

图2-5　Android程序执行序列图

2.3 Android应用程序的主要组件

Android是一种基于Linux的自由及开源代码的操作系统，有许多优点。Linux平台上一些编程语言的复杂性和高难度往往令许多编程初学者望而却步，尤其是如何在Linux平台中设计出良好的人机交互界面和进行信息正确的传递令程序员颇耗费精力。虽然Android是一种基于Linux的操作系统，相对而言，Android程序员可以较方便、简单地实现所需Android应用程序及功能。下面将向读者展示Android应用程序的主要组件以及在Android应用程序中如何进行信息的传递。

与其他计算机编程语言生成的程序不同，Android应用程序没有唯一的启动入口［如C语言中main()函数入口］，一个Android应用程序是由多个不同的组件组合而成，组件之间通过Intent来实现通信和传递信息。Android系统的基本组件包括Activity、Service、BroadcastReceiver、ContentProvider等，此外还包括专门负责在基本组件之间传递消息的Intent组件。所有这些组件都必须在AndroidManifest.xml文件中声明。这些组件是如何协调工作的呢？下面我们用一个简单的用户与应用程序交互的例子来说明Android程序中上述组件是如何配合的，这里包含了两个Activity，如图2-6所示。

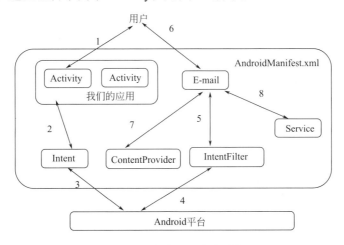

图2-6 示例应用程序的结构

首先，用户通过Activity与我们的应用程序交互，如图2-6中步骤1。

我们的应用程序中的Activity通过Intent来向Android平台请求启动一个能处理打开E-mail的应用程序，如图2-6中步骤2和步骤3。

Android系统通过AndroidManifest.xml中声明的IntentFilter找到能处理打开E-mail的应用程序，如图2-6中步骤4和步骤5。

接下来，用户与E-mail应用程序进行交互，如图2-6中步骤6。

E-mail应用程序通过ContentProvider来使用另一个录音应用程序产生的音频文件，如图2-6中步骤7。

用户播放刚才的音频文件，并返回到之前的应用程序，此时音频文件仍然会继续播放，因为Service将在后台工作，如图2-6中的步骤8。

当然,并不是每个Android应用程序都必须包含这些组件,但是一旦确定了用户的应用程序中需要的组件,那么就应该在AndroidManifest.xml中声明它们。接下来我们就对这些基本组件做简单的介绍,使读者对这些组件建立一个大致的认识,后面章节还会对这些组件做更详细的介绍。

2.3.1 Activity

Activity是应用程序的表示层。应用程序中的每个屏幕显示都通过集成和扩展基类Activity来实现。Activity利用View来实现应用程序的GUI,而我们的手机用户则直接通过GUI和应用程序做交互,如应用程序通过GUI向用户显示信息,用户通过GUI向应用程序发出指令和响应。

例如一个短信应用程序,我们需要一个Activity来显示联系人列表,同时需要另一个Activity显示用户输入的短信的内容,甚至还可能需要第三个Activity显示已收到短信的内容。虽然这些Activity整体形成了一个完整的短信程序用户界面,但实际上每个Activity是独立的。当然,它们也有共同点——每个Activity都是继承自Activity的子类。

应用程序往往由多个Activity组成。一个应用程序需要多少个Activity?每个Activity表示什么样的用户界面?这些问题都取决于具体的应用程序设计。通常的原则是,程序启动后显示的第一幅画面是应用程序的第一个Activity,以后根据应用程序的需要从一个Activity跳转到一个新的Activity。下面这段代码展示了我们之前的项目HelloWorld创建Activity的方法。

```java
public class MainActivity extends Activity {

    @Override
    protected void onCreate(Bundle savedInstanceState) {
        super.onCreate(savedInstanceState);
        setContentView(R.layout.activity_main);
    }
}
```

对于每个Activity,系统会分配一个默认的窗口。一般情况下,窗口将占满整个屏幕。改变默认属性,窗口大小也是可调整的。窗口的显示位置也可以悬浮在其他窗口之上。Activity同时也能使用别的窗口,为了提醒用户,可以在一个Activity中使用弹出对话框。

Activity窗口内的可见内容通过View提供,View对象继承自View类,每个View对象控制这窗口内的一个巨型空间。View是一种层次结构,父View包含的布局属性会被子View继承。位于View层次关系最底部的子View对象所代表的矩形空间就是跟用户进行交互的地方。例如,我们可以用一个View对象来显示图片,并在用户点击图片时产生相应的动作。Android自带了很多不同的View供开发者使用,比如按钮、文本框、滚动条、菜单项等,这些内容会在后续章节进行详细介绍。

既然Activity的内容通过View来显示,那么如何才能将View对象放入Activity中呢?我们可以调用Activity.setContentView(),如前文代码中的最后一行。

2.3.2 Service

Service与Activity的地位是并列的,它也代表一个单独的Android组件。但与Activity相反,Service没有可见的界面,它的特点是能长时间在后台运行,也可以这样理解,Service

是具有一段较长生命周期且没有用户界面的程序。

为什么我们需要长时间在后台运行的 Service？想想音乐播放器。用户可能在播放音乐的同时去编辑短信或者浏览网页，这种情况下用户的音乐播放器不可能一直处于前台。为了让音乐一直播放下去，需要将播放音乐的任务放在后台。这样，即使音乐播放器已经不再显示了，用户仍然可以听到音乐。所以，我们需要这样的机制——长时间在后台运行的 Service。与 Activity 组件需要集成 Activity 基类相似，Service 组件需要继承 Service 基类。一个 Service 组件被运行起来之后，它将拥有自己独立的生命周期。

2.3.3　BroadcastReceiver

BroadcastReceiver 是用户接收广播通知的组件。广播是一种同时通知多个对象的事件通知机制。Android 中的广播通知要么来自系统，要么来自普通应用程序。很多事件都可能导致系统广播，比如手机所在的时区发生变化、电池电量低、用户改变系统语言设置等。当然也有广播来自应用程序，比如一个应用程序通知其他应用程序某些数据已经下载完毕。

为了响应不同的事件通知，应用程序可以注册不同的 BroadcastReceiver，而所有的 BroadcastReceiver 都继承自基类 BroadcastReceiver。

说明：BroadcastReceiver 自身并不实现图形用户界面，但是当它收到某个通知消息后，BroadcastReceiver 可以启动 Activity 作为响应，或者通过 NotificationManager 提醒用户。

2.3.4　ContentProvider

在 Android 中，每个应用程序都使用自己的用户 id 并在自己的进程中运行。这样做的好处是，可以保护系统及应用程序，避免被其他不正常的应用程序影响，每个进程都拥有独立的进程地址空间和虚拟内存。当应用程序彼此间需要共享资源时，这样的架构必须要有一个妥善的解决方案。例如，Contacts 应用程序内存中保存使用者的联系资料，当用户在 E-mail 中要填写收信人时，希望读取 Contacts 内的联系人资料。由于 Contacts 和 E-mail 这两个应用程序运行在不同的进程中，因此它们无法直接通过内存共享联系人资料。为了解决应用程序间数据通信、共享的问题，Android 提供了 ContentProvider 机制。

ContentProvider 能将应用程序特定的数据提供给另一个应用程序使用。数据的存储方式可以是 Android 文件系统，也可以是 SQLite 数据库，或者别的合理的方式。

ContentProvider 继承自父类 ContentProvider，并且实现了一组标准的接口，通过这组接口，其他应用程序能对数据进行读写。然而，需要使用数据的应用程序并不是直接调用这组方法，而是通过调用 ContentResolver 对象的方法来完成的。ContentResolver 对象可以与任意 ContentProvider 通信。

2.3.5　Intent 和 IntentFileter

严格地说，Intent 并不是 Android 应用的组件，但是它对于 Android 应用的作用非常大——它是 Android 应用程序内不同组件之间通信的载体。当 Android 运行时需要连接不同的组件时，通常就需要借助于 Intent 来实现。Intent 可以启动应用中另一个 Activity 组件，也可以启动一个 Service 组件，还可以发送一条广播消息来触发系统中的 BroadcastReceiver。也就是说，Activity、Service、BroadcastReceiver 三种组件之间的通信都以 Intent 作为载体，只是不同组件使用 Intent 的机制略有区别而已。具体组件间如何通过 Intent 进行通信，后续章节会详细介绍，在此不再赘述。

强化训练

本章主要对 Android 应用程序进行了深入剖析，首先介绍了 Android 应用程序的构成及程序的内部执行流程，并对一个 Android 应用程序所需要的基本组件的功能及其作用进行简单的介绍，使读者对 Android 应用程序的内部执行有一个初步的认识。

学完本章知识后，不妨通过以下习题来巩固所学的内容吧。

一、填空题

1. Android Studio 项目中，strings.xml 用来定义字符串和数值，在 Activity 中使用 _____ 或 _____ 取得资源。

2. Android Studio 项目中，drawable 和 mipmap 区别其实不大，只是使用 _____ 会在图片缩放时提供一定的性能优化，分辨率不同系统会根据屏幕分辨率来选择 ____、____、____、____、____ 下的对应图片。

3. Android Studio 项目中，_____ 与 Activity 的地位是并列的，它也代表一个单独的 Android 组件。

4. Android Studio 项目中，要生成 APK，Android 程序的编译过程包括：_____、_____、_____、_____、_____、_____、_____。

5. 与在网页布局中使用 HTML 文件一样，Android 在 _____ 文件中使用 _____ 元素来设定屏幕布局。

6. 在 Android Studio 中，_____ 的真实含义是工作空间，相当于 Eclipse 中的 Workspace；_____ 为一个具体的项目，相当于 Eclipse 中的 _____，一个 _____ 可以包含多个 _____。

二、选择题

1. Android 应用程序需要打包成（　　）文件格式在手机上安装运行。
   A. .apk
   B. .class
   C. .xml
   D. .dex

2. 中等分辨率的图像文件一般放在 res 的（　　）子目录下。
   A. mipmap-hdpi
   B. mipmap-mdpi
   C. mipmap-xhdpi
   D. mipmap-xxhdpi

3. Activity、Service、BroadcastReceiver 三种组件之间的通信都以（　　）作为载体。
   A. Activity
   B. Intent
   C. ContentProvider
   D. Interface

4. Android Studio 项目中的文件夹中（　　）存放 Android Studio 生成的配置文件。
   A. app
   B. gradle
   C. build
   D. .idea
5. 下列（　　）是 Android 应用程序 app 的表示层。
   A. Service
   B. Activity
   C. BroadcastReceiver
   D. ContentProvider
6. 每个应用程序都有一个功能清单文件（　　）。
   A. activity_main.xml
   B. AndroidManifest.xml
   C. strings.xml
   D. Android Asset Packaging Tool

三、操作题

1. 在 Android Studio 中一个 Android 应用程序项目包含有相对比较复杂的目录结构和文件。练习利用 Android Studio 开发环境创建一个新的 Android 应用程序项目，浏览并学习项目的目录、文件夹和文件，然后对项目进行编译、执行，观察项目编译、执行等过程。

2. Android 应用程序的图标和程序名称是 Android 应用程序最常被用到的内容。练习利用 Android Studio 开发环境创建一个新的 Android 应用程序项目，尝试着修改 Android 应用程序的图标和程序名称，再次编译并运行，验证是否修改成功。

第3章

界面布局精讲

内容导读

应用程序界面（UI）是人与机器之间传递和交换信息的媒介。它实现信息的内部形式与用户可以接受的形式之间的转换。如何将Android应用程序交互界面设计得美观、操作简单、逻辑性强，甚至让程序变得更有个性更有品位，这是Android程序开发人员必须要面对的实际问题之一。在Android系统中，程序开发人员可以开发出绚丽多彩的UI，而且是在开放式环境中，这是Android程序开发的魅力，也是吸引更多优秀开发人员加入的原因。

本章将着重介绍Android UI的基础知识和几种常见的Android布局格式。

学习目标

- 了解Android UI的基础知识；
- 掌握使用线性布局构建复杂UI的方法；
- 掌握使用相对布局构建复杂UI的方法；
- 掌握使用约束布局构建复杂UI的方法。

3.1 UI概述

UI是User Interface(用户界面)的缩写。UI设计往往是设计程序时需要面对的第一个难题，甚至直接影响到整个软件系统开发的成败。界面美观的Android程序不仅可以大大增加用户的黏性，还可以吸引更多的新用户。Android程序往往运行在移动触摸屏设备（尤其是

第3章 界面布局精讲

手机或平板电脑）上，所以Android程序的UI设计又有其独特的一面。

在Android程序开发中，一个用户界面是由View和ViewGroup对象来构建的。虽然它们有很多的种类，不过都是View类的子类，View类是Android系统平台上用户界面表示的基本单元。View类的一些子类被统称为"widgets（工具）"，它们提供了诸如文本输入框和按钮之类的UI对象的完整实现。ViewGroup是View的一个扩展，它可以容纳多个子View。通过扩展ViewGroup类，用户可以创建由相互联系的子View组成的复合控件。ViewGroup类同样可以被扩展用作layout（布局）管理器，如LinearLayout（线性布局）、TableLayout(表格布局)以及RelativeLayout(相对布局)等布局架构，并且用户可以通过用户界面与程序进行交互。通过使用这些布局模型的组合、嵌套并设置子控件的布局参数，我们完全可以构建出各种复杂的用户界面。下面将对这几种布局模型分别进行详细介绍。

读者还可以通过查看官方文档来了解所有的布局信息，对于每个Android开发者而言，Android提供的官方文档是必看的。这里简单给读者介绍一下该如何查看Android文档——这实际上是一种学习方法（笔者常常觉得掌握学习方法比记住几个知识点更重要）。首先定位到Android Studio SDK的安装目录，找到docs子目录，打开docs子目录下的index.html页面，按照Develop/API Guides的标签路径点击，用户将看到如图3-1所示页面。在此页面，可以通过点击链接查看自己感兴趣的内容，如果具有良好的英文阅读能力加上扎实的Java基础，完全可以通过阅读此指南开发出各种Android应用程序。

图3-1 Android开发指南

Android开发中一般分为以下几种布局：线性布局（LinearLayout）、相对布局（RelativeLayout）、绝对布局（AbsoluteLayout）、表格布局（TableLayout）和帧布局（FrameLayout）。本章我们主要学习LinearLayout、RelativeLayout、TableLayout和AbsoluteLayout，本章案例的布局主要是在src/main/res/layout/ activity_main.xml中编写。

3.2 线性布局LinearLayout

线性布局LinearLayout是Android程序开发中最基本、最常见的布局形式，这种布局形式相对而言简单易学、容易上手，初学者往往从学习线性布局开始。本节将对线性布局LinearLayout进行详细分析，并用实例来展示线性布局LinearLayout的应用。

线性布局由LinearLayout类来代表。线性布局有点像AWT编程里的FlowLayout，它们会将容器里的组件一个挨着一个地排列起来。LinearLayout不仅可以控制各组件横向排列，也可以控制纵向排列。LinearLayout以它的垂直或水平的属性值来排列所有的子元素。所有的子元素都被堆放在其他元素之后，因此一个垂直列表的每一行只会有一个元素，而不管它们有多宽；一个水平列表将会只有一个行高（高度为最高子元素的高度加上边框高度）。

Android的线性布局不会换行，当组件一个挨着一个排到头之后，剩下的组件将不会被显示出来。LinearLayout的常用XML属性及相关方法见表3-1。

表3-1 LinearLayout常用属性说明

XML属性	相关方法	说明
android:orientation	setOrientation(int)	设置布局管理器内组件的排列方式，vertical为垂直，默认horizontal（水平）
android:gravity	setGravity(int)	设置布局管理器内组件对齐方式，该属性支持top、bottom、left、right、center_vertical等，可以同时指定多种对齐方式，多个属性值间竖线隔开，竖线前后不能有空格
android:layout_weight	setLayoutParams()	设置控件占屏幕的比例，在垂直布局时，代表行距；水平的时候代表列宽；weight值越大就越大

　　LinearLayout支持为其包含的widget或者是container指定填充权值。允许其包含的widget或者container可以填充屏幕上的剩余空间。剩余的空间会按这些widget或者container指定的权值比例分配屏幕。默认的weight值为0，表示按照widget或者container实际大小来显示；若大于0，则将container剩余可用空间分割，分割大小具体取决于每一个widget或者container的layout_weight及该权值在所有widget或者container中的比例。

　　Android中定义的长度单位有以下几种。
- px（pixel，像素）：对应屏幕上的实际像素点。
- in（inch，英寸）：屏幕物理长度单位。
- mm（millimeter，毫米）：屏幕物理长度单位。
- pt（point，磅）：屏幕物理长度单位，1/72英寸。
- dp（与密度无关的像素）：逻辑长度单位，在每英寸160点的屏幕上，1dp=1px=1/160英寸。随着密度变化，对应的像素数量也变化，但并没有直接的变化比例。
- dip：与dp相同，多用于Google示例中。
- sp（与密度和字体缩放度无关的像素）：与dp类似，但是可以根据用户的字体大小首选项进行缩放。

　　在实际应用开发中，尽量使用dp作为空间大小单位，sp作为和文字相关大小单位。

☞【实例3-1】通过实例演示LinearLayout的用法。

扫一扫 看视频

　　在Android Studio中创建一个Android工程，工程名为"Chapter03"，在该工程下创建一个Module，命名为"LinearLayout_demo"，打开其目录下src/main/res/layout/activity_main.xml文件，修改代码并输入一些代码，代码如下所示。

```xml
<?xml version="1.0" encoding="utf-8"?>
<LinearLayout xmlns:android="http://schemas.android.com/apk/res/android"
    xmlns:tools="http://schemas.android.com/tools"
    android:id="@+id/activity_main"
    android:layout_width="match_parent"
```

```xml
        android:layout_height="match_parent"
        android:orientation="vertical"
        tools:context="cn.androidstudy.linearlayout_demo.MainActivity">
    <LinearLayout
        android:layout_width="match_parent"
        android:layout_height="match_parent"
        android:orientation="horizontal"
        android:id="@+id/ll_hor"
        android:layout_weight="3" >
        <TextView
            android:layout_width="wrap_content"
            android:layout_height="match_parent"
            android:layout_weight="1"
            android:background="#aa0000"
            android:text="red"
            android:textSize="30dp"/>
        <TextView
            android:layout_width="wrap_content"
            android:layout_height="match_parent"
            android:layout_weight="2"
            android:background="#00aa00"
            android:text="green"
            android:textSize="30dp"/>
        <TextView
            android:layout_width="wrap_content"
            android:layout_height="match_parent"
            android:layout_weight="1"
            android:background="#0000aa"
            android:text="blue"
            android:textSize="30dp"
            android:textColor="#aaaaaa" />
    </LinearLayout>
    <LinearLayout
        android:layout_width="match_parent"
        android:layout_height="match_parent"
        android:orientation="vertical"
        android:layout_weight="1" >
        <TextView
            android:layout_width="match_parent"
            android:layout_height="wrap_content"
            android:text="row1"
```

```xml
            android:background="#aaaaaa"
            android:textSize="30dp"
            android:layout_weight="1" />
        <TextView
            android:layout_width="match_parent"
            android:layout_height="wrap_content"
            android:background="#00aa00"
            android:layout_weight="1"
            android:textSize="30dp"
            android:text="row2" />
        <TextView
            android:layout_width="match_parent"
            android:layout_height="wrap_content"
            android:layout_weight="1"
            android:background="#0000aa"
            android:textSize="30dp"
            android:text="row3"
            android:textColor="#aaaaaa"/>
    </LinearLayout>
</LinearLayout>
```

运行该项目，效果如图3-2所示。该示例的布局效果是通过使用布局嵌套实现的，最外层是一个垂直排布的LinearLayout；在最外层的LinearLayout中再嵌套两个（上、下）LinearLayout子布局；在嵌套的子LinearLayout布局中，上部分的LinearLayout使用水平排布，里面放3个TextView，下部分的LinearLayout使用垂直排布，里面放3个TextView。

在上述的activity_main.xml布局文件中，给2个LinearLayout子布局和6个TextView控件都设置了layout_weight属性，指定了在分配各自父容器的剩余空间时各自所占的比例。我们能发现2个子布局在显示效果中垂直方向的占比与layout_weight属性值相反，而两组TextView在各自不居中水平或者垂直方向的占比与layout_weight属性值相同，区别在于子布局或者TextView控件这些元素自身的layout_width、layout_height两个属性值。当同排布方向的高度或宽度值为"wrap_content"时，父容器剩余空间为正值，各子元素按layout_weight属性值分配后，显示效果同layout_weight属性值同比例。而当同排布方向的高度或宽度值为"match_parent"时，父容器剩余空间为负值，各子元素按layout_weight属性值分配后，显示效果同layout_weight属性值成反比。

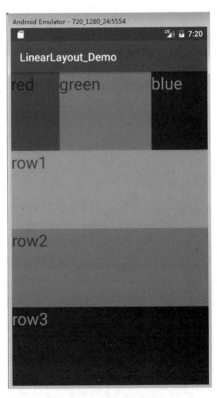

图3-2　LinearLayout线性布局示例效果图

3.3 相对布局 RelativeLayout

相对布局 RelativeLayout 与线性布局 LinearLayout 比较而言复杂一些，需要编程者有一定的空间感。因为相对布局 RelativeLayout 容器内子组件的位置总是相对兄弟组件、父容器来决定的。如果 A 组件的位置是由 B 组件的位置来决定的，Android 要求先定义 B 组件，再定义 A 组件。当编程者对相对布局 RelativeLayout 熟悉后，会有部分编程者习惯并喜欢这种布局形式。

RelativeLayout 常见的一些重要属性，可以分为两组：一组用来设置控件与给定控件之间的关系属性，如表3-2所示；另一组用来设置控件与父容器之间的关系属性，如表3-3所示。

RelativeLayout 允许子元素指定它们相对于其他元素或父元素的位置（通过ID指定）。因此，用户可以左右对齐，或上下对齐，或置于屏幕中央的形式来排列两个元素。元素按顺序排列，因此如果第一个元素在屏幕的中央，那么相对于这个元素的其他元素将以屏幕中央的相对位置来排列。如果使用 XML 来指定这个 layout，在用户定义它之前，被关联的元素必须定义。

表3-2 RelativeLayout 中设置控件与给定控件关系的属性

属性	说明	备注
android:layout_above	将该控件置于给定ID的控件之上	
android:layout_below	将该控件置于给定ID的控件之下	
android:layout_toLeftOf	将该控件置于给定ID的控件之左	
android:layout_toRightOf	将该控件置于给定ID的控件之右	
Android:layout_alignBaseline	该控件的baseline和给定ID控件的baseline对齐	属性值为某个给定控件的ID，例如：android:layout_above="@id/×××"
android:layout_alignBottom	将该控件的底部边缘与给定ID控件的底部边缘对齐	
android:layout_alignLeft	将该控件的左边边缘与给定ID控件的左边边缘对齐	
android:layout_alignTop	将该控件的顶部边缘与给定ID控件的顶部边缘对齐	
android:layout_alignRight	将该控件的右边边缘与给定ID控件的右边边缘对齐	

表3-3 RelativeLayout 中设置控件与父容器关系的属性

属性	说明	备注
android:layout_alignParentBottom	该控件的底部与父控件的底部对齐	属性值为true或false，这里假设值为true。如果不指定，默认是false，表示的都是控件自身的上下左右边缘与父控件的对应边缘是否对齐。由于控件是在父控件的内部，所以是内对齐
android:layout_alignParentLeft	该控件的左边缘与父控件的左边缘对齐	
android:layout_alignParentRight	该控件的右边缘与父控件的右边缘对齐	
android:layout_alignParentTop	该控件的顶部与父控件的顶部对齐	
android:layout_marginLeft	该控件距离父类容器左边的距离	属性值为像素值，例如android:layout_marginLeft = "30dp"，表示该控件距离父类容器左边的距离为30dp

续表

属性	说明	备注
android:layout_marginRight	该控件距离父类容器右边的距离	属性值为像素值，例如android:layout_marginLeft = "30dp"，表示该控件距离父类容器左边的距离为30dp
android:layout_marginTop	该控件距离父类容器顶部的距离	
android:layout_marginBottom	该控件距离父类容器底部的距离	
android:layout_margin	该控件距离父类容器上、下、左、右四周的距离	
android:layout_centerHorizontal	该控件在其父控件范围内水平居中	属性值为true或false，这里假设值为true。如果不指定，默认是false，表示的都是控件自身相对于父控件范围内的居中情况
android:layout_centerInparent	该控件在其父控件范围内垂直且水平居中	
android:layout_centerVertical	该控件在其父控件范围内垂直居中	

下面通过一个简单的案例来学习上面所讲相对布局以及设置控件的属性。运用RelativeLayout(相对布局)制作一幅由9幅小图片组合而成的大图片。

【实例3-2】通过实例演示RelativeLayout的用法。

在工程"Chapter03"下创建一个Module,命名为"RelativeLayout.demo"，打开其目录下src/main/res/layout/activity_main.xml文件，修改代码并输入一些代码，代码如下所示。

```xml
<?xml version="1.0" encoding="utf-8"?>
<RelativeLayout xmlns:android="http://schemas.android.com/apk/res/android"
    xmlns:tools="http://schemas.android.com/tools"
    android:id="@+id/activity_main"
    android:layout_width="match_parent"
    android:layout_height="match_parent"
    android:paddingBottom="@dimen/activity_vertical_margin"
    android:paddingLeft="@dimen/activity_horizontal_margin"
    android:paddingRight="@dimen/activity_horizontal_margin"
    android:paddingTop="@dimen/activity_vertical_margin"
    tools:context="cn.androidstudy.relativelayoutdemo.MainActivity">
    <ImageView
        android:layout_width="wrap_content"
        android:layout_height="wrap_content"
        android:id="@+id/iv_5"
        android:src="@mipmap/five"
        android:layout_centerInParent="true"
        android:layout_marginTop="3dp"
        android:layout_marginLeft="3dp"/>
    <ImageView
        android:layout_width="wrap_content"
        android:layout_height="wrap_content"
        android:id="@+id/iv_1"
        android:src="@mipmap/one"
```

第3章 界面布局精讲

```
        android:layout_above="@id/iv_5"
        android:layout_toLeftOf="@id/iv_5" />
<ImageView
        android:layout_width="wrap_content"
        android:layout_height="wrap_content"
        android:id="@+id/iv_2"
        android:src="@mipmap/two"
        android:layout_above="@id/iv_5"
        android:layout_toRightOf="@id/iv_1"
        android:layout_marginLeft="3dp"/>
<ImageView
        android:layout_width="wrap_content"
        android:layout_height="wrap_content"
        android:id="@+id/iv_3"
        android:src="@mipmap/three"
        android:layout_above="@id/iv_5"
        android:layout_toRightOf="@id/iv_2"
        android:layout_marginLeft="3dp"/>
<ImageView
        android:layout_width="wrap_content"
        android:layout_height="wrap_content"
        android:id="@+id/iv_4"
        android:src="@mipmap/four"
        android:layout_below="@id/iv_1"
        android:layout_toLeftOf="@id/iv_5"
        android:layout_marginTop="3dp"/>
<ImageView
        android:layout_width="wrap_content"
        android:layout_height="wrap_content"
        android:id="@+id/iv_6"
        android:src="@mipmap/six"
        android:layout_below="@id/iv_3"
        android:layout_toRightOf="@id/iv_5"
        android:layout_marginTop="3dp"
        android:layout_marginLeft="3dp"/>
<ImageView
        android:layout_width="wrap_content"
        android:layout_height="wrap_content"
        android:id="@+id/iv_7"
        android:src="@mipmap/seven"
        android:layout_below="@id/iv_4"
        android:layout_toLeftOf="@id/iv_5"
        android:layout_marginTop="3dp"/>
```

```xml
<ImageView
    android:layout_width="wrap_content"
    android:layout_height="wrap_content"
    android:id="@+id/iv_8"
    android:src="@mipmap/eight"
    android:layout_below="@id/iv_5"
    android:layout_toRightOf="@id/iv_7"
    android:layout_marginTop="3dp"
    android:layout_marginLeft="3dp"/>
<ImageView
    android:layout_width="wrap_content"
    android:layout_height="wrap_content"
    android:id="@+id/iv_9"
    android:src="@mipmap/nine"
    android:layout_below="@id/iv_6"
    android:layout_toRightOf="@id/iv_8"
    android:layout_marginTop="3dp"
    android:layout_marginLeft="3dp"/>
</RelativeLayout>
```

运行该项目，其效果如图3-3所示。

图3-3　RelativeLayout相对布局示例效果图

3.4 绝对布局AbsoluteLayout

与前面两种布局形式不同，绝对布局AbsoluteLayout会采用另外一种不同的布局思路设计程序UI，Android为不同风格的程序开发人员提供多种布局设计方式。与Java AWT中的空布局类似，Android不提供任何布局控制，而是由编程者自己通过X、Y两个坐标来控制组件的位置，进而实现程序UI设计。当使用AbsoluteLayout作为布局容器时，布局容器不再管理组件的位置、大小，都需要开发人员自己控制。

绝对布局AbsoluteLayout可以让子元素指定准确的X、Y坐标值，并显示在屏幕上。(0, 0)为左上角，当向下或向右移动时，坐标值将变大。AbsoluteLayout没有页边框，允许元素之间互相重叠（尽管不推荐）。一般不建议使用绝对布局，因为运行Android应用的手机往往千差万别，屏幕大小、分辨率都可能存在差异，使用绝对布局会很难兼顾不同屏幕大小、分辨率的问题。

使用绝对布局时，每个子组件都可指定如下两个XML属性。

- layout_x：指定该子组件的X坐标；
- layout_y：指定该子组件的Y坐标。

【实例3-3】通过实例演示如何使用绝对布局创建一个简单登录页面。

在工程"Chapter03"下创建一个Module，命名为"AbsoluteLayout_Demo"，打开其目录下src/main/res/layout/activity_main.xml文件，修改代码并输入一些代码，代码如下所示。

```xml
<?xml version="1.0" encoding="utf-8"?>
<AbsoluteLayout xmlns:android="http://schemas.android.com/apk/res/android"
    xmlns:tools="http://schemas.android.com/tools"
    android:id="@+id/activity_main"
    android:layout_width="match_parent"
    android:layout_height="match_parent"
    android:orientation="vertical"
    tools:context="cn.androidstudy.absolutelayout_demo.MainActivity">
    <!-- 定义一个文本框，使用绝对布局 -->
    <TextView
        android:layout_width="wrap_content"
        android:layout_height="wrap_content"
        android:layout_x="20dip"
        android:layout_y="20dip"
        android:text="用户名：" />
    <!-- 定义一个文本编辑框，使用绝对布局 -->
    <EditText
        android:layout_width="wrap_content"
```

```xml
        android:layout_height="wrap_content"
        android:layout_x="80dip"
        android:layout_y="10dip"
        android:width="700px" />
    <!-- 定义一个文本框,使用绝对布局 -->
    <TextView
        android:layout_width="wrap_content"
        android:layout_height="wrap_content"
        android:layout_x="20dip"
        android:layout_y="80dip"
        android:text="密    码:" />
    <!-- 定义一个文本编辑框,使用绝对布局 -->
    <EditText
        android:layout_width="wrap_content"
        android:layout_height="wrap_content"
        android:layout_x="80dip"
        android:layout_y="70dip"
        android:password="true"
        android:width="700px" />
    <!-- 定义一个按钮,使用绝对布局 -->
    <Button
        android:layout_width="wrap_content"
        android:layout_height="wrap_content"
        android:layout_x="130dip"
        android:layout_y="135dip"
        android:text="登    录" />
</AbsoluteLayout>
```

运行该项目,其效果如图3-4所示。

图3-4　AbsoluteLayout绝对布局示例效果图

3.5 表格布局TableLayout

表格布局TableLayout是网页程序设计人员所熟悉的一种UI布局形式,其优点是结构位置简单、布局简单、容易上手。表格布局TableLayout采用行列的形式来管理UI组件,TableLayout并不需要明确地声明需要多少行、多少列,而是通过添加TableRow并在其中添加其他组件来控制表格的行数和列数。当然,由于表格的嵌套使用可能会造成代码的复杂度更高。

每次向TableLayout中添加一个TableRow,该TableRow就是一个表格行,TableRow也是容器,因此也可以向它添加其他组件,每添加一个组件该表格就增加一列。如果直接向TableLayout中添加组件,那么此组件将直接占用一行。

在表格布局中,列的宽度由该列中最宽的那个单元格决定,整个表格布局的宽度则取决于父容器的宽度(默认总是占满父容器本身)。

TableLayout常见的一些重要属性,可以分为两组:一组是TableLayout布局容器本身所使用的属性,称为全局属性,如表3-4所示;另一组是TableLayout布局容器中包含的控件所使用的属性,称为局部属性,如表3-5所示。

表3-4 TableLayout的全局属性

属性	说明	备注
android:collapseColumns	隐藏索引列,可隐藏指定的列	属性值为列号,从0开始编号,可同时赋值多列,例如:android:collapseColumns="1,2,6"
android:shrinkColumns	收缩索引列,可收缩因内容太多而过宽的列,使其不会被挤出屏幕	属性值为列号,从0开始编号,可同时赋值多列,例如:android:shrinkColumns="1,2,6",可以通过"*"代替收缩所有列
android:stretchColumns	拉伸索引列,可通过拉伸指定列来填满剩下的多余空白空间	属性值为列号,从0开始编号,可同时赋值多列,例如:android:stretchColumns="1,2,6",可以通过"*"代替拉伸所有列

注:一列可以同时收缩和拉伸。

表3-5 TableLayout的局部属性

属性	说明	备注
android:layout_column	表示该控件显示在第几列	属性值为列号,从0开始编号
android:layout_span	表示该控件占据几列	属性值为大于0小于总列数的整数

☞【实例3-4】演示使用表格布局实现一个简单的人员信息表。

在工程"Chapter03"下创建一个Module,命名为"TableLayout_Demo",打开其目录下src/main/res/layout/activity_main.xml文件,修改代码并输入一些代码,代码如下所示。

```
<?xml version="1.0" encoding="utf-8"?>
<TableLayout xmlns:android="http://schemas.android.com/apk/res/
android"
    xmlns:tools="http://schemas.android.com/tools"
    android:id="@+id/activity_main
```

```xml
        android:layout_width="match_parent"
        android:layout_height="match_parent"
        android:stretchColumns="0,1,2"
        tools:context="cn.androidstudy.tablelayout_demo.MainActivity">
    <TableRow
        android:layout_width="wrap_content"
        android:layout_height="wrap_content" >
        <TextView
            android:gravity="center"
            android:padding="3dip"
            android:text="姓名" />
        <TextView
            android:gravity="center"
            android:padding="3dip"
            android:text="性别" />
        <TextView
            android:gravity="center"
            android:padding="3dip"
            android:text="电话" />
    </TableRow>
    <TableRow
        android:layout_width="wrap_content"
        android:layout_height="wrap_content" >
        <TextView
            android:gravity="center"
            android:padding="3dip"
            android:text="杨过" />
        <TextView
            android:gravity="center"
            android:padding="3dip"
            android:text="男" />
        <TextView
            android:gravity="center"
            android:padding="3dip"
            android:text="1234567" />
    </TableRow>
    <TableRow
        android:layout_width="wrap_content"
        android:layout_height="wrap_content" >
        <TextView
            android:gravity="center"
```

```
            android:padding="3dip"
            android:text="小龙女" />
    <TextView
            android:gravity="center"
            android:padding="3dip"
            android:text="女" />
    <TextView
            android:gravity="center"
            android:padding="3dip"
            android:text="7654321"/>
    </TableRow>
</TableLayout>
```

- 表格布局的风格跟HTML中的表格比较接近，只是所采用的标签不同。
- <TableLayout>是顶级元素，说明采用的是表格布局。
- <TableRow>定义一行。
- <TextView>定义一个单元格的内容。
- android:stretchColumns="0,1,2"指定每行都由"0、1、2"列占满空白空间。
- gravity指定文字对齐方式，本例都设为居中对齐。
- padding指定视图与视图间的内容空隙，单位为像素。

运行该项目，结果如图3-5所示。

图3-5　TableLayout表格布局示例效果图

3.6 约束布局ConstraintLayout

约束布局ConstraintLayout是Android Studio 2.2版本带给用户的新增功能之一，也是给编程人员带来的一个惊喜和另一种布局思路。简而言之，约束布局ConstraintLayout是相对布局的升级版，不过比相对布局更加强调约束。这里的约束，是控件之间的关系。

约束布局ConstraintLayout有两大优点：一是非常适合使用可视化的方式来编写界

面，可视化操作的背后还是使用XML代码来实现的，只不过这些代码是由Android Studio根据我们的操作自动生成的；另一方面，它还可以有效地解决布局嵌套过多的问题，ConstraintLayout通过使用约束的方式来指定各个控件的位置和关系，这样可以避免在复杂约束布局中使用多层的嵌套，而嵌套越多，程序的性能也就越差。

☞【实例3-5】实例演示约束布局的简单使用。

在工程Chapter03下创建一个Module，命名为"ConstraintLayout_Demo"，打开其目录下src/main/res/layout/activity_main.xml文件，查看布局标签名，如果是ConstraintLayout则可直接修改编写布局；如果不是，则先在预览界面中找到Component Tree/activity_main，单击右键，选择Convert RelativeLayout to ConstraintLayout，如图3-6所示，确认后再修改布局。

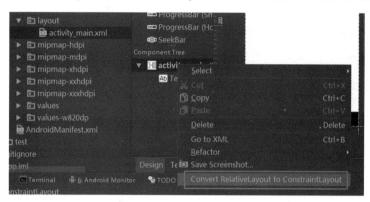

图3-6　RelativeLayout转ConstraintLayout示例图

在ConstraintLayout主操作区域内有两个类似于手机屏幕的界面，如图3-7所示，左边是设计预览界面，右边是蓝图界面。两者都可以用于进行布局编辑工作，区别是左边部分主要用于预览最终的界面效果，右边部分主要用于观察界面内各个控件的约束情况。

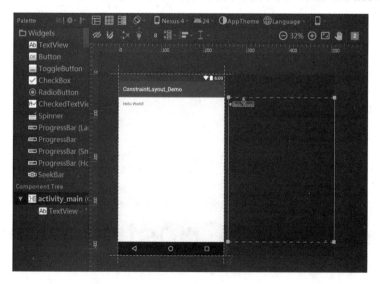

图3-7　ConstraintLayout主操作区域

下面通过在主操作区域中拖动添加控件，并给控件添加约束，实现一个简单的登录页面。

先选中初始布局中的TextView，使用Delete将其删除，然后给界面中拖动添加一个Button按钮，并给其添加约束。每个控件的约束都分为垂直和水平两类，一共可以在四个方向上给控件添加约束,在控件的上下左右各有一个圆圈,这圆圈就是用来添加约束的,我们可以将约束添加到ConstraintLayout,也可以将约束添加到另一个控件。例如,要让添加的Button按钮居中显示,只需要依次拖动按钮控件上下左右四个圆圈分别指向布局界面的上下左右边界即可,如图3-8所示。

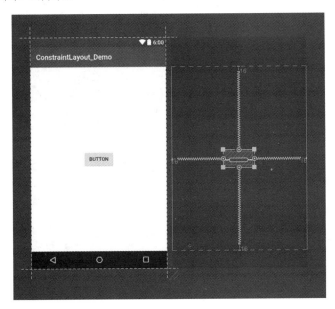

图3-8 ConstraintLayout添加约束使控件居中

还可以将约束添加到另外一个控件。例如,再拖动添加一个Button按钮,通过添加约束将其放置到第一个按钮的右侧,水平间距为96dp,如图3-9所示。

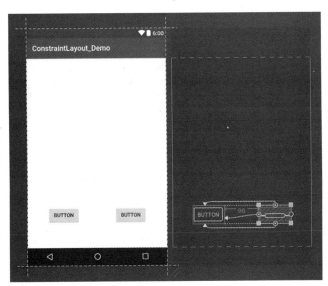

图3-9 ConstraintLayout添加约束使控件与控件按一定间距水平排布

有时候,因布局设计调整需要将添加的约束删除掉。要删除约束有三种方式:第一种用于删除一个单独的约束,将鼠标悬浮在某个约束的圆圈上,然后该圆圈会变成红色,这个时候单击一下就能删除该约束;第二种用于删除某一个控件的所有约束,选中一个控件,然后它的左下角会出现一个删除约束的图标，点击该图标就能删除当前控件的所有约束；第三种用于删除当前界面中的所有约束，点击工具栏中的删除约束图标即可，如图3-10所示。

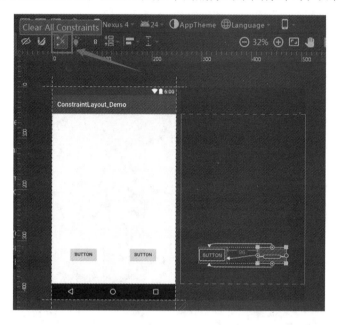

图3-10 ConstraintLayout删除界面中的所有约束

如果要使添加的两个控件同时水平或者垂直居中，可以使用Android Studio提供的Guidelines功能。首先点击通知栏中的Guidelines图标，可以添加一个垂直或水平方向上的Guideline，这里我们需要的是垂直方向上的。而Guideline默认使用dp尺，我们需要选中Guideline，并点击一下最上面的箭头图标将它改成百分比尺，然后将垂直方向上的Guideline调整到50%的位置，这样准备工作就做好了。接下来以该居中线为依据，给控件添加约束即可，如图3-11所示。

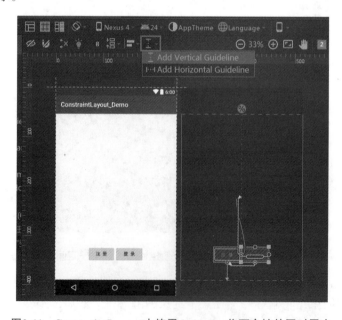

图3-11 ConstraintLayout中使用Guideline将两个控件同时居中

如果界面中的内容变得复杂起来，给每个控件一个个地添加约束也是一件很烦琐的事情。为此，ConstraintLayout中支持自动添加约束的功能，可以很大程度上简化那些烦琐的操作。自动添加约束的方式主要有两种：一种叫Autoconnect；另一种叫Inference。默认情况下Autoconnect是不启用的，想要使用Autoconnect，首先需要在工具栏中将这个功能启用，如图3-12所示。

图3-12　ConstraintLayout中启用Autoconnect的按钮

Autoconnect可以根据我们拖放控件的状态自动判断应该如何添加约束。比如我们将Button放到界面的正中央，那么它的上下左右都会自动地添加上约束。Autoconnect会判断用户的意图，并自动给控件添加约束，不过Autoconnect是无法保证百分百准确判断出用户的意图的，如果自动添加的约束并不是用户想要的话，还可以在任意时刻进行手动修改。可以把它当成一个辅助工具，但不能完全靠它去添加控件的约束。

Inference也是用于自动添加约束的，但它比Autoconnect的功能更为强大，因为Autoconnect只能给当前操作的控件自动添加约束，而Inference会给当前界面中的所有元素自动添加约束。因而Inference适合用于复杂程度比较高的界面，它可以一键自动生成所有的约束。

例如，在一个页面中添加2个Button、2个TextView和2个EditText，先将各个控件按照设计摆放到界面中相应的位置，这个位置不是精确的，然后点击一下工具栏中的Inter Constraints按钮，就能给界面汇总的所有的控件自动添加约束。预览效果如图3-13所示。

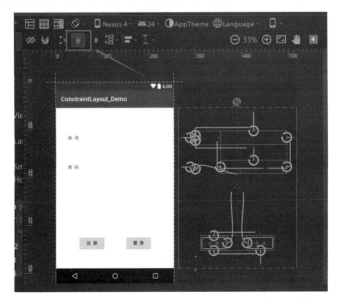

图3-13　ConstraintLayout中使用Inference自动添加约束

运行该案例，运行结果如图3-14所示。

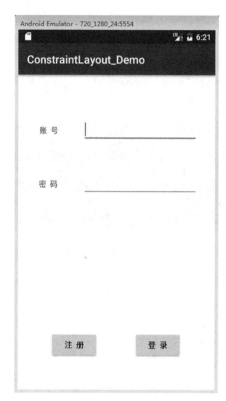

图3-14　ConstraintLayout案例运行结果

强化训练

为了给用户带来交互良好、操作简便、具有个性化的Android应用程序界面，程序设计人员必须对界面中的组件进行合理的设计。

本章主要介绍了Android常用的线性布局（LinearLayout）、相对布局（RelativeLayout）、绝对布局（AbsoluteLayout）、表格布局（TableLayout）、约束布局（ConstraintLayout），并分别通过实例展示了各种布局的使用方法。

通过本章的学习，读者对于Android的界面设计将有一定的了解。学完本章知识后，不妨通过以下习题来巩固所学的内容吧。

一、填空题

1. 一个Android应用的用户界面是由_____和_____对象构建的。
2. Android开发中一般分为以下几种布局：_____、_____、_____、_____和_____。
3. Android中定义的长度单位有以下几种：_____，_____，_____，_____，_____，_____，_____。
4. 相对布局RelativeLayout容器内子组件的位置总是相对_____、_____来决定的，因此这种布局方式被称为相对布局。
5. RelativeLayout常见的一些重要属性，可以分为两组；一组是

_____；另一组是_____。

6. 使用绝对布局时，每个子组件都可指定如下两个 XML 属性：_____，_____。

二、选择题

1. RelativeLayout 中设置控件与给定控件关系的属性中，将该控件的底部边缘与给定 ID 控件的底部边缘对齐的属性是（   ）。

   A. Android:layout_alignBaseline

   B. android:layout_toRightOf

   C. android:layout_alignBottom

   D. android:layout_toLeftOf

2. 以下为 TableLayout 的局部属性的是（   ）。

   A. android:collapseColumns

   B. android:layout_column

   C. android:stretchColumns

   D. android:shrinkColumns

3. Android 不提供任何布局控制，而是开发人员自己通过 X、Y 两个坐标来控制组件的位置。当使用 AbsoluteLayout 作为布局容器时，布局容器不再管理组件的位置、大小，需要开发人员自己控制的布局是（   ）。

   A. TableLayout

   B. ConstraintLayout

   C. RelativeLayout

   D. AbsoluteLayout

4. （   ）为逻辑长度单位，在每英寸 160 点的屏幕上，1dp=1px=1/160 英寸。随着密度变化，对应的像素数量也变化，但并没有直接的变化比例。

   A. dp

   B. px

   C. in

   D. sp

5. （   ）有两大优点：一方面是非常适合使用可视化的方式来编写界面，可视化操作的背后仍然还是使用的 XML 代码来实现；另一方面，它还可以有效地解决布局嵌套过多的问题，（   ）通过使用约束的方式来指定各个控件的位置和关系，这样可以避免在复杂约束布局中使用多层的嵌套，而嵌套越多，程序的性能也就越差。

   A. TableLayout

   B. ConstraintLayout

   C. RelativeLayout

   D. AbsoluteLayout

三、操作题

1. 使用 Android Studio 创建项目进行线性布局练习，构建如图 3-2 的效果图。

2. 使用 Android Studio 创建项目进行相对布局练习，构建如图 3-3 的效果图。

3. 使用 Android Studio 创建项目，利用线性布局、相对布局和约束布局分别构建如图 3-14 所示的登录界面。通过构建同一个用户界面，加深对每种布局的特点的理解；根据布局文件的复杂程度，分析这些布局的优缺点；这样在实际开发中，就可以根据需要选择合适的布局方式。

第4章

常见资源和控件精讲

内容导读

在Android开发中，编程人员经常会遇到资源和控件的使用。在Android Studio IDE中进行Android开发可以使用各式各样的资源和控件，不过Android开发的初学者刚开始可能会为这么多的资源和控件的使用所困扰。为了使读者能快速熟悉资源和控件的使用方法，本章将从最常见的资源和控件开始，带领读者学习Android中资源和基本控件的使用。

学习目标

- 熟悉Android中常见资源；
- 掌握Android基本控件的使用；
- 掌握使用Toast的显示提示信息。

4.1 常见资源

Android中常见的资源有字符串资源（string）、颜色资源（color）、尺寸资源（dimension）、形状（shape）、按钮背景及点击状态（selector）等；常用的基本控件有TextView（文本框）、EditText（编辑框）、ImageView（图片控件）、Button（按钮）、ImageButton（图片按钮）、RadioButton（单选框）、ChekBox（复选框）、Toast（提示）等。

编程人员开发一个Android应用程序，除了包含实现各种应用功能的代码外，还有与代码相独立的各种资源，例如常用的字符串、颜色、尺寸、形状、按钮背景及状态等资源，以及图像、音频、动画、布局、菜单等。在Android开发中使用资源有两大好处：首先，使用资源可以有效地支持多国语言，有利于国际化，如果是在代码中直接定义，当手机更换语言的时候，直接定义的文字是不会自动变化的，而使用定义资源的方式可以根据手机系统的语言来自动变更语言；其次，使用资源还有助于适应应用程序在不同尺寸屏幕的移动设备上使用。本节主要简单介绍常见的字符串、颜色等资源的定义和使用，其他资源会在后续章节中使用到相应资源时加以介绍。

Android Studio中基本延续了Eclipse中res资源目录结构，有一点主要的区别是在保留drawable资源文件夹的同时，增添了一组mipmap文件夹，包括mipmap-hdpi、mipmap-mdpi、mipmap-×hdpi、mipmap-××hdpi、mipmap-×××hdpi，该组文件夹的使用与Eclipse中drawable系列文件夹类似，但是功能更强大。把图片放到mipmap，可以提高系统渲染图片的速度，提高图片质量，减少GPU压力。另外，使用mipmap系统会在图片缩放上提供一定的性能优化，性能更好，占用内存更少，系统会根据屏幕分辨率来选择mipmap-hdpi、mipmap-mdpi、mipmap-×mdpi、mipmap-××hdpi、mipmap-×××hdpi下的对应图片。在drawable文件夹中也可以放置图片资源，还可以放一些自定义形状和按钮背景及点击状态之类的XML资源文件。

在Android工程文件中专门有个res目录用于存放资源，该目录下的资源可以进行可视化的编辑，编写好的资源通过aapt（Android AssetPackaging Tool）工具自动生成gen目录下的R.java资源索引文件，之后在Java代码和XML资源文件中就可以利用索引来调用资源了。

4.1.1 字符串资源

【实例4-1】在res/values目录下包含一个strings.xml文件（如果没有可以自定义，文件名随意，但是目录是固定的），该文件中主要定义的就是应用程序需要用到的字符串资源。

字符串资源的定义：

```
<resources>
    <string name="app_name">Chapter04</string>
</resources>
```

其中，"app_name"为该字符串资源的名称，Chapter04是该字符串资源对应的内容值。使用该字符串资源的调用方法有两种。

- 在XML资源文件中引用：android:text="@string/app_name"
- 在Java代码中引用：R.string.app_name;

假设有个资源文件为res/values/strings.xml，其内容如下：

```
<?xml version="1.0"encoding="utf-8"?>
<resources>
  <string name="hello">Hello Android Studio!</string>
</resources>
```

那么这个hello字符串资源在其他XML资源文件中的调用如下所示：

```
<TextView
 android:layout_width="wrap_content"
 android:layout_height="wrap_content"
```

```
android:text="@string/hello"/>
```

在Java代码中的调用如下：

```
String str_hello = getString(R.string.hello);
```

4.1.2 颜色资源

在res/values目录下包含一个colors.xml文件（如果没有可以自定义，文件名随意，但是目录是固定的），该文件中主要定义的就是应用程序需要用到的颜色资源。在该资源文件中定义的颜色资源一般使用RGB值和alpha通道指定颜色值。可以在应用程序中任何接受十六进制颜色值的地方使用color资源；也可以在XML里用到drawable资源时使用color 资源（例如：android:drawable="@color/red"）。

颜色值总是以（#）字符开头，后面跟着Alpha-红-绿-蓝信息，格式如下之一：
- #RGB
- #ARGB
- #RRGGBB
- #AARRGGBB

例如，#f00是#RGB例子中省略了alpha值的不透明位红色，而#90ff0000是#AARRGGBB例子中透明度（alpha值，16进制）为90的红色。

颜色资源的定义：

```
<resources>
    <color name="red">#f00</color>
</resources>
```

color是简单类型资源，是用名称（name）属性（而非XML文件名）来直接引用的。颜色资源的引用同样有两种方式。
- 在XML资源文件中引用：android:textColor="@color/red"
- 在Java代码中引用：int color = res.getColor(R.color.red);

假设有个资源文件为res/values/colors.xml，其内容如下：

```
<?xml version="1.0"encoding="utf-8"?>
<resources>
    <color name="red">#f00</color>
    <color name="green">#00ff00</color>
    <color name="blue">#860000ff</color>
</resources>
```

那么这些颜色资源在其他XML资源文件中的调用如下所示：

```
<TextView
    android:layout_width="wrap_content"
    android:layout_height="wrap_content"
    android:text="@string/hello
```

```
android:textColor="@color/red"/>
```

在Java代码中的调用如下：

```
tv= (TextView) findViewById(R.id.tv_hello);
Resources res = getResources();
int color = res.getColor(R.color.red);
tv.setTextColor(color);
```

或者使用如下代码也可以实现同样效果：

```
tv.setTextColor(getResources().getColor(R.color.red));
```

4.1.3 尺寸资源

在res/values目录下包含一个dimens.xml文件（如果没有可以自定义，文件名随意，但是目录是固定的），该资源定义跟屏幕显示相关的一些尺寸常量。

常见的具体度量单位如下：
- px（像素）：屏幕实际的像素，常说的分辨率1024×768pixels就是横向1024px，纵向768px，不同设备显示效果相同。
- dp/dip：与密度无关的像素，一种基于屏幕密度的抽象单位。在每英寸160点的显示器上，1dp =1px。但dp和px的比例会随着屏幕密度的变化而改变，不同设备有不同的显示效果。
- sp：与刻度无关的像素，主要用于字体显示 "best for textsize"，作为和文字大小相关的单位。
- in（英寸）：屏幕的物理尺寸，1in等于2.54cm。
- mm（毫米）：屏幕的物理尺寸。
- pt（点）：屏幕的物理尺寸，1/72in。

尺寸资源的定义：

```
<resources>
    <dimen name="horizontal_margin">16dp</dimen>
</resources>
```

尺寸资源的引用同样有两种方式：
- 在XML资源文件中引用：android:paddingTop="@dimen/vertical_margin"
- 在Java代码中引用：float paddingTopSize = res.getDimension(R.dimen.vertical_margin)

资源文件为res/values/dimens.xml，其内容如下：

```
<?xml version="1.0" encoding="utf-8"?>
<resources>
    <!-- Default screen margins, per the Android Design guidelines. -->
    <dimen name="horizontal_margin">16dp</dimen>
    <dimen name="vertical_margin">26dp</dimen>
```

```
</resources>
```

那么这些颜色资源在其他XML资源文件中的调用如下所示：

```xml
<RelativeLayout xmlns:android="http://schemas.android.com/apk/res/android"
    xmlns:tools="http://schemas.android.com/tools"
    android:id="@+id/activity_main"
    android:layout_width="match_parent"
    android:layout_height="match_parent"
    android:paddingBottom="@dimen/vertical_margin"
    android:paddingLeft="@dimen/horizontal_margin"
    android:paddingRight="@dimen/horizontal_margin"
    android:paddingTop="@dimen/vertical_margin"
    tools:context="cn.androidstudy.chapter04.MainActivity">
</RelativeLayout>
```

在Java代码中的调用如下：

```java
iv_background= (ImageView) findViewById(R.id.iv_background);
Resources res = getResources();
int paddingTopSize = (int) res.getDimension(R.dimen.vertical_margin);
int paddingBottomSize = (int) res.getDimension(R.dimen.vertical_margin);
int paddingLeftSize = (int) res.getDimension(R.dimen.horizontal_margin);
int paddingRightSize = (int) res.getDimension(R.dimen.horizontal_margin);
iv_background.setPadding(paddingLeftSize,paddingTopSize,
                    paddingRightSize,paddingBottomSize);
```

4.1.4 形状

在res/drawable目录下可以创建shape类型的XML资源文件，用来定义应用程序中使用到的一些形状，基本的形状类型有水平直线（line）、椭圆形（oval）、矩形（rectangle）、环形（ring）等。shape常用的属性及其作用如下：

• shape：形状，使用指定定义的具体形状，有水平直线（line）、椭圆形（oval）、矩形（rectangle）、环形（ring），不指定的话默认值为矩形。

```xml
<shape xmlns:android="http://schemas.android.com/apk/res/android"
    android:shape="rectangle">
</shape>
```

• solid：实心，就是填充的意思，使用android:color指定形状填充的颜色。

• gradient：渐变，使用android:startColor和android:endColor分别为形状设置渐变的起始和结束颜色，android:angle用来指定渐变角度，其值必须为45°的整数倍。另外，渐变默认的模式为android:type="linear"，即线性渐变。可以指定渐变为径向渐变android:type="radial"，径向渐变需要指定半径，例如android:gradientRadius="50"。

• stroke：描边，使用android:width="2dp"设置描边的宽度，使用android:color设置描边

的颜色。还可以把描边弄成虚线的形式，设置方式为：

```
android:dashWidth="5dp"
android:dashGap="3dp"
```

其中，android:dashWidth 表示'-'这样一个横线的宽度，android:dashGap 表示之间隔开的距离。

- corners：圆角，使用 android:radius 为形状设置圆角的弧度，值越大角越圆。

我们还可以把四个角设定成不同的角度，方法为：

```
<corners
     android:topRightRadius="20dp"        //右上角
     android:bottomLeftRadius="20dp"      //右下角
     android:topLeftRadius="1dp"          //左上角
     android:bottomRightRadius="0dp"      //左下角
/>
```

- padding：间隔，用来设置形状里的内容文字至形状边界的距离。

```
<padding
  android:left="10dp"
  android:top="10dp"
  android:right="10dp"
  android:bottom="10dp" />
```

下面来学习 shape 的定义和使用。新建一个 shape 类型的 XML 文件，命名为 shape_demo，设置形状为矩形，并为其设置圆角、填充颜色、边框颜色及宽度等属性值，代码如下所示：

```
<?xml version="1.0" encoding="utf-8"?>
<shape xmlns:android="http://schemas.android.com/apk/res/android"
   android:shape="rectangle">
   <!-- 矩形的圆角弧度 -->
   <corners android:radius="10dp" />
   <!-- 矩形的填充色 -->
   <solid android:color="@color/blue" />
   <!-- 矩形边框的宽度和颜色 -->
   <stroke
       android:width="2dp"
       android:color="@color/red" />
</shape>
```

在其他 XML 文件中引用自定义形状 shape 的方法如下所示：

```
<Button
    android:layout_width="wrap_content"
```

```
    android:layout_height="wrap_content"
    android:id="@+id/btn"
    android:text="形状示例按钮"
    android:background="@drawable/shape_demo"/>
```

4.1.5 按钮背景及点击状态

在Android应用程序中常会将按钮的状态分为被点击状态和未被点击状态。为了将按钮的这两个状态可视化地区分开来，可以使用自定义selector。同shape一样定义在res/drawable目录下，定义selector类型的XML文件即可。在使用selector时通常会用item标签分别设置按钮被点击和未被点击时的相关属性值，主要用到的属性如下。

• state_pressed：点击，使用android:state_pressed来设置区分按钮被点击或者未被点击的状态，属性值为true或false。

• drawable：背景，用来设置在某个状态下按钮的背景，可通过引用自定义的shape资源为其赋值。

下面来学习selector的定义和使用。新建一个selector类型的XML文件，命名为selector_btn_text，并新建两个shape类型的XML文件shape_btn_bg_pressed和shape_btn_bg_normal，分别为按钮设置被点击和未被点击状态下的背景。selector_btn_text.xml文件的代码如下所示：

```
<?xml version="1.0"encoding="utf-8"?>
<selector xmlns:android="http://schemas.android.com/apk/res/android">
    <!-- 指定按钮被点击时的背景 -->
    <item android:state_pressed="true"
        android:drawable="@drawable/shape_btn_bg_pressed"/>
    <!-- 指定按钮未被点击时的背景 -->
    <item android:state_pressed="false"
        android:drawable="@drawable/shape_btn_bg_normal"/>
</selector>
```

4.2 TextView和EditText控件

文字处理在Android编程中会经常遇到，而Android编程人员经常会用TextView控件和EditText控件来实现文字处理功能。Android 所有的控件和布局都是类，并且都有着一个父类View。TextView、ImageView、ViewGroup均继承自View，TextView有显示文本的特性，只要带有文字的控件基本都继承自 TextView，常见的有 Button、EditText、CheckBox、RadioButton等。ImageView 带有显示图片的特性，其子类相对较少，常见的有ImageButton。ViewGroup是Android中最基本的容器类，绝大部分布局都继承自ViewGroup，其中的LinearLayout、RelativeLayout、GridLayout等都是比较常见的，同时ViewGroup还有一些其他可以包含控件的子类，如RecyclerView、AdapterView等。常用的ListView、GridView都继承自AdapterView。Android通过继承的方式，对父类控件进行继承，由此在父类基础上创造新的控件。

绝大部分属性都会包含一些基本的属性，例如id、layout_width、layout_height、layout_weight、background、padding、gravity等。在正式学习Android中的基本控件之前，我们先来了解一下这些控件的属性，如表4-1所示。

表4-1 控件的基本属性及对应方法说明

属性名称	对应方法	说明
android:id	setId(CharSequence)	设置控件的id
android:layout_height	setHeight(int)	设置控件的高度，单位为px
android:layout_width	setWidth(int)	设置控件的宽度，单位为px
android:layout_weight		设置布局中控件所占的权重
android:background	setBackground(drawable)	设置控件的背景
android:padding	setPadding(int)	设置控件的内边距，单位为dip
android:gravity	setGravity(int)	设置控件包含内容的对齐方式

TextView是Android中最常用的控件，它直接继承自View父类，同时它也是EditText、Button等UI组件类的父类。它的作用就是在界面上显示文本。从这点上看，它类似于Swing中的JLable，不过它比JLable的功能更加强大。

从功能上看，TextView其实就是一个文本编辑器，只是Android关闭了它的文字编辑功能。如果想要定义一个可编辑的文本框，则可以使用它的子类EditText。

TextView提供了大量的XML属性，这些XML属性大部分既可适用于TextView，也可适用于EditText，但有少量XML只能适用于其中之一。表4-2列出了TextView的常见属性及对应方法说明。

表4-2 TextView的常见属性及对应方法说明

属性名称	对应方法	说明
android:text	setText(CharSequence)	设置文本框内文本的内容
android:textColor	setTextColor(ColorStateList)	设置文本框内文本的颜色
android:textSize	setTextSize(float)	设置文本框内字体的大小
android:autoLink	setAutoLinkMask(int)	设置是否将指定格式的文本转换为可单击的超链接显示
android:hint	setHint(int)	设置文本框内容为空时，默认显示的提示文本
android:typeface	setTypeFace(Typeface)	设置文本框内文本的字体
android:phoneNumber	setKeyListener(KeyListener)	设置文本框只能接受电话号码
android:singleLine	setSingleLine(boolean)	如果设置为true，则控件当中的内容在同一行显示，多余内容省略号表示

表4-2中android:autoLink属性值是如下几个属性值的一个或几个，多个属性值之间用竖线"|"隔开。

- none：不设置任何超链接。
- web（对应于Linkfy.WEB_URLS）：将文本中的URL地址转换为超链接。
- email（对应于Linkfy.EMAL_ADRESSES）：将文本中的E-mail地址转换为超链接。
- phone（对应于Linkfy.PHONE_NUMBERS）：将文本中的电话号码转换为超链接。
- map（对应于Linkfy.MAP_ADDRESSES）：将文本中的街道地址转换为超链接。

- all：相当于指定web|email|phone|map。

Android:ellipsize属性可支持如下几个属性值。
- none:不进行任何处理。
- start:在文本开头部分进行省略。
- middle:在文本中间部分进行省略。
- end:在文本结尾部分进行省略。
- marquee:在文本结尾处以淡出的方式省略。

用户可以通过以下两种方法创建TextView。

① 在程序中创建TextView对象，代码段如下：

```
TextView tv_hello=new TextView (this);
tv_hello.setText("hello");
setContentView(tv_hello);
```

② 在XML布局文件中创建，代码段如下：

```
<TextView
android:id="@+id/tv_hello"
android:layout_width="fill_parent"
android:layout_height="wrap_content"
android:text="你好"
/>
```

EditText是TextView的一个子类，可以将其视为一个可编辑的文本框，主要用来在界面上编辑输入。在表4-1所介绍的属性当中，大部分都是EditText和TextView的共有属性，比较特殊的是android:hint属性，该属性一般只在EditText中使用，用来显示输入框的输入提示信息。另外，EditText不使用android:text属性。

同样也可以通过以下两种方法创建EditText。

① 在程序中创建EditText对象，代码段如下：

```
EditText et_name=new EditText(this);
et_name.setHint("请输入姓名！");
setContentView(et_name);
```

② 在XML布局文件中创建，代码段如下：

```
<EditText
android:id="@+id/et_name"
android:layout_width="fill_parent"
android:layout_height="wrap_content"
android:hint="请输入姓名！"
/>
```

【实例4-2】通过一个简单的案例来学习TextView（文本显示控件）和EditText（文本编辑框控件，即可以在此文本框中输入内容），运用TextView和EditText控件编写应用程序登录的界面，如图4-1所示。

第4章 常见资源和控件精讲

图4-1 登录界面效果图

案例分析

本案例中采用线性布局来实现。在src/main/res/layout/activity_main.xml文件中布局，使用线性布局，添加2个TextView控件和2个EditText控件。

实现步骤

① 创建一个Android工程，工程名为"Chapter04"，在该工程下创建一个Module,命名为"TV_ET_demo"。

② 打开"Chapter04"工程中"TV_ET_demo"项目下的src/main/res/layout/activity_main.xml文件，修改代码并输入一些代码（本案例布局具体代码见配套资源）。

4.3 Button和ImageButton控件

按钮是Android用户最经常使用的控件之一。由于其使用简单、功能强大，Android编程人员经常使用按钮来实现其所需功能。在Android Studio中，Button控件和ImageButton控件是经常使用的两种按钮。Button继承自TextView，ImageButton继承自Button，它们的主要功能就是在界面上生成一个按钮，用户可以单击，从而触发onClick事件。关于事件处理机制在后续章节中会有详细讲解。Button和ImageButton的主要区别在于：前者生成的按钮上显示文字，而后者上面显示图片。

按钮的用法比较简单，可以通过设置相关的属性值为按钮增加背景颜色或图片，但这都是固定的，不会随着用户的动作而改变。如果使用图片按钮，可以通过指定属性值指定图片，但是却不能指定文字。在实际应用中，用户可能会对按钮有更高的要求，例如按钮的形状、颜色、图片和文字等，这就需要对按钮进行更高级的定制。下面通过实例来演示按钮的复杂设计。

☞【实例4-3】为了定义图片随用户动作改变的按钮，可以使用XML资源文件或者图片来定义drawable对象，再将其设为Button的android:background属性值，或者设为ImageButton的android:src属性值。

在工程"Chapter04"下新建一个Module,命名为"Btn_Ibtn_demo"，在其res/drawable下添加图片btn_red和btn_blue，并创建selector_ibtn.xml；在res/drawable下创建selector_btn——text.xml、shape_btn_bg_pressed、shape_btn_bg_normal；修改项目"Btn_Ibtn_demo"的布局文件activity_main.xml的内容。本案例布局具体代码见配套资源。

运行该项目，效果如图4-2所示。

图4-2　按钮和图片按钮示例运行效果图

图4-2所示界面中，上方两个按钮的背景色和图片都是固定的，当用户单击按钮时不会产生任何变化；用户按下下面两个按钮时，能看到按钮的图片切换成红色按钮图片或者是按钮的背景切换成红色蓝边圆角矩形。

ImageView控件

仅仅靠文字已不能满足用户日益增长的需求。图片能带给用户更加直观形象的感受，也会给用户带来愉悦的心情。图片处理也就成了Android编程人员经常解决的事情。ImageView控件是Android中的基础图片显示控件。该控件有两个重要的属性是src和scaleType：src属性用来给ImageView控件赋值，可以通过引用res/mipmap系列文件夹下的图片指定相应图片；scaleType属性用以表示显示图片的方式。如表4-3所示。

表4-3　scaleType的值

scaleType的值	描述
scaleType.CENTER	默认值,图片大小为原始大小，如果图片大小大于ImageView控件，则截取图片中间部分，若小于，则直接将图片居中显示

续表

scaleType的值	描述
scaleType.CENTER_CROP	将图片等比例缩放，让图像的短边与ImageView的边长度相同，即不能留有空白，缩放后截取中间部分进行显示
scaleType.CENTER_INSIDE	将图片大小大于ImageView的图片进行等比例缩小，直到整幅图能够居中显示在ImageView中，小于ImageView的图片不变，直接居中显示
scaleType.FIT_CENTER	ImageView的默认状态，大图等比例缩小，使整幅图能够居中显示在ImageView中，小图等比例放大，同样要整体居中显示在ImageView中
scaleType.FIT_END	缩放方式同FIT_CENTER，只是将图片显示在右方或下方，而不是居中
scaleType.FIT_START	缩放方式同FIT_CENTER，只是将图片显示在左方或上方，而不是居中
scaleType.FIT_XY	将图片非等比例缩放到大小与ImageView相同
scaleType.MATRIX	根据一个3×3的矩阵对其中图片进行缩放

☞【实例4-4】下面通过一个简单的案例来学习ImageView控件以及其属性的使用。使用ImageView设计一个界面，效果如图4-3所示。

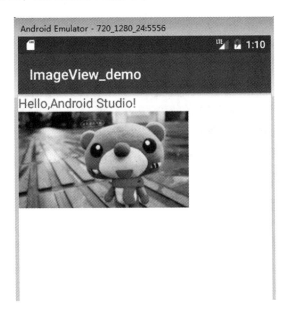

图4-3　ImageView案例效果图

实现步骤

① 在工程"Chapter04"下创建一个Module,命名为"IamgeView_demo"。

② 把图片导入到工程项目资源文件中：将图片拖拽或者复制粘贴到项目res\mipmap开头的5个文件夹下，它们分别代表了高、中、低分辨度的图片。Android读取图片时自动优化，选用合适分辨率的图片显示。

③ 打开src/main/res/layout/activity_main.xml文件，修改代码并输入相关代码（本案例布局具体代码见配套资源）。

4.5 RadioButton和ChekBox控件

选择框是Android程序员在设计UI时会经常用到的控件，通过用户操作选择框可以实现用户与系统之间的信息交流。RadioButton（单选框）和ChekBox（复选框）是所有用户界面中最普通的UI组件。由于Android中这两种选择组件都继承自Button按钮，因此它们可以直接使用Button支持的属性和方法，不过它们也有自己不同的属性和方法。本节将通过实例来讲解和说明。

RadioButton和ChekBox与普通按钮的区别是它们多了一个可选中的功能，它们具有属性android:checked，该属性用于指定组件初始是否被选中。RadioButton与ChekBox的区别在于，一组RadioButton只能选中其中一个，因此，RadioButton通常与RadioGroup一起使用，用于定义一组单选按钮。下面通过实例介绍RadioButton和ChekBox的用法。

☞【实例4-5】在工程"Chapter04"下新建一个Module，命名为"RB_CB_demo"，修改其布局文件activity_main.xml（本案例布局具体代码见配套资源）。

在RadioGroup的属性中有一个比较特殊的属性android:checkedButton，通过指定id设置单选按钮的默认选定项，本案例将id为woman的RadioButton设置为默认选定项。注意，"@"后有"+"，被指定为默认选中的RadionButton的id不用"+"，直接应用前面定义过的id标识即可。

运行项目，效果如图4-4所示。

图4-4　单选框和复选框示例运行效果图

性别选项是使用RadioGroup和RadioButton实现的单选框，只能选中一项，男或者女；爱好选项是使用CheckBox实现的多选框，可以同时选中多项，并且可以取消选中。

4.6 AnalogClock和DigitalClock控件

由于智能手机中附带有强大的时钟功能，用户对手表的需求越来越弱。当然，Android Studio中也提供了时钟UI组件，以供Android程序开发者使用。AnalogClock和DigitalClock是两个非常简单的时钟UI组件，前者继承了View组件，它重写了View的onDraw方法，会在View上显示模拟时钟；后者继承了TextView，即它本身就是文本框，只不过它里面显示的内容是当前时间。这两个组件都会显示当前时间，不同的是，DigitalClock显示数字时钟，可以显示当前的秒数；而AnalogClock显示模拟时钟，不能显示当前秒数。

☞【实例4-6】下面将演示这两种组件的用法。在工程"Chapter04"下新建一个Module，命名为"AC_DC_demo"，修改其布局文件activity_main.xml（本案例布局具体代码见配套资源）。

运行项目，效果如图4-5所示。

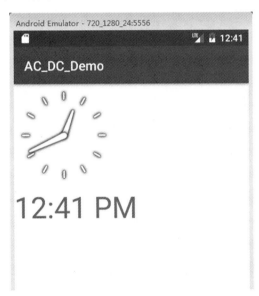

图4-5 数字时钟和模拟时钟示例运行效果图

4.7 Toast控件

在人机交互中，轻量级信息提示框的使用会经常遇到，Android编程平台提供了一种简单易用的消息提示框。当程序有大量消息、图片需要向用户提示时，可以考虑使用前面介绍的对话框，但如果程序只有少量信息要向用户呈现，则可以考虑使用更轻量级的对话框，即Android提供的消息提示。本节主要讲解Toast（吐司）消息提示框的功能及用法。

Toast是一种非常方便的消息提示框，它会在程序界面上显示一条简单的提示信息。这个提示框用于向用户生成简单的提示信息。它具有两个特点：

① Toast提示信息不会获得焦点。
② Toast提示信息过一段时间会自动消失。
使用Toast来生成提示消息非常简单，只要按照如下步骤进行即可。
① 调用Toast的构造器或makeText方法创建一个Toast对象。
② 调用Toast的方法来设置该消息提示的对齐方式、页边距、显示内容等。
③ 调用Toast的show()方法将它显示出来。

Toast的功能和用法都比较简单，大部分时候它只能显示简单的文本提示，如果程序需要显示图片、列表之类的复杂提示，一般用对话框完成比较适合。

下面实例讲解Toast的用法。

☞【实例4-7】在工程"Chapter04"下新建一个Module,命名为"toast_demo"，修改其布局文件activity_main.xml（本案例布局具体代码见配套资源）。

打开src/main/java/cn.androidstudy.toast_demo/MainActivity文件，修改并输入部分代码，如下所示：

```java
package cn.androidstudy.toast_demo;
import android.support.v7.app.AppCompatActivity;
import android.os.Bundle;
import android.view.View;
import android.widget.Button;
import android.widget.Toast;
public class MainActivity extends AppCompatActivity {
    Button btn_eclipse;
    Button btn_studio;
    @Override
    protected void onCreate(Bundle savedInstanceState) {
        super.onCreate(savedInstanceState);
        setContentView(R.layout.activity_main);
        btn_eclipse= (Button) findViewById(R.id.btn_eclipse);
        btn_studio= (Button) findViewById(R.id.btn_studio);
        btn_eclipse.setOnClickListener(new View.OnClickListener()
{
            @Override
            public void onClick(View v) {
                Toast toast=Toast.makeText(MainActivity.this,
"点击了Eclipse！", Toast.LENGTH_SHORT);//定义Toast
                toast.show();//显示Toast
            }    });
        btn_studio.setOnClickListener(new View.OnClickListener() {
            @Override
            public void onClick(View v) {
                Toast toast=Toast.makeText(MainActivity.this,
"点击了Andriod Studio！", Toast.LENGTH_SHORT);
                toast.show();
```

```
        } });
    }}
```

运行该项目，效果如图4-6所示。界面上有两个文本信息不同的按钮，当点击其中任意一个按钮时，会弹出一个Toast，显示的内容为所点击按钮上的文本。

图4-6　Toast示例运行效果图

强化训练

各种资源和控件是Android编程的重点之一，如何利用好Android提供的各种资源和控件是每个Android程序设计人员必须面对的问题之一。

本章以案例的形式讲解了字符串资源（string）、颜色资源（color）、尺寸资源（dimension）、形状（shape）、按钮背景及点击状态（selector）等常见资源和TextView（文本框）、EditText（编辑框）、Button（按钮）、ImageButton（图片按钮）、ImageView（图片控件）、RadioButton（单选框）、CheckBox（多选框）、Toast（提示）等基本控件，介绍了这些资源和控件的常用属性以及使用方法，对于初学者来说有一定帮助。请在开发工具中多调试本章的案例。

学完本章知识后，不妨通过以下习题来巩固所学的内容吧。

一、填空题

1. Android中常见的资源有_____、_____、尺寸资源（dimension）、形状（shape）、

按钮背景及点击状态（selector）等。

2. 在Android编程中常用的基本控件有_____、_____、_____、_____、ImageButton（图片按钮）、RadioButton（单选框）、ChekBox（复选框）、Toast（提示）等。

3. Android Studio项目中，在/res/values目录下包含一个_____文件（如果没有可以自定义，文件名随意，但是目录是固定的），该文件中主要定义的就是应用程序需要用到的颜色资源。在该资源文件中定义的颜色资源一般使用_____值和_____通道指定颜色值。

4. 在Android应用程序中常会将按钮的状态分为_____和_____，为了将按钮的这两个状态可视化地区分开来，可以使用自定义selector。

5. 在Android编程中文字处理控件TextView、ImageView、ViewGroup均继承自View，_____有显示文本的特性，只要带有文字的控件基本都继承自_____。

6. 在Android编程的按钮控件中_____带有显示图片的特性，其子类相对较少，常见的有ImageButton。_____是Android中最基本的容器类，绝大部分布局都继承自_____，其中的LinearLayout、RelativeLayout、GridLayout等都是比较常见的。

二、选择题

1. 大部分控件共有的属性中，(   ) 设置控件的内边距，单位为dip。
   A. android:padding
   B. android:gravity
   C. android:layout_height
   D. android:layout_width

2. TextView的常见属性中，设置文本框内容为空时，默认显示的提示文本是(   )。
   A. android:autoLink
   B. android:typeface
   C. android:hint
   D. android:singleLine

3. shape常用的属性及其作用中，(   ) 是描边，使用android:width="2dp" 设置描边的宽度，使用android:color 设置描边的颜色。
   A. gradient
   B. solid
   C. stroke
   D. corners

4. 在使用selector时通常会用item标签分别设置按钮被点击和未被点击时的相关属性值，主要用到的属性中(   ) 代表背景。
   A. drawable
   B. state_pressed
   C. android:state_pressed
   D. solid android:color

5. ImageButton继承自(   )。
   A. TextView
   B. OnClick
   C. Drawable
   D. Button

6. ImageView 控件是Android中的基础图片显示控件。该控件有两个重要的属性是src

070

和 scaleType（   ）。
   A. CENTER 和 scaleType
   B. CENTER 和 src
   C. src 和 scaleType
   D. CENTER_CROP 和 CENTER

三、操作题

1. 练习利用 Android Studio 开发环境创建 Android 应用程序，练习使用 Android 的常见资源，例如字符串资源、颜色资源、尺寸资源、形状资源等。

2. TextView 和 EditText 都是 Android 中最常用的控件，它们的作用就是在界面上显示文本、编辑文本。练习利用 Android Studio 开发环境创建 Android 应用程序，练习使用 TextView 和 EditText 控件编写应用程序登录的界面。

3. Button 继承自 TextView，ImageButton 继承自 Button，它们的主要功能就是在界面上生成一个按钮，用户可以单击，从而触发 onClick 事件。练习利用 Android Studio 开发环境创建 Android 应用程序，练习各种不同类型的 Button 按钮，并根据按钮的事件处理机制定义按钮图片随用户动作改变。

第5章

事件处理机制与多线程

内容导读

多线程是实现Android程序并发执行的技术基础，可以极大地提高手机或移动设备资源的利用率，可以在有限的资源空间中给用户以更好的体验。Android多线程机制的使用可以有效地加强应用程序处理状态与用户之间的互动，提升用户在有限资源空间中的实际体验效果。Android程序利用的是标准的图形化交互界面，Android程序通过事件处理来对用户的动作做出响应，因此事件处理机制也就成了Android编程的重点之一。为了高效对用户指令做出响应，Android提供了强大的事件处理机制。本章介绍事件处理机制与多线程以及Handler的相关知识。

学习目标

- 了解Android的事件处理机制；
- 熟悉Android的多线程机制；
- 掌握Android多线程的使用。

5.1 事件处理机制

在Android程序中经常要对用户或系统的事件做出响应，这种响应就是事件处理。在Android中提供了强大的事件处理机制，支持两种处理机制：基于监听的事件处理机制和基于回调的事件处理机制。对于Android基于监听的事件处理，事件源和事件监听器是分离的，

第5章 事件处理机制与多线程

当事件源上发生特定事件时，该事件交给事件监听器处理。对于Android基于回调的事件处理，事件源和事件监听器是统一的，当事件源发生特定事件时，该事件还是由事件源本身负责处理。

5.1.1 基于监听接口的事件处理

对于一个Android应用程序来说，事件处理是必不可少的，用户与应用程序之间的交互便是通过事件处理来完成的。关于Android事件处理模型应该注意以下几点。

① 事件源与事件监听器。当用户与应用程序交互时，一定是通过触发某些事件来完成的，让事件来通知程序应该执行哪些操作，在这个繁杂的过程中主要涉及两个对象：事件源与事件监听器。

② 事件源指的是事件所发生的控件，各个控件在不同情况下触发的事件不尽相同，而且产生的事件对象也可能不同。

③ 监听器则是用来处理事件的对象，实现了特定的接口，根据事件的不同重写不同的事件处理方法来处理事件。

④ 将事件源与事件监听器联系到一起，就需要为事件源注册监听器，当事件发生时，系统才会自动通知事件监听器来处理相应的事件。

事件处理的过程一般分为三步。

步骤一：首先为事件源对象添加监听，这样当某个事件被触发时，系统才会知道通知谁来处理该事件。

步骤二：当事件发生时，系统会将事件封装成相应类型的事件对象，并发送给注册到事件源的事件监听器。

步骤三：当监听器对象接收到事件对象之后，系统会调用监听器中相应的事件处理方法来处理事件并给出响应。

主要的监听器接口如下。

（1）onClickListener接口

功能	该接口处理的是点击事件。在触控模式下，是在某个View上按下并抬起的组合动作；在键盘模式下，是某个View获得焦点后点击确定键或者按下轨迹球事件
回调方法	public void onClick(View v) 说明：参数v为事件发生的事件源

（2）onLongClickListener接口

功能	onLongClickListener接口与之前介绍的onClickListener接口原理基本相同，只是该接口为View长按事件的捕捉接口，即当长时间按下某个View时触发的事件
回调方法	public boolean onLongClick(View v) 说明：参数v为事件源控件，当长时间按下此控件时才会触发该方法。 返回值：该方法的返回值为一个boolean类型的变量。当返回true时，表示已经完整地处理了这个事件，并不希望其他的回调方法再次进行处理；当返回false时，表示并没有完全处理完该事件，希望其他方法继续对其进行处理

（3）onFocusChangeListener接口

功能	onFocusChangeListener接口用来处理控件焦点发生改变的事件。如果注册了该接口，当某个控件失去焦点或者获得焦点时都会触发该接口中的回调方法

续表

回调方法	public void onFocusChange(View v, Boolean hasFocus) 说明：参数v为触发该事件的事件源； 　　　参数hasFocus表示v的新状态，即v是否获得焦点

（4）onKeyListener接口

功能	onKeyListener是对手机键盘进行监听的接口，通过对某个View注册该监听，当View获得焦点并有键盘事件时，便会触发该接口中的回调方法
回调方法	public boolean onKey(View v, int keyCode, KeyEvent event) 说明：参数v为事件源控件； 　　　参数keyCode为手机键盘的键盘码； 　　　参数event为键盘事件封装类的对象，其中包含了事件的详细信息，例如发生的事件、事件的类型等

（5）onTouchListener接口

功能	onTouchListener接口是用来处理手机屏幕事件的监听接口，当在View的范围内触摸、按下、抬起或滑动等动作时都会触发该事件
回调方法	public boolean onTouch(View v, MotionEvent event) 说明：参数v为事件源对象； 　　　参数event为事件封装类的对象，其中封装了触发事件的详细信息，同样包括事件的类型、触发时间等信息

（6）onCreateContextMenuListener接口

功能	onCreateContextMenuListener接口是用来处理上下文菜单显示事件的监听接口。该方法是定义和注册上下文菜单的另一种方式
回调方法	public void onCreateContextMenu(ContextMenu menu, View v, ContextMenuInfo info) 说明：参数menu为事件的上下文菜单； 　　　参数v为事件源View，当该View获得焦点时才可能接收该方法的事件响应； 　　　参数info：info对象中封装了有关上下文菜单额外的信息，这些信息取决于事件源View

【实例5-1】下面以onClickListener接口为例，介绍按钮控件基于监听接口的事件处理的方法。创建一个Android工程，工程名为"Chapter05"，在该工程下创建一个Module，命名为"onClickListener_Demo"。打开src/main/res/layout/activity_main.xml文件，修改代码并输入相关代码（本案例布局具体代码见配套资源）。

扫一扫　看视频

打开src/main/java/cn.androidstudy.onclicklistener_demo路径下的MainActivity.java文件，修改并输入部分代码，如下所示：

```
package cn.androidstudy.onclicklistener_demo;
import android.support.v7.app.AppCompatActivity;
import android.os.Bundle;
import android.view.View;
import android.widget.Button;
import android.widget.Toast;
public class MainActivity extends AppCompatActivity {
    Button btn_Onclick;
    @Override
    protected void onCreate(Bundle savedInstanceState) {
        super.onCreate(savedInstanceState);
```

```
        setContentView(R.layout.activity_main);
        btn_Onclick= (Button) findViewById(R.id.btn);
        btn_Onclick.setOnClickListener(new View.OnClickListener() {
            @Override
            public void onClick(View v) {
                //定义Toast（吐司）
                Toast toast=Toast.makeText(MainActivity.this,"点击了监听示例按钮！", Toast.LENGTH_SHORT);
                toast.show();//显示Toast
            }
        });
    }}
```

运行项目，效果如图5-1所示。当点击界面中的"点击监听示例"按钮时，会执行指定操作，弹出一个Toast提示信息，告诉用户点击了监听示例按钮。

图5-1 点击监听示例运行效果图

【实例5-2】下面将以onCheckedChangeListener接口为例，介绍单选、多选按钮选择监听的事件处理的方法。在工程"Chapter05"下创建一个Module,命名为"OnCheckChanged_Demo"。打开src/main/res/layout/activity_main.xml文件，修改并输入相关代码（本案例布局具体代码见配套资源）。

打开src/main/java/cn.androidstudy.oncheckchanged_demo路径下的MainActivity.java文件，修改并输入部分代码，如下所示：

```java
package cn.androidstudy.oncheckchanged_demo;
import android.support.v7.app.AppCompatActivity;
import android.os.Bundle;
import android.view.View;
import android.widget.Button;
import android.widget.CheckBox;
import android.widget.CompoundButton;
import android.widget.RadioGroup;
import android.widget.Toast;
public class MainActivity extends AppCompatActivity {
    //声明RadioGroup和CheckBox对象
    private RadioGroup RG_sex;
    private CheckBox CB_sport;
    private CheckBox CB_shopping;
    private CheckBox CB_game;
    private CheckBox CB_movie;
    private Button btn_CB;
    private String multi_list[]={"运动","购物","玩游戏","看电影"};
    @Override
    protected void onCreate(Bundle savedInstanceState) {
        super.onCreate(savedInstanceState);
        setContentView(R.layout.activity_main);
        RG_sex= (RadioGroup) findViewById(R.id.rg_sex);
        RG_sex.setOnCheckedChangeListener(new RadioGroup.OnCheckedChangeListener() {
            @Override
            public void onCheckedChanged(RadioGroup group, int checkedId) {
                if(checkedId==R.id.rb_woman){
                    Toast.makeText(MainActivity.this, "你是女生！", Toast.LENGTH_LONG).show();      }
                else if(checkedId==R.id.rb_man){
                    Toast.makeText(MainActivity.this, "你是男生！", Toast.LENGTH_LONG).show();           }     });
        CB_sport= (CheckBox) findViewById(R.id.cb_sport);
        CB_shopping= (CheckBox) findViewById(R.id.cb_shopping);
        CB_game= (CheckBox) findViewById(R.id.cb_game);
        CB_movie= (CheckBox) findViewById(R.id.cb_movie);
        //设置默认情况下，每个选项是否被勾选的状态
        final boolean b[]=new boolean[]{false,false,false,false};
        CB_sport.setOnCheckedChangeListener(new CompoundButton.OnCheckedChangeListener() {
            @Override
            public void onCheckedChanged(CompoundButton buttonView,
```

第5章 事件处理机制与多线程

```
boolean isChecked)
            {         b[0]-isChecked;           } });
         CB_shopping.setOnCheckedChangeListener(new CompoundButton.
OnCheckedChangeListener() {
             @Override
         public void onCheckedChanged(CompoundButton buttonView,
boolean isChecked)
            {         b[1]=isChecked;           } });
          CB_game.setOnCheckedChangeListener(new CompoundButton.
OnCheckedChangeListener() {
             @Override
         public void onCheckedChanged(CompoundButton buttonView,
boolean isChecked)
            {         b[2]=isChecked;      } });
          CB_movie.setOnCheckedChangeListener(new CompoundButton.
OnCheckedChangeListener() {
             @Override
         public void onCheckedChanged(CompoundButton buttonView,
boolean isChecked)
            {         b[3]=isChecked;           } });
        btn_CB= (Button) findViewById(R.id.btn_CheckBox);
        btn_CB.setOnClickListener(new View.OnClickListener() {
             @Override
            public void onClick(View v) {
                String itemChecked="";
                //取出被选中的内容
                for (int i = 0; i < b.length; i++) {
                    if (b[i]) {
                        itemChecked+=multi_list[i]+",";       } }
                   Toast.makeText(MainActivity.this,"你选择的爱好是：
"+itemChecked, Toast.LENGTH_LONG).show();
                } });    }}
```

运行项目，当点击界面中的性别单选按钮时，会弹出相应的Toast提示所选择的项的内容，如图5-2所示。对于多项选择来说，每一项都是既可以选中也可以取消，一般都需要获取最终选择的一个或多个选项内容，因此在页面最终添加了"提交多选"按钮，当选择好多选项后点击按钮提交，则会用Toast的方式显示出所选中的选项内容，如图5-3所示。

5.1.2 基于回调机制的事件处理

所谓的回调机制就是在A类中定义一个方法，这个方法中用到了一个接口和该接口中的抽象方法，但是抽象方法没有具体的实现，需要B类去实现，B类实现该方法后其本身不会调用该方法，而是传递给A类去调用。也就是将事件的处理绑定在控件上，由图形用户界面

图5-2 选择监听示例单选运行效果图

图5-3 选择监听示例多选运行效果图

控件自己处理事件，回调机制需要自定义View来实现。Android平台中，每个View都有自己的处理事件的回调方法，开发人员可以通过重写View中的这些回调方法来实现需要的响应事件。当某个事件没有被任何一个View处理时，便会调用Activity中相应的回调方法。

回调方法如下。

（1）onKeyDown方法

功能	onKeyDown方法是接口KeyEvent.Callback中的抽象方法，所有的View全部实现了该接口并重写了该方法。该方法用来捕捉手机键盘被按下的事件
方法声明	public boolean onKeyDown (int keyCode, KeyEvent event) 参数keyCode：该参数为被按下的键值即键盘码，手机键盘中每个按钮都会有其单独的键盘码，应用程序都是通过键盘码才知道用户按下的是哪个键； 参数event：该参数为按键事件的对象，其中包含了触发事件的详细信息，例如事件的状态、事件的类型、事件发生的时间等，当用户按下按键时，系统会自动将事件封装成KeyEvent对象供应用程序使用； 返回值：该方法的返回值为一个boolean类型的变量，当返回true时，表示已经完整地处理了这个事件，并不希望其他的回调方法再次进行处理，而当返回false时，表示并没有完全处理完该事件，希望其他回调方法继续对其进行处理，例如Activity中的回调方法

（2）onKeyUp方法

功能	onKeyUp方法同样是接口KeyEvent.Callback中的一个抽象方法，并且所有的View同样全部实现了该接口并重写了该方法。onKeyUp方法用来捕捉手机键盘按键抬起的事件
方法声明	public boolean onKeyUp (int keyCode, KeyEvent event) 参数keyCode：参数keyCode为触发事件的按键码，需要注意的是，同一个按键在不同型号的手机中的按键码可能不同； 参数event：参数event为事件封装类的对象，其含义与onKeyDown方法中的完全相同，在此不再赘述； 返回值：该方法返回值表示的含义与onKeyDown方法相同，同样通知系统是否希望其他回调方法再次对该事件进行处理

（3）onTouchEvent方法

功能	onTouchEvent方法在View类中定义，并且所有的View子类全部重写了该方法，应用程序可以通过该方法处理手机屏幕的触摸事件
方法声明	public boolean onTouchEvent (MotionEvent event) 参数event：参数event为手机屏幕触摸事件封装类的对象，其中封装了该事件的所有信息，例如触摸的位置、触摸的类型以及触摸的时间等，该对象会在用户触摸手机屏幕时被创建； 返回值：该方法的返回值机制与键盘响应事件的相同，同样是当已经完整地处理了该事件且不希望其他回调方法再次处理时返回true，否则返回false。 该方法并不像之前介绍过的方法只处理一种事件，一般情况下以下三种情况的事件全部由onTouchEvent方法处理，只是三种情况中的动作值不同。 ● 屏幕被按下：当屏幕被按下时，会自动调用该方法来处理事件，此时MotionEvent.getAction的值为MotionEvent.ACTION_DOWN，如果在应用程序中需要处理屏幕被按下的事件，只需重新调用该方法，然后在方法中进行动作的判断即可。 ● 屏幕被抬起：当触控笔离开屏幕时触发的事件，该事件同样需要onTouchEvent方法来捕捉，然后在方法中进行动作判断。当MotionEvent.getAction的值为MotionEvent.ACTION_UP时，表示屏幕被抬起的事件。 ● 在屏幕中拖动：该方法还负责处理触控笔在屏幕上滑动的事件，调用MotionEvent.getAction方法来判断动作值是否为MotionEvent.ACTION_MOVE再进行处理

（4）onTrackBallEvent方法

功能	onTrackBallEvent方法为手机中轨迹球的处理方法。所有的View全部实现了该方法
方法声明	public boolean onTrackballEvent (MotionEvent event) 参数event：为手机轨迹球事件封装类的对象，其中封装了触发事件的详细信息，包括事件的类型、触发时间等，一般情况下，该对象会在用户操控轨迹球时被创建； 返回值：该方法的返回值与前面介绍的各个回调方法的返回值机制完全相同

☞【实例5-3】下面通过一个简单的案例来了解回调机制的使用。在工程"Chapter05"下创建一个Module，命名为"CallBack_Demo"。打开src/main/res/layout/activity_main.xml文件，修改代码并输入相关代码（本案例布局具体代码见配套资源）。

定义一个接口ChangeTitle，定义一个方法onChangeTitle，参数为一个字符串，代码如下所示。

```
package cn.androidstudy.callback_demo;
public interface ChangeTitle {
    void onChangeTitle(String title);}
```

定义一个class类，名为MyTask，继承自AsyncTask，把接口作为构造方法参数，在doInBackground方法中判断，如果有数据则接口回调，代码如下所示。

```
package cn.androidstudy.callback_demo;
import android.content.Context;
import android.os.AsyncTask;
public class MyTask extends AsyncTask<String,Void,String>{
    private ChangeTitle changeTitle;
    public MyTask(ChangeTitle changeTitle) {
        this.changeTitle = changeTitle;         }
    @Override
    protected String doInBackground(String... strings) {
        if (strings[0]!=null){
            changeTitle.onChangeTitle(strings[0]);          }
```

```
            return null;      }}
```

打开 src/main/java/cn.androidstudy.callback_demo 路径下的 MainActivity.java 文件，给异步任务参数传递 this，即接口回调方法在此类中执行，那么就需要实现 ChangeTitle 接口，重写接口中 onChangeTitle 方法，代码如下所示。

```
package cn.androidstudy.callback_demo;
import android.support.v7.app.AppCompatActivity;
import android.os.Bundle;
import android.widget.TextView;
public class MainActivity extends AppCompatActivity implements ChangeTitle {
    private TextView textView;
    @Override
    public void onCreate(Bundle savedInstanceState) {
        super.onCreate(savedInstanceState);
        setContentView(R.layout.activity_main);
        textView = (TextView) findViewById(R.id.tv_title);
        new MyTask(this).execute("我是标题");        }
// 重写接口方法，执行相应操作
    @Override
    public void onChangeTitle(String title) {
        textView.setText(title);       }}
```

运行项目，效果如图 5-4 所示。通过回调机制将制定的标题文本传递赋值给界面中的文本框控件 tv_title，从而显示到界面中。

图 5-4　回调机制示例运行效果图

5.1.3　回调方法应用案例

【实例 5-4】下面将通过介绍一个回调的应用案例介绍 onTouchEvent 方法，加强对回调方法的理解和掌握。其他的回调方法的使用方式类似，读者可以自行练习。本实例实现了在用户点击的位置绘制一个矩形，然后监测用户触控笔的状态，当用户在屏幕上移动触控笔时，使矩形随之移动，而当用户触控笔离开手机屏幕时，停止绘制矩形。

在工程"Chapter05"下创建一个Module，命名为"DrawRactangle_Demo"。打开src/main/java/cn.androidstudy.drawractangle_demo路径下的MainActivity.java文件，修改并输入部分代码，如下所示。

```java
package cn.androidstudy.drawractangle_demo;
import android.content.Context;
import android.graphics.Canvas;
import android.graphics.Color;
import android.graphics.Paint;
import android.support.v7.app.AppCompatActivity;
import android.os.Bundle;
import android.view.MotionEvent;
import android.view.View;
public class MainActivity extends AppCompatActivity {
    private MyView myView;          //自定义View的引用
    public void onCreate(Bundle savedInstanceState)
    {   //重写的onCreate方法，该方法会在此Activity创建时被系统调用，在方法中先初始化自定义的View，然后将当前的用户界面设置成该View
        super.onCreate(savedInstanceState);
        myView = new MyView(this);      //初始化自定义的View
        setContentView(myView);         //设置当前显示的用户界面          }
    @Override
    public boolean onTouchEvent(MotionEvent event)
    {// onTouchEvent回调方法重写的屏幕监听方法，在该方法中，根据事件动作的不同执行不同的操作
        switch(event.getAction()){
//当前事件为屏幕被按下的事件，通过调用MotionEvent的getX()和getY()方法得到事件发生的坐标，然后设置给自定义View的x与y成员变量
            case MotionEvent.ACTION_DOWN:    //按下
                myView.x = (int) event.getX();   //改变x坐标
                myView.y = (int) event.getY()-52;   //改变y坐标
                myView.postInvalidate();     //重绘
                break;
// 表示在屏幕上滑动时的事件，同样是得到事件发生的位置并设置给View的x、y。需要注意的是，因为此时手机屏幕并不是全屏模式，所以需要对坐标进行调整
            case MotionEvent.ACTION_MOVE:    //移动
                myView.x = (int) event.getX(); //改变x坐标
                myView.y = (int) event.getY()-52;   //改变y坐标
                myView.postInvalidate();     //重绘
                break;
// 以下是屏幕被抬起的事件，此时将View的x、y成员变量设成-100，表示并不需要在屏幕中绘制矩形
```

```
            case MotionEvent.ACTION_UP:         //抬起
                myView.x = -100;             //改变x坐标
                myView.y = -100;             //改变y坐标
                myView.postInvalidate();     //重绘
                break;                }
        return super.onTouchEvent(event);   }
    class MyView extends View {         //自定义的View
        Paint paint;              //画笔
        int x = 50;             //x坐标
        int y = 50;             //y坐标
        int w = 80;         //矩形的宽度
        public MyView(Context context)
        { //构造器
            super(context);
            paint = new Paint();           //初始化画笔   }
        @Override
        protected void onDraw(Canvas canvas)
        { //绘制方法
            canvas.drawColor(Color.GRAY);        //绘制背景色
            canvas.drawRect(x, y, x+w, y+w, paint);    //根据成员变量
绘制矩形
            super.onDraw(canvas);
        } }}
```

自定义的View并不会自动刷新,所以每次改变数据模型时都需要调用postInvalidate方法进行屏幕的刷新操作。运行该项目,效果如图5-5所示。

图5-5 回调方法应用案例运行效果图

案例运行的初始界面中会有一个绘制的黑色矩形,当点击屏幕时,初始的黑色矩形会取消绘制,并且会在点击的位置绘制一个矩形,当触控笔在屏幕中滑动时,该矩形会随之移动,当触控笔离开屏幕时,便会取消绘制矩形。

5.2 Android多线程机制

线程与进程是两个不同的概念。进程指正在运行的程序,当一个程序进入内存运行,就变成一个进程,进程是处于运行过程中的程序,并且具有一定独立功能。线程是进程中的一个实体,是被系统独立调度和分派的基本单位,一个进程中是可以有多个线程的。线程本身不拥有系统资源,或只拥有少量运行中必不可少的资源,但它可与同属一个进程的其他线程共享进程所拥有的全部资源。一个线程可以创建和撤销另一个线程,同一进程中的多个线程之间可以并发执行。线程之间相互制约,在运行中呈现间断性。

线程有就绪、阻塞和运行三种基本状态。每一个程序都至少有一个线程,若程序只有一个线程,那就是程序本身。线程是程序中一个单一的顺序控制流程。在单个程序中同时运行多个线程完成不同的工作,称为多线程。

Android程序开发人员为什么要使用多线程呢?这是因为使用多线程可以提高资源的使用效率以及系统的执行效率。主要体现在以下几方面:

① 使用线程可以合理分配"时间片",把占用时间长的任务放置到后台处理;

② 实现更人性化的用户界面设计,例如用户通过点击按钮触发了某些事件的处理,可以弹出一个进度条来显示处理的进度;

③ 在一些需要等待的任务实现上,如用户输入、文件读写和网络收发数据等,利用多线程可以释放一些珍贵的资源(如内存占用等)。

5.2.1 多线程机制的特点

Andriod的多线程是基于Linux本身的多线程机制,而多线程之间的同步又是通过Java本身的线程同步。在Androidk中使用多线程的优势主要体现在以下三个方面。

(1)避免ANR,提升用户体验

在Android中,如果应用程序在某段时间内响应不够灵敏,系统会向用户显示一个对话框,这个对话框称为应用程序无响应(Application Not Responding ANR)对话框。用户可以选择"等待"而让程序继续运行,也可以选择"强制关闭"。 默认情况下,在Android中Activity的最长执行时间是5s,BroadcastReceiver的最长执行时间是10s。一个流畅合理的应用程序中不应该出现ANR,因为这样会带来令人不满意的用户体验,使用多线程可以避免ANR。例如在访问网络服务端时返回过慢、数据过多导致滑动屏幕不流畅,或者I/O读取大资源时,可以通过开启一个新线程来处理比较耗时的操作。这里需要提到我们开发时的一个事件处理的原则:把所有可能耗时的操作都放到其他线程去处理。参考代码:

```
new Thread() {
    public void run() {
        //耗时操作代码         }}.start();
```

Android中的Main线程在时间处理时若无法在5s内得到响应,就会弹出ANR对话框。在Main线程中执行的方法有以下几种:

① Activity的生命周期方法,例如onCreate、onStart、onResume等;

② 事件处理方法，例如 onClick、onItemClick 等。

一般来说，Activity 的 onCreate、onStart、onResume 方法的执行时间决定了应用首页打开的时间，要尽量把不必要的操作放到其他线程去处理，如果仍然很耗时，可以使用动态的 SplashScreen 告知用户应用正在运行。

（2）实现异步处理

当用户与应用交互时，事件处理方法的执行效率决定了应用程序的响应性能，分为两种情况。

① 同步，需要等待返回结果。例如用户点击了"登录"按钮，需要等待服务端返回结果，那么需要有一个进度条来提示用户程序的运行情况。

② 异步，不需要等待返回结果。异步的概念和同步相对。当一个异步过程调用发出后，调用者不能立刻得到结果。调用的部件在完成后，通过状态、通知和回调来通知调用者。例如微博中的收藏功能，点击完"收藏"按钮后，会通知"收藏成功"而无需用户等待，这里就需要异步实现。

无论同步还是异步，事件处理都有可能比较耗时，此时需要放到其他线程中处理，处理完成后，再通知界面刷新。有一点需要注意，不是所有的界面刷新行为都需要放到 Main 线程中处理，例如 TextView 的 setText 方法需要在 Main 线程中，否则会抛出 "Called From Wrong Thread Exception" 异常，而 ProgressBar 的 setProgress 方法则不需要在 Main 线程中处理。

（3）实现多任务

多任务是一个操作系统可以同时执行多个程序的能力。基本上，操作系统使用一个硬件时钟为同时运行的每个进程分配"时间片"。如果时间片足够小，并且机器负荷不重，那么在用户看来，所有的程序似乎在同时运行。程序使用多线程在后台执行长作业，从而使用户仍然可以使用计算机进行其他工作。例如，我们向打印机发出打印的命令，假如此时计算机停止响应了，那我们岂不是必须停止手上的工作来等待低速的打印机工作？所幸的是，我们在打印机工作的同时可以进行听音乐、画图、看电影等应用。这是因为每一个程序被分成了独立不同的任务，使用多线程，即使某一部分任务失败了，也不会对其他的造成影响，不会导致整个程序崩溃。

总之，使用多线程可以获得更高效的 CPU 利用率、更好的系统可靠性，改善多处理器计算机的性能等，在许多应用中，可以同步地调用资源。尽管多线程具有以上诸多的优势，切不可过多地使用多线程。这是因为过多使用多线程会出现数据同步的问题，需特别处理，在使用多线程的时候必须尽量保持每个线程的独立性不被其他线程干预。另外，如果一个程序有多个线程，那么其他程序的线程必然只能占用更少的 CPU 时间，这样还需要大量的 CPU 时间做线程调度，大量操作系统的内存空间维护每个线程的上下文信息，反而会降低系统的运行效率。

5.2.2 多线程的实现

Android 中的线程是基于 Java 定义的线程。一个应用程序中可能会包含多个线程（Thread），每个线程中都有一个 run 方法，run 方法内部的程序执行完毕后，所在的线程就自动结束。每个线程都有一个消息队列，用于不同的线程之间传递消息。

（1）线程定义

Android 中定义线程的方法和 Java 是一样的，有两种方式：一种是 Thread；另一种是 Runnable。Thread 是一个类，而根据 Java 的继承要求，一个类只能有一个父类，所以继

第5章 事件处理机制与多线程

承了 Thread 的子类不能再继承其他类，限制了这种方法的使用。而 Runnable 是一个接口（interface），同样可以启动一个线程，不同的是它可以被多继承。参见代码．

```
package org.thread.demo;
class MyThread extends Thread{
    private String name;
    public MyThread(String name) {
        super();
        this.name = name; }
    public void run(){
        for(int i=0;i<10;i++){
            System.out.println("线程开始:"+this.name+",i="+i); } }}
package org.thread.demo;
public class ThreadDemo01 {
    public static void main(String[] args) {
        MyThread mt1=new MyThread("线程a");
        MyThread mt2=new MyThread("线程b");
        mt1.run();
        mt2.run(); }};
```

程序的运行很有规律，先执行第一个对象，然后执行第二个对象，并没有相互运行。在 JDK 的文档中可以发现，一旦调用 start 方法，则会通过 JVM 找到 run 方法。下面用 start 方法启动线程：

```
package org.thread.demo;
public class ThreadDemo01 {
    public static void main(String[] args) {
        MyThread mt1=new MyThread("线程a");
        MyThread mt2=new MyThread("线程b");
        mt1.start();
        mt2.start();}};
```

这样程序可以正常完成交互式运行。那么为什么要使用 start 方法启动多线程呢？这是因为在 JDK 的安装路径下，src.zip 是全部的 Java 源程序，通过此代码找到 Thread 中的 start 方法的定义，可以发现此方法中使用了"private native void start0();"。其中 native 关键字表示可以调用操作系统的底层函数，那么这样的技术称为 JNI 技术（Java Native Interface）。

在实际开发中，一个多线程的操作很少使用 Thread 类，而是通过 Runnable 接口完成。

```
public interface Runnable{
    public void run();}
```

例子：
```
package org.runnable.demo;
class MyThread implements Runnable{
    private String name;
    public MyThread(String name) {
```

```
            this.name = name;}
        public void run(){
            for(int i=0;i<100;i++){
                System.out.println("线程开始："+this.name+",i="+i);
            }}};
```

但是在使用 Runnable 定义的子类中没有 start 方法，只有 Thread 类中才有。此时观察 Thread 类，有一个构造方法：public Thread(Runnable targer)，此构造方法接受 Runnable 的子类实例，也就是说可以通过 Thread 类来启动 Runnable 实现多线程［start 可以协调系统的资源］：

```
package org.runnable.demo;
import org.runnable.demo.MyThread;
public class ThreadDemo01 {
    public static void main(String[] args) {
        MyThread mt1=new MyThread("线程a");
        MyThread mt2=new MyThread("线程b");
        new Thread(mt1).start();
        new Thread(mt2).start();}};
```

在实际程序开发中，多线程的实现多以 Runnable 接口为主，因为实现 Runnable 接口相比继承 Thread 类有如下好处：避免点继承的局限；一个类可以继承多个接口；适合于资源的共享等。

（2）Handler、Message 和 Looper

Handler 主要接受子线程发送的数据，并用此数据配合主线程更新 UI。一个线程中只能有一个 Handler 对象，可以通过该对象向所在线程发送消息。Handler 主要有两种用途：一是实现一个定时任务，类似于 Windows 程序中的定时器功能，可以通过 Handler 对象向所在线程发送一个延时消息，当消息指定的时间到达后，通过 Handler 的消息处理函数完成指定的任务；二是在线程间传递数据。

① 完成定时任务　在一个 Activity 内部实现定时器的功能，需要通过 Handler 对象的延迟发送消息方法来实现。Handler 有两类发送消息的方式：一类是 post×××方法，用于把一个 Runnable 对象发送到消息队列，从而当消息被处理时，能够执行 Runnable 对象；另一类是 send×××方法，用于发送一个 Message 类型的消息到消息队列，当消息被处理时，系统会调用 Handler 对象定义的 handleMessage 方法处理该信息。

send×××类包含以下方法。
- sendEmptyMessage(int)：空消息；
- sendMessage(Message)：发送 Message 指定的消息；
- sendMessageAtTime(Message,long)：在指定的时间点发送该消息；
- sendMessageDelayed(Message,long)：在指定的时间后发送该消息。

② 在线程之间传递数据　如果一个进程获得了另一个进程的 Handler，那么这个进程就可以通过 handler.sendMessage(Message) 方法向那个进程发送数据。基于这个机制，我们在处理多线程的时候可以新建一个 Thread，这个 Thread 拥有 UI 线程中的一个 Handler。当 Thread 处理完一些耗时的操作后通过传递过来的 Handler 像 UI 线程发送数据，由 UI 线程去更新界面。

Thread 在默认的情况下，只要 run 函数执行完毕，线程就结束。但有时新建的线程需要接收消息并处理，因此，在新线程中，除了需要添加一个 Handler 对象外，还需要从线程的消息队列中取出消息，并负责分发消息，这就需要用 Looper 来实现了。事实上，Activity 内部就只有一个 Looper，只是 Activity 是一个特殊的 Thread，操作系统已经将其封装了而已。

在 Android 中，Handler 和 Message、Thread 有着很密切的关系。Handler 主要是负责 Message 的分发和处理。Message 由一个消息队列进行管理，消息队列又是由一个 Looper 进行管理的。Android 系统中，Looper 负责管理线程的消息队列和消息循环。通过 Loop.myLooper 可以得到当前线程的 Looper 对象，通过 Loop.getMainLooper 可以获得当前进程的主线程的 Looper 对象。Android 系统的消息队列和消息循环都是针对具体线程的，一个线程可以存在（也可以不存在）一个消息队列和一个消息循环（Looper），特定线程的消息只能分发给本线程，不能进行跨线程、跨进程通信。但是创建的工作线程默认是没有消息循环和消息队列的，如果想让该线程具有消息队列和消息循环，需要在线程中首先调用 Looper.prepare 来创建消息队列，然后调用 Looper.loop 进入消息循环。先来看一下 Looper.prepare。

若现在线程中有一个 Looper 对象，其内部维护了一个消息队列 MQ（Message Queue）。一个 Thread 只能有一个 Looper 对象，参照如下源代码：

```
public class Looper {
    // 每个线程中的 Looper 对象其实是一个 ThreadLocal，即线程本地存储(TLS)对象
    private static final ThreadLocal sThreadLocal = new ThreadLocal();
    // Looper 内的消息队列
    final MessageQueue mQueue;
    // 当前线程
    Thread mThread;
    //其他属性
    // 每个 Looper 对象中有它的消息队列和它所属的线程
    private Looper() {
        mQueue = new MessageQueue();
        mRun = true;
        mThread = Thread.currentThread();   }
    // 我们调用该方法会在调用线程的 TLS 中创建 Looper 对象
    public static final void prepare() {
        if (sThreadLocal.get() != null) {
            // 试图在有 Looper 的线程中再次创建 Looper 将抛出异常
         throw new RuntimeException("Only one Looper may be created per thread");    }
        sThreadLocal.set(new Looper());}}
```

调用 loop() 方法后，Looper 线程就开始真正工作了，它不断从自己的 MQ 中取出队头的消息(也叫任务)执行。其源代码分析如下：

```
public static final void loop() {
        Looper me = myLooper();   //得到当前线程Looper
        MessageQueue queue = me.mQueue;  //得到当前looper的MessageQueue
        Binder.clearCallingIdentity();
        final long ident = Binder.clearCallingIdentity();
        // 开始循环
```

```java
            while (true) {
                Message msg = queue.next(); // 取出message
                if (msg != null) {
                    if (msg.target == null) {
                        // message没有target为结束信号，退出循环
                        return;    }
            // 日志
            if (me.mLogging!= null) me.mLogging.println(
                    ">>>>> Dispatching to " + msg.target + " "
                    + msg.callback + ": " + msg.what    );
            msg.target.dispatchMessage(msg);
            //将处理工作交给message的target，即handler
            if (me.mLogging!= null) me.mLogging.println(
                    "<<<<< Finished to    " + msg.target + " "
                    + msg.callback);
            final long newIdent = Binder.clearCallingIdentity();
            if (ident != newIdent) {
                Log.wtf("Looper", "Thread identity changed from 0x"
                    + Long.toHexString(ident) + " to 0x"
                    + Long.toHexString(newIdent) + " while dispatching to "
                    + msg.target.getClass().getName() + " "
                    + msg.callback + " what=" + msg.what);     }
            msg.recycle();    // 回收message资源    } } }
```

除了prepare和loop方法，Looper类还提供如下的方法，比如：Looper.myLooper得到当前线程Looper对象；getThread得到Looper对象所属线程；quit方法结束Looper循环等。

Handler扮演了往MQ上添加消息和处理消息的角色（只处理由自己发出的消息），即通知MQ它要执行一个任务(sendMessage)，并在loop到自己的时候执行该任务(handleMessage)，整个过程是异步的。Handler创建时会关联一个Looper，默认的构造方法将关联当前线程的Looper，不过这也是可以设置的。LooperThread类加入Handler的实现代码如下：

```java
    public class LooperThread extends Thread {
        private Handler handler1;
        private Handler handler2;
        @Override
        public void run() { // 将当前线程初始化为Looper线程
            Looper.prepare();
            // 实例化两个handler
            handler1 = new Handler();
            handler2 = new Handler();
            // 开始循环处理消息队列
            Looper.loop();    } }
```

第5章 事件处理机制与多线程

一个线程可以有多个Handler，但是只能有一个Looper。Handler可以在任意线程发送消息，它首先创建消息，根据Looper找到相关联的消息队列，然后这些消息会被添加到关联的消息队列上。Handler是在它关联的Looper线程中处理消息的。Looper首先取出消息队列的头消息，对应的Handler执行handlerMessage，最后返回Looper继续执行。

这就解决了Android不能在其他非主线程中更新UI的问题。Android的主线程也是一个Looper线程，我们在其中创建的Handler默认将关联主线程消息队列。因此，利用Handler的一个Solution就是在Activity中创建Handler并将其引用传递给worker thread，worker thread执行完任务后使用Handler发送消息通知Activity更新UI。

（3）线程间的消息传递

虽说特定线程的消息只能分发给本线程，不能进行跨线程通信，但是可以通过获得线程的Looper对象来实现不同线程间消息的传递，代码如下：

```
package com.mytest.handlertest;
import android.app.Activity;
import android.graphics.Color;
import android.os.Bundle;
import android.os.Handler;
import android.os.Looper;
import android.os.Message;
import android.util.Log;
import android.view.View;
import android.view.View.OnClickListener;
import android.view.ViewGroup.LayoutParams;
import android.widget.Button;
import android.widget.LinearLayout;
import android.widget.TextView;
public class HandlerTest extends Activity implements OnClickListener{
    private String TAG = "HandlerTest";
    private boolean bpostRunnable = false;
    private NoLooperThread noLooperThread = null;
    private OwnLooperThread ownLooperThread = null;
    private ReceiveMessageThread receiveMessageThread =null;
    private Handler  mOtherThreadHandler=null;
    private EventHandler mHandler = null;
    private Button btn1 = null;
    private Button btn2 = null;
    private Button btn3 = null;
    private Button btn4 = null;
    private Button btn5 = null;
    private Button btn6 = null;
    private TextView tv = null;
```

```java
/** Called when the activity is first created. */
@Override
public void onCreate(Bundle savedInstanceState) {
    super.onCreate(savedInstanceState);
    LinearLayout layout = new LinearLayout(this);
    LinearLayout.LayoutParams params = new LinearLayout.LayoutParams(250, 50);
    layout.setOrientation(LinearLayout.VERTICAL);
    btn1 = new Button(this);
    btn1.setId(101);
    btn1.setText("message from main thread self");
    btn1.setOnClickListener(this);
    layout.addView(btn1, params);
    btn2 = new Button(this);
    btn2.setId(102);
    btn2.setText("message from other thread to main thread");
    btn2.setOnClickListener(this);
    layout.addView(btn2, params);
    btn3 = new Button(this);
    btn3.setId(103);
    btn3.setText("message to other thread from itself");
    btn3.setOnClickListener(this);
    layout.addView(btn3, params);
    btn4 = new Button(this);
    btn4.setId(104);
    btn4.setText("message with Runnable as callback from other thread to main thread");
    btn4.setOnClickListener(this);
    layout.addView(btn4, params);
    btn5 = new Button(this);
    btn5.setId(105);
    btn5.setText("main thread's message to other thread");
    btn5.setOnClickListener(this);
    layout.addView(btn5, params);
    btn6 = new Button(this);
    btn6.setId(106);
    btn6.setText("exit");
    btn6.setOnClickListener(this);
    layout.addView(btn6, params);
    tv = new TextView(this);
    tv.setTextColor(Color.WHITE);
    tv.setText("");
```

```java
        params=new LinearLayout.LayoutParams(LayoutParams.FILL_
PARENT,LayoutParams.WRAP_CONTENT);
        params.topMargin=10;
        layout.addView(tv, params);
        setContentView(layout);
        receiveMessageThread = new ReceiveMessageThread();
        receiveMessageThread.start();        }
    class EventHandler extends Handler{
        public EventHandler(Looper looper){
            super(looper);        }
        public EventHandler(){
            super();        }
        @Override
        public void handleMessage(Message msg) {
            // TODO Auto-generated method stub
            super.handleMessage(msg);
            Log.e(TAG, "CurrentThread id:----------+>" +
Thread.currentThread().getId());
            switch(msg.what){
            case 1:   tv.setText((String)msg.obj);
                break;
            case 2:   tv.setText((String)msg.obj);
                noLooperThread.stop();
                break;
            case 3://不能在非主线程的线程里面更新UI，这里通过Log打印
信息
                Log.e(TAG,(String)msg.obj);
                ownLooperThread.stop();
                break;
            default:   Log.e(TAG,(String)msg.obj);
                break;}}}
    //ReceiveMessageThread has his own message queue by execute
Looper.prepare();
    class ReceiveMessageThread extends Thread {
        @Override
        public void run(){
            Looper.prepare();
            mOtherThreadHandler= new Handler(){
                @Override
                public void handleMessage(Message msg) {
                    // TODO Auto-generated method stub
                    super.handleMessage(msg);
```

```java
                        Log.e(TAG,"-------+>"+(String)msg.obj);
                        Log.e(TAG, "CurrentThread id:----------+>" + Thread.currentThread().getId());
                } };
            Log.e(TAG, "ReceiveMessageThread id:--------+>" + this.getId());
            Looper.loop();}}
    class NoLooperThread extends Thread {
        private EventHandler mNoLooperThreadHandler;
        @Override
        public void run() {
            Looper myLooper = Looper.myLooper();
            Looper mainLooper= Looper.getMainLooper();

            String msgobj;
            if(null == myLooper){
                //这里获得的是主线程的Looper,由于NoLooperThread没有自己的Looper,所以这里肯定会被执行
                mNoLooperThreadHandler = new EventHandler(mainLooper);
                msgobj = "NoLooperThread has no looper and handleMessage function executed in main thread!";
            } else{
                mNoLooperThreadHandler = new EventHandler(myLooper);
                msgobj = "This is from NoLooperThread self and handleMessage function executed in NoLooperThread!";    }

            mNoLooperThreadHandler.removeMessages(0);

            if(bpostRunnable == false){
                //send message to main thread
                Message msg = mNoLooperThreadHandler.obtainMessage(2, 1, 1, msgobj);
                mNoLooperThreadHandler.sendMessage(msg);
                Log.e(TAG, "NoLooperThread id:--------+>" + this.getId());
            }else{
                //后面实现了Runnable接口的对象中run函数是在主线程中执行,不是在NoLooperThread进程中执行
                //注意Runnable是一个接口,它里面的run函数被执行时不会再新建一个线程
```

第5章 事件处理机制与多线程

```java
                    mNoLooperThreadHandler.post(new Runnable(){
                        public void run() {
                            // TODO Auto-generated method stub
                            tv.setText("update UI through handler post runnalbe mechanism!");
                            Log.e(TAG, "update UI id:--------+>" + Thread.currentThread().getId());
                            noLooperThread.stop();   }}}}
    class OwnLooperThread extends Thread{
        private EventHandler mOwnLooperThreadHandler = null;

        @Override
        public void run() {
            Looper.prepare();
            Looper myLooper = Looper.myLooper();
            Looper mainLooper= Looper.getMainLooper();
            String msgobj;
            if(null == myLooper){
                mOwnLooperThreadHandler = new EventHandler(mainLooper);
    msgobj = "OwnLooperThread has no looper and handleMessage function executed in main thread!"; }
    else{
                mOwnLooperThreadHandler = new EventHandler(myLooper);
     msgobj = "This is from OwnLooperThread self and handleMessage function executed in NoLooperThread!";   }
            mOwnLooperThreadHandler.removeMessages(0);

            //给自己发送消息
    Message msg = mOwnLooperThreadHandler.obtainMessage(3,1,1,msgobj);
            mOwnLooperThreadHandler.sendMessage(msg);
            Looper.loop();     }}
    public void onClick(View v) {
        // TODO Auto-generated method stub
        switch(v.getId()){
        case 101:
            //主线程发送消息给自己
    Looper looper = Looper.myLooper();//get the Main looper related with the main thread
    //如果不给任何参数的话会用当前线程对应的Looper(这里就是Main Looper)为Handler里面的成员mLooper赋值
```

```java
                mHandler = new EventHandler(looper);
                // 清除整个MessageQueue里的消息
                mHandler.removeMessages(0);
                String obj = "This main thread's message and received by itself!";
                Message msg = mHandler.obtainMessage(1,1,1,obj);
                // 将Message对象送入到主线程的MessageQueue里面
                mHandler.sendMessage(msg);
                break;
            case 102:
                //其他线程发送消息给主线程。首先bpostRunnable设为false,表示不通过Runnable方式进行相关的操作,然后建立并启动线程noLooperThread可发送消息给主线程
                bpostRunnable = false;
                noLooperThread = new NoLooperThread();
                noLooperThread.start();
                break;
            case 103:
                //其他线程建立并启动ownLooperThread后可获取自己发送的消息
                tv.setText("please look at the error level log for other thread received message");
                ownLooperThread = new OwnLooperThread();
                ownLooperThread.start();
                break;
            case 104: //其他线程通过Post Runnable方式发送消息给主线程
                bpostRunnable = true;
                noLooperThread = new NoLooperThread();
                noLooperThread.start();
                break;
            case 105:    //主线程发送消息给其他线程
                if(null!=mOtherThreadHandler){
                    tv.setText("please look at the error level log for other thread received message from main thread");
                    String msgObj = "message from mainThread";
                    Message mainThreadMsg = mOtherThreadHandler.obtainMessage(1, 1, 1, msgObj);
                    mOtherThreadHandler.sendMessage(mainThreadMsg);
                }
                break;
            case 106: finish();
                break;
        }}}
```

第5章 事件处理机制与多线程

 强化训练

事件处理机制和多线程机制是Android的两个经常使用的重要机制，Android程序开发人员需要熟悉并掌握。

本章主要介绍了Android的事件处理机制和多线程机制，事件处理机制是实现用户与系统有效交互的必要手段，而多线程机制则是系统开发时有效配置资源和提高执行效率的重要手段。

学完本章知识后，不妨通过以下习题来巩固所学的内容吧。

一、填空题

1. Android提供了强大的事件处理机制，它包括两套处理机制：一个是_____，另一个是_____。
2. 对于一个Android应用程序来说，_____是必不可少的，用户与应用程序之间的交互便是通过_____来完成的。
3. 所谓的回调机制就是在_____中定义一个方法，这个方法中用到了一个接口和该接口中的抽象方法，但是抽象方法没有具体的实现，需要_____去实现，_____实现该方法后其本身不会调用该方法，而是传递给_____去调用。
4. 线程是_____中的一个实体，是被系统独立调度和分派的_____，线程本身不拥有_____，或只拥有少量运行中必不可少的资源，但它可与_____共享进程所拥有的全部资源。
5. 一个线程可以_____另一个线程，同一进程中的多个线程之间可以_____。线程之间相互制约，在运行中呈现_____。
6. 线程有____、____和____三种基本状态。

二、选择题

1. (　　) 不是多线程机制的特点。
   A. 避免ANR，提升用户体验
   B. 实现异步处理
   C. 实现多任务
   D. 计算量大
2. (　　) 接口的功能为在触控模式下，是在某个View上按下并抬起的组合动作，而在键盘模式下，是某个View获得焦点后点击确定键或者按下轨迹球事件。
   A. OnClickListener接口
   B. OnFocusChangeListener接口
   C. OnKeyListener接口
   D. OnCreateContextMenuListener接口
3. 回调方法(　　) 的功能为接口KeyEvent.Callback中的一个抽象方法，并且所有的View全部实现了该接口并重写了该方法，onKeyUp方法用来捕捉手机键盘按键抬起的事件。
   A. onTouchEvent方法
   B. onTrackBallEvent方法
   C. onKeyUp方法

D. onTrackBallEvent 方法
4. 下列哪个不属于多线程的实现过程。(　　)
A. 线程定义
B. 定义子类
C. Handler、Message 和 Looper
D. 线程间的消息传递

三、操作题

1. 练习利用 Android Studio 开发环境创建一个新的 Android 应用程序项目，使用按钮实现基于监听接口的事件处理点击监听。

2. 练习利用 Android Studio 开发环境创建一个新的 Android 应用程序项目，使用回调方法实现下列功能：在用户点击的位置绘制一个矩形，然后监测用户触控笔的状态，当用户在屏幕上移动触控笔时，使矩形随之移动，而当用户触控笔离开手机屏幕时，停止绘制矩形。

3. Android 中的线程是基于 Java 定义的线程。一个应用程序中可能会包含多个线程（Thread），每个线程都有一个消息队列，用于不同的线程之间传递消息。练习利用 Android Studio 开发环境创建一个新的 Android 应用程序项目，实现跨线程通信。

第6章

Android 控件进阶

内容导读

本书前面为读者讲解了一部分基本UI组件，本章将继续介绍UI的高级组件。使用基本UI组件可以开发出普通的UI界面，但是不一定会满足用户的多样性需求。一般程序员可以开发出强大完备的功能，但审美方面不一定非常好。Android提供了大量的UI控件以供程序员选择，只要合理地利用它们，也可以设计出各种漂亮的UI界面。系统提供的组件越多，越有利于程序开发人员构造满足用户需求的UI界面。

学习目标

- 熟悉使用Android的进度条组件；
- 熟悉使用Android的列表视图组件；
- 熟悉使用Android的下拉列表组件；
- 熟悉使用Android的网格视图组件；
- 熟悉使用Android的日期和时间选择器组件。

6.1 进度条 ProgressBar

进度条是UI界面中一种经常使用的组件。进度条可以在处理任务时，实时地以图形方式显示处理任务的速度、完成度、剩余未完成任务量的多少或者可能需要处理时间，一般以长方形条状显示。进度条可以为用户动态显示进度,避免长时间地执行某个耗时操作时,让用户感觉程序失去了响应,从而更好地提高用户界面的友好性。

Android支持几种风格的进度条,最常见的两种进度条是"环形进度条"和"水平进度条",可以通过style属性指定风格,水平进度条对应的属性为:style="?android:attr/progressBarStyleHorizontal"。

ProgressBar类中常用的方法如下:
- ProgressBar.setMax(int max):设置总长度;
- ProgressBar.setProgress(int progress):设置已经开启长度为0,假设设置为50,进度条将进行到一半停止。

☞【实例6-1】下面通过实例学习ProgressBar的使用。创建一个Android工程,工程名为"Chapter06",在该工程下创建一个Module,命名为"ProgressBar_Demo"。

扫一扫 看视频

打开/res/values目录下的strings.xml文件,修改编写代码如下:

```xml
<resources>
    <string name="app_name">ProgressBar_Demo</string>
    <string name="hello_AS">Hello Android Studio! progressBarActivity</string>
    <string name="progressbar_ring">环形进度条</string>
    <string name="progressbar_horizontal">水平进度条</string>
    <string name="menu_settings">Settings</string>
</resources>
```

打开"Chapter06"工程中"ProgressBar_Demo"项目下res/layout/activity_main.xml文件,修改代码并输入相关代码(本案例布局具体代码见配套资源)。

打开src/main/java/cn.androidstudy.progressbar_demo/MainActivity文件,修改并输入部分代码,如下所示:

```java
package cn.androidstudy.progressbar_demo;
import android.os.Handler;
import android.support.v7.app.AppCompatActivity;
import android.os.Bundle;
import android.widget.ProgressBar;
public class MainActivity extends AppCompatActivity {
    private ProgressBar mProgress;
    private int mProgressStatus=0;
    //创建一个Handler对象
    private Handler mHandler=new Handler();
    @Override
    protected void onCreate(Bundle savedInstanceState) {
        super.onCreate(savedInstanceState);
        setContentView(R.layout.activity_main);
        mProgress=(ProgressBar)findViewById(R.id.progressBar_Ring);
        //设置进度条的最大值,其将为该进度条显示的基数
        mProgress.setMax(10000);
```

第6章 Android控件进阶

```
        //新开启一个线程
        new Thread(new Runnable(){
            public void run(){
                //循环10000次，不停地更新mProgressStatus的值
                while(mProgressStatus++<10000){
                    //将一个Runnable对象添加到消息队列中
                    //并且当执行到该对象时执行run()方法
                    mHandler.post(new Runnable(){
                        public void run(){
                            //重新设置进度条当前值
                            mProgress.setProgress(mProgressStatus);
                        } }); } }
        }).start(); }}
```

运行该项目，效果如图6-1所示。

图6-1　ProgressBar_Demo示例运行效果图

6.2 列表视图ListView

列表视图控件ListView是Android软件开发中经常用到的重要组件之一，它以垂直列表的形式展示所有列表项（例如清单、联系人等），并且能够根据数据的长度自适应显示。很多Android软件里面都会使用ListView列表视图控件。本节将以实例来展示列表视图控件ListView的使用。

列表的显示需要以下三个元素。

① ListVeiw：用来展示列表的View。

② 适配器：用来把数据映射到ListView上的中介。一般有三种：ArrayAdapter、SimpleAdapter

和SimpleCursorAdapter。其中，以ArrayAdapter最为简单，只能展示一行字。SimpleAdapter有最好的扩充性，可以自定义出各种效果。SimpleCursorAdapter可以认为是SimpleAdapter对数据库的简单结合，可以方便地把数据库的内容以列表的形式展示出来。

③ 数据：即具体的将被映射的字符串、图片或者基本组件等。

ListView类的常用方法有以下几个。

- setAdapter(ListAdapter adapter)：为ListView绑定一个Adapter；
- setChoiceMode（int choiceMode）：为ListView指定一个显示模式，可选值有三个：CHOICE_MODE_NONE（默认值，没有单选或多选效果）、CHOICE_MODE_SINGLE（单选效果）、CHOICE_MODE_MULTIPLE（多选框效果）；
- setOnItemClickListener（AdapterView.onItemClickListener listener）：为其注册一个元素被点击事件的监听器，当其中某一项被点击时调用其参数listener中的onItemClick方法。

【实例6-2】下面通过两个案例分别实现一个简单的纯文本列表控件和一个图文并茂相对复杂的列表控件。先以ArrayAdapter数组适配器为例，使用Android提供的自带的布局文件，实现简单的纯文本列表控件，当点击某个列表项时，用Toast显示相应信息。

扫一扫 看视频

扫一扫 看视频

在工程"Chapter06"下新建一个Module，命名为"ListView_demo"，修改其布局文件activity_main.xml（本案例布局具体代码见配套资源）。

打开src/main/java/cn.androidstudy.listview_demo/MainActivity文件，修改并输入部分代码，如下所示：

```java
package cn.androidstudy.listview_demo;
import android.support.v7.app.AppCompatActivity;
import android.os.Bundle;
import android.view.View;
import android.widget.AdapterView;
import android.widget.ArrayAdapter;
import android.widget.ListView;
import android.widget.Toast;
public class MainActivity extends AppCompatActivity {
    private ListView listView;
    private String[] className={"计科一班","计科二班","软件一班","软件二班","移动一班","移动二班"};
    protected void onCreate(Bundle savedInstanceState) {
        super.onCreate(savedInstanceState);
        setContentView(R.layout.activity_main);
        listView=(ListView)findViewById(R.id.lv_className);
        //创建一个ArrayAdapter
        ArrayAdapter adapter=new ArrayAdapter(this,
                android.R.layout.simple_list_item_1,className);
        listView.setAdapter(adapter);
        //为listView注册一个元素点击事件监听器
        listView.setOnItemClickListener(new AdapterView.
```

```
OnItemClickListener() {
            public void onItemClick(AdapterView<?> arg0,
                                    View arg1, int arg2, long arg3) {
                Toast.makeText(MainActivity.this, "您选择的是："+className[arg2],
                    Toast.LENGTH_LONG).show();
            } }); }}
```

运行项目，效果如图6-2所示。

图6-2 纯文本ListView示例运行效果图

以SimpleAdapter数组适配器为例，使用自定义的布局文件，实现相对复杂的图文并茂的列表控件，当点击某个列表项时，用Toast显示相应信息。

【实例6-3】在工程"Chapter06"下新建一个Module,命名为"ListView_Image_Demo"，修改其布局文件activity_main.xml。本案例布局具体代码见光盘中相关代码。

在res/layout文件夹下新建一个XML布局文件，命名为"listview_layout"，用于自定义列表项的布局（本案例布局具体代码见配套资源）。

打开src/main/java/cn.androidstudy.listview_image_demo/MainActivity文件，修改并输入部分代码，如下所示：

```
package cn.androidstudy.listview_image_demo;
import android.support.v7.app.AppCompatActivity;
import android.os.Bundle;
import android.view.View;
```

```java
import android.widget.AdapterView;
import android.widget.ListView;
import android.widget.SimpleAdapter;
import android.widget.Toast;
import java.util.ArrayList;
import java.util.HashMap;
import java.util.Map;
public class MainActivity extends AppCompatActivity {
    private String[] listFruitName = { "苹果", "西瓜", "橙子", "桃子","草莓","葡萄"};
    private String[] listFruitPrice = { "13.6￥/kg", "3.6￥/kg", "15.0￥/kg", "5.8￥/kg","19.8￥/kg","6.0￥/kg"};
    private String[] listFruitExplain = { "红红苹果，满满能量！", "旱地西瓜，甘甜解暑！", "不催熟、不打蜡、不上色！", "原生态，无公害！","口口流汁，粒粒甜美！","自然生长，甜蜜悠长！"};
    private ListView listView_Fruit;
    ArrayList<Map<String,Object>> listData= new ArrayList<Map<String,Object>>();
    @Override
    protected void onCreate(Bundle savedInstanceState) {
        super.onCreate(savedInstanceState);
        setContentView(R.layout.activity_main);
        listView_Fruit = (ListView) findViewById(R.id.lv_fruit);
        int length = listFruitName.length;
        for(int i =0; i < length; i++) {
            Map<String,Object> item = new HashMap<String,Object>();
            item.put("image", R.mipmap.fruit);
            item.put("name", listFruitName[i]);
            item.put("price", listFruitPrice[i]);
            item.put("explain", listFruitExplain[i]);
            listData.add(item);          }
        SimpleAdapter myAdapter = new SimpleAdapter(this,listData, R.layout.listview_layout,new String[]{"image","name","price","explain"},new int[]{R.id.image_fruit,R.id.tv_name,R.id.tv_price,R.id.tv_explain});
        listView_Fruit.setAdapter(myAdapter);
        listView_Fruit.setOnItemClickListener(new AdapterView.OnItemClickListener() {
            @Override
            public void onItemClick(AdapterView<?> adapterView, View view, int position, long id) {
                Toast.makeText(MainActivity.this,"您点击的水果是："
```

```
+ listFruitName[position] + ",价格是:"+listFruitPrice[position]+","+l
istFruitExplain[position], Toast.LENGTH_SHORT).show();      } }); }}
```

运行项目,效果如图6-3所示。

图6-3 含图片ListView示例运行效果图

本书后续章节会介绍从数据库及内容提供者处获取数据的方法,届时再对利用SimpleAdapter、CursorAdapter绑定数据进行详细介绍。

6.3 下拉列表Spinner

由于受到实际操控性的限制,手机屏幕一般不会太大。当需要用户选择的时候,可以提供一个下拉列表将所有可选项列出来,供用户选择,以此提高用户的体验。下拉列表控件Spinner与列表视图控件ListView一样,也是AdapterView的一个间接子类,是一个显示数据的窗口。本节中将以实例来展示下拉列表控件Spinner的使用。

Spinner的数据源和ListView一样,可以通过数组给适配器提供数据,也可以通过资源文件给适配器提供数据,下面通过一个选择血型的案例,讲解通过数组的方式获取数据的方法,使用资源文件的方式会在后续章节中进一步介绍。

☞【实例6-4】在工程"Chapter06"下新建一个Module,命名为"Spinner_Demo",修改其布局文件activity_main.xml(本案例布局具体代码见配套资源,后续案例代码均见同样位置)。

打开src/main/java/cn.androidstudy.spinner_demo/MainActivity文件,修改并输入部分代码,如下所示:

```java
package cn.androidstudy.spinner_demo;
import android.support.v7.app.AppCompatActivity;
import android.os.Bundle;
import android.view.View;
import android.widget.AdapterView;
import android.widget.ArrayAdapter;
import android.widget.Spinner;
import android.widget.TextView;
public class MainActivity extends AppCompatActivity {
    private static final String[] bloodType={"A型","B型","AB型","O型","其他"};
    private TextView tv_Spinner;
    private Spinner spinner_BT;
    private ArrayAdapter<String> arr_adapter;
    @Override
    protected void onCreate(Bundle savedInstanceState) {
        super.onCreate(savedInstanceState);
        setContentView(R.layout.activity_main);

        tv_Spinner=(TextView)findViewById(R.id.tv_spinner);
        spinner_BT=(Spinner)findViewById(R.id.spinner_bloodType);
        //将选项数据与ArrayAdapter适配器连接起来
        arr_adapter=new ArrayAdapter<String>(this,android.R.layout.simple_spinner_item,bloodType);
        //设置下拉列表的风格
        arr_adapter.setDropDownViewResource(android.R.layout.simple_spinner_dropdown_item);
        //将adapter添加到Spinner中
        spinner_BT.setAdapter(arr_adapter);
        //为Spinner添加监听事件
        spinner_BT.setOnItemSelectedListener(new SpinnerSelectdeListener());
        //设置默认值
        spinner_BT.setVisibility(View.VISIBLE);    }
    //使用数组形式操作数据
    class SpinnerSelectdeListener implements AdapterView.OnItemSelectedListener {
        public void onItemSelected(AdapterView<?> arg0, View arg1, int arg2,
                                   long arg3) {
            tv_Spinner.setText("您选择的血型是："+bloodType[arg2]);    }
```

```
        public void onNothingSelected(AdapterView<?> arg0) {
        } }}
```

运行该项目，效果如图6-4所示。点击某一下拉选项，则该选项的值会添加在TextView的文本后面显示出来。

图6-4　Spinner示例运行效果图

6.4　网格视图GridView

网格视图控件GridView是按照行、列分布的方式来显示内容的，通常用于显示图片等内容。比如实现九宫格图，用GridView是首选，也是最简单的。在使用网格视图控件GridView时，经常使用Adapter为网格视图控件GridView提供数据来源。本节中将以实例来展示网格视图控件GridView的使用。

（1）Context

Context提供了关于应用环境全局信息的接口。它是一个抽象类，它的执行由Android系统提供。它允许获取以应用为特征的资源和类型，同时启动应用级的操作，如启动Activity、broadcasting和接收intents。

（2）public void setAdapter (ListAdapter adapter)

用于设置GridView的数据，其参数adapter为GridView提供数据的适配器。

（3）public View getView(int position, View convertView, ViewGroup parent)

其参数的含义如下。
- position：该视图在适配器数据中的位置；
- convertView：旧视图；
- parent：此视图最终会被附加到的父级视图。

（4）ImageView

用于显示任意图像，例如图标。ImageView类可以加载各种来源的图片（如资源或图片库），需要计算图像的尺寸，以便它可以在其他布局中使用，并提供例如缩放和着色（渲染）等显示选项。

（5）public void setAdjustViewBounds (boolean adjustViewBounds)

当需要在ImageView调整边框时保持可绘制对象的比例，将该值设为真。

（6）public void setScaleType (ImageView.ScaleType scaleType)

控制图像应该如何缩放和移动，以使图像与ImageView一致。参数scaleType是需要的缩放方式。

☞【实例6-5】下面通过一个简单的案例，演示如何使用GridView实现一个九宫格图片效果。在工程"Chapter06"下新建一个Module，命名为"gridview_demo"，修改其布局文件activity_main.xml。

打开src/main/java/cn.androidstudy.gridview_demo/MainActivity文件，修改并输入部分代码，如下所示：

```
package cn.androidstudy.gridview_demo;
import android.os.Bundle;
import android.app.Activity;
import android.content.Context;
import android.view.Menu;
import android.view.View;
import android.view.ViewGroup;
import android.widget.BaseAdapter;
import android.widget.GridView;
import android.widget.ImageView;
public class MainActivity extends Activity {
    private GridView gv;
    @Override
    public void onCreate(Bundle savedInstanceState) {
        super.onCreate(savedInstanceState);
        setContentView(R.layout.activity_main);
        gv=(GridView)findViewById(R.id.GridViewOne);
        gv.setNumColumns(3);//设置GridView的列数
        gv.setAdapter(new MyAdapter(this));  //为GridView设置适配器  }
    //自定义适配器
```

```java
class MyAdapter extends BaseAdapter{        ////图片ID数组
    private Integer[] imgs = {
            R.mipmap.one,
            R.mipmap.two,
            R.mipmap.three,
            R.mipmap.four,
            R.mipmap.five,
            R.mipmap.six,
            R.mipmap.seven,
            R.mipmap.eight,
            R.mipmap.nine        };
    Context context;////上下文对象
    MyAdapter(Context context){//构造方法
        this.context = context;         }
    public int getCount() {//获得数量
        // TODO Auto-generated method stub
        return imgs.length;         }
    public Object getItem(int position) {//获得当前选项
        // TODO Auto-generated method stub
        return position;            }
    public long getItemId(int position) {//获得当前选项ID
        // TODO Auto-generated method stub
        return position;            }
        //创建View方法
    public View getView(int position, View convertView, ViewGroup parent) {
        ImageView imageView;
        if (convertView == null) {
            imageView = new ImageView(context);//实例化ImageView对象
            //设置ImageView对象布局
            imageView.setLayoutParams(new GridView.LayoutParams(240,240));
            imageView.setAdjustViewBounds(false);//设置边界对齐
            //设置刻度类型
            imageView.setScaleType(ImageView.ScaleType.CENTER_CROP);
            imageView.setPadding(8, 8, 8, 8);//设置间距
        } else {    imageView = (ImageView) convertView;        }
        //为ImageView设置图片资源
        imageView.setImageResource(imgs[position]);
        return imageView;       } }}
```

运行该项目，效果如图6-5所示，可以得到一个使用GridView控件实现的九宫格图片。

图6-5 GridView示例运行效果图

6.5 日期和时间选择器DatePicker和TimePicker

在Android Studio中，提供了一些与日期和时间相关的组件以供Android开发人员使用。例如DatePicker和TimePicker是两个比较易用的控件，它们都是从FrameLayout派生出来的。DatePicker控件供用户选择日期，TimePicker控件供用户选择时间。这两个控件使用非常方便，DatePicker控件和TimePicker控件都可以从Android Studio的可视化界面设计器中拖拽到布局界面中。本节将以实例来展示DatePicker控件和TimePicker控件的使用。

DatePicker和TimePicker在FrameLayout的基础上提供了一些方法来获取当前用户所选择的日期、时间；如果程序需要获取用户选择的日期、时间，则可以通过为它们添加相应的监听器来实现。

（1）DatePicker

日期选择器常见方法如下。

- public int getDayOfMonth ()：获取选择的天数。
- public int getMonth ()：获取选择的月份（注意：返回数值为0~11，需要自己+1来显示）。
- public int getYear ()：获取选择的年份。
- public void init (int year, int monthOfYear, int dayOfMonth, DatePicker.OnDate

ChangedListener onDateChangedListener)：初始化状态（初始化年月日），其中各参数含义如下：

　　year：初始年［注意：使用new Date()初始化年时，需要+1900，例如：date.getYear() + 1900］；

　　monthOfYear：初始月；

　　dayOfMonth：初始日。

　　onDateChangedListener：日期改变时通知用户的事件监听器，可以为空(null)。

　•　public void setEnabled (boolean enabled)：设置视图的启用状态。该启用状态随子类的不同而有不同的解释。其中enabled设置为true表示启动视图，反之禁用。

　•　public void updateDate (int year, int monthOfYear, int dayOfMonth)：更新日期。

（2）TimePicker

时间选择器的常用方法如下：

　•　public int getBaseline ()：返回窗口空间的文本基准线到其顶边界的偏移量。如果这个部件不支持基准线对齐，这个方法返回−1。

　•　public Integer getCurrentHour ()：获取当前时间的小时部分（0～23）。

　•　public Integer getCurrentMinute ()：获取当前时间的分钟部分。

　•　public boolean is24HourView ()：获取当前系统设置是否是24小时制。

　•　public void setCurrentHour (Integer currentHour)：设置当前小时。

　•　public void setCurrentMinute (Integer currentMinute)：设置当前分钟（0～59）。

　•　public void setEnabled (boolean enabled)：设置可用的视图状态。可用的视图状态的解释在子类中改变。

　•　public void setIs24HourView (Boolean is24HourView)：设置是24小时还是上午/下午制。

　•　public void setOnTimeChangedListener (TimePicker.OnTimeChangedListener onTimeChangedListener)：设置时间调整事件的回调函数，其中参数onTimeChangedListener不能为空。

☞【实例6-6】下面通过一个具体实例演示这两种控件的用法。在"Chapter06"下新建一个Module，命名为"DP_TP_Demo"，修改其布局文件activity_main.xml。

　　打开src/main/java/cn.androidstudy.dp_tp_demo/MainActivity文件，修改并输入部分代码，如下所示：

```
package cn.androidstudy.dp_tp_demo;
import android.support.v7.app.AppCompatActivity;
import android.os.Bundle;
import android.widget.DatePicker;
import android.widget.EditText;
import android.widget.TextView;
import android.widget.TimePicker;
import java.util.Calendar;
public class MainActivity extends AppCompatActivity {
    //定义5个记录当前时间的变量
    private int year;
    private int month;
```

```java
    private int day;
    private int hour;
    private int minute;
    @Override
    protected void onCreate(Bundle savedInstanceState) {
        super.onCreate(savedInstanceState);
        setContentView(R.layout.activity_main);
        DatePicker datePicker = (DatePicker)findViewById(R.id.datePicker);
        TimePicker timePicker = (TimePicker)findViewById(R.id.timePicker);
        //获取当前的年、月、日、小时、分钟
        Calendar c = Calendar.getInstance();
        year = c.get(Calendar.YEAR);
        month = c.get(Calendar.MONTH);
        day = c.get(Calendar.DAY_OF_MONTH);
        hour = c.get(Calendar.HOUR);
        minute = c.get(Calendar.MINUTE);
        //初始化DatePicker组件，初始化时指定监听器
        datePicker.init(year , month ,day
            , new DatePicker.OnDateChangedListener()
            {    @Override
                public void onDateChanged(DatePicker arg0, int year
                    , int month, int day)
                {    MainActivity.this.year = year;
                    MainActivity.this.month = month;
                    MainActivity.this.day = day;
                    //显示当前日期、时间
                    showDate(year, month+1 , day , hour, minute);
                } });
        //为TimePicker指定监听器
         timePicker.setOnTimeChangedListener(new TimePicker.OnTimeChangedListener()
        {    @Override
            public void onTimeChanged(TimePicker arg0, int hour, int minute)
            {    MainActivity.this.hour = hour;
                MainActivity.this.minute = minute;
                //显示当前日期、时间
                showDate(year, month+1 , day , hour, minute);
            } }); }
    //定义在TextView中显示当前日期、时间的方法
    private void showDate(int year, int month , int day
        , int hour , int minute)
```

```
{   TextView tv_show = (TextView) findViewById(R.id.TV_show);
    tv_show.setText("您的购买日期为:" + year + "年" + month + "月"
        + day + "日 " + hour + "时" + minute + "分");   }}
```

运行该项目,效果如图6-6所示。可以得到一个使用DatePicker和TimePicker控件实现的简单的日期时间选择器,可以选择设置购买产品的日期时间,可区分上午或者下午,并通过TextView将选择设置的时间给显示出来。

图6-6 日期、时间选择器示例运行效果图

6.6 控件的综合应用案例

控件是Android程序开发的基本单位,通过灵活地使用控件可以高效地开发Android应用程序。本书在前面章节中已经详细介绍了Android开发中常用的基本控件,加上本章前面所讲解的部分高级组件,这里已向读者展示了Android中的大部分控件,熟练掌握控件的使用还需要大量的实战练习。本节我们就利用之前所学习的这些控件,编写一个相对较完善的注册界面。

【实例6-7】本案例中会使用到TextView、EditText、RadioButton、Button、ToggleButton、CheckBox、Spinner、ImageButton、ImageView、Spinner等控件,采用的布局方式是相对布局。首先在"Chapter06"下新建一个Module,命名为"Widget_Synthetical_Demo",修改其布局文件activity_main.xml。

打开src/main/java/cn.androidstudy.widget_synthetical_demo/MainActivity文件,修改并输

入部分代码，如下所示：

```
package cn.androidstudy.widget_synthetical_demo;
import android.support.v7.app.AppCompatActivity;
import android.os.Bundle;
import android.widget.ArrayAdapter;
import android.widget.Spinner;
public class MainActivity extends AppCompatActivity {
    private Spinner zwxz;
    @Override
    protected void onCreate(Bundle savedInstanceState) {
        super.onCreate(savedInstanceState);
        setContentView(R.layout.activity_main);
        zwxz = (Spinner) findViewById(R.id.zwxz);
        String[] a = { "CEO", "CFO", "PM" };
        ArrayAdapter A = new ArrayAdapter(this,
                android.R.layout.simple_spinner_item, a);
        zwxz.setAdapter(A);
    }}
```

运行该项目，效果如图6-7所示，可以得到一个符合预期的相对完善的注册界面。

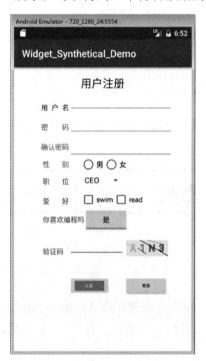

图6-7　控件综合示例运行效果图

强化训练

Android提供了大量的UI控件如何合理选择并搭配相关控件组成UI界面是程序员要通过大量实践练习才能熟练掌握的。

本章主要以案例的形式讲解了ProgressBar（进度条）、ListView（列表视图）、Spinner（下拉列表）、GridView（网格视图）等控件，介绍了这些控件的常用属性以及使用方法。最后通过一个控件的综合应用案例进一步加深对这些控件使用方法的理解和掌握。对于初学者来说有一定帮助，请在开发工具中多调试本章的案例。

学完本章知识后，不妨通过以下习题来巩固所学的内容吧。

一、填空题

1. ＿＿＿＿是UI界面中一种经常使用的组件，通常用于向用户显示某个耗时操作完成的百分比。

2. Android支持几种风格的进度条，最常见的两种进度条是"＿＿＿＿"和"＿＿＿＿"，可以通过style属性指定风格。

3. 每个软件基本上都会使用ListView，列表的显示需要三个元素：＿＿＿＿，＿＿＿＿，＿＿＿＿。

4. Spinner的数据源和＿＿＿＿一样，可以通过＿＿＿＿给适配器提供数据，也可以通过＿＿＿＿给适配器提供数据。

5. ＿＿＿＿是按照行列的方式来显示内容的，一般用于显示图片等内容，比如实现九宫格图，用＿＿＿＿是首选，也是最简单的。一般主要用于设置＿＿＿＿。

6. ＿＿＿＿和＿＿＿＿是两个比较易用的控件，它们都是从FrameLayout派生出来的，前者供用户选择日期，后者供用户选择时间。

二、选择题

1. 若要实现九宫格图，在Android Studio中需要（　　）视图。
   A. ListView
   B. Spinner
   C. GridView
   D. DatePicker

2. 在Android编程中日期选择器常见方法中，利用（　　）选择的天数。
   A. public int getMonth ()
   B. public void init
   C. onDateChangedListener
   D. public int getDayOfMonth ()

3. 在Android编程中时间选择器的常用方法中，（　　）功能是设置可用的视图状态。
   A. public void setEnabled
   B. public Integer getCurrentMinute ()
   C. public void setIs24HourView
   D. public int getBaseline ()

4. 在Android Studio中能够提高用户界面友好性的是（　　）。
   A. ListView
   B. Spinner
   C. ProgressBar
   D. TimePicker

三、操作题

1. 进度条是UI界面中一种经常使用的组件。练习利用Android Studio开发环境创建一个新的Android应用程序，练习使用实例演示学习ProgressBar（进度条）的应用。

2. ListView（列表视图）是Android软件开发中非常重要组件之一，它以列表的形式展示具体内容。练习利用Android Studio开发环境创建一个新的Android应用程序，练习实现一个简单的纯文本列表控件和一个相对复杂的列表控件。

3. 手机屏幕较小，当需要用户选择的时候，可以提供一个下拉列表将所有可选项列出来供用户选择，以此提升用户的体验。练习利用Android Studio开发环境创建一个新的Android应用程序，练习使用实例演示学习Spinner（下拉列表）的应用。

第7章

菜单和对话框的应用

内容导读

在应用程序中,菜单(Menu)的使用非常广泛。由于特殊情况,Android应用程序中菜单的使用在一定程度上有所减少。不过Android的SDK还是提供了三种类型的菜单供程序员使用:Options Menu(选项菜单)、Context Menu(上下文菜单)和Sub Menu(子菜单)。对话框是用户经常使用的控件之一,一般用来显示一些提示性信息,让用户自主选择,或者在一些操作不可逆的情况下提示用户是否继续操作。本章将通过实际案例介绍常见的对话框,分别为提示对话框、单选对话框、多选对话框、列表对话框、自定义对话框、进度对话框、日期和时间选择对话框等。

学习目标

- 了解Android中的菜单处理机制;
- 掌握Android中选项菜单和子菜单的使用;
- 掌握Android中上下文菜单的使用;
- 掌握Android中常用对话框的使用。

7.1 选项菜单和子菜单

由于屏幕等资源的限制,在Android中一个菜单只会显示不多的菜单项,超出部分会自动隐藏，而且会出现一个"更多"的菜单项提示用户。在Android开发中，基于相关资源的限制，如何使用好菜单控件是Android程序开发人员要仔细思考的。

Android开发中，菜单的创建有两种方式。

第一种是通过动态创建，也就是通过在Activity中重写onCreateOptionMenu方法来创建菜单，一般使用menu.add("选项名")添加菜单选项，menu.getItem(int itemIndex)则是根据选项的索引获取该菜单选项，并对其进行操作，例如添加单击事件onMenuItemClickListener，用addSubMenu添加子菜单等。这种方法的优点是创建方便，不需要建立资源文件，而缺点是对菜单条目过多或多级菜单操作不方便，按键多的时候添加单击事件不方便。

另一种方式是通过XML资源文件创建，也就是在res文件夹下创建名为menu的文件夹，通过该文件夹下的menu.xml创建菜单样式，然后再利用getMenuInflater().inflate(menuRes, menu)将创建的菜单样式压入（inflate）到程序中。这种方法大体分为三步：首先，建立XML资源文件；其次，在onCreateOptionMenu回调函数完成XML资源的压入，把菜单资源文件压入到程序中，不同菜单使用不同的回调函数，例如ContextMenu使用onCreateContextMenu；最后，如果需要给菜单添加单击事件，一般是通过重写onMenuItemSelected来设置的，不同菜单使用不同的单击回调函数，例如ContextMenu使用onContextItemSelected。

7.1.1 选项菜单Options Menu

Options Menu是通过按Menu键来显示，Context Menu通过在View（视图控件上）上按2秒之后才会显示，这两种菜单都可以嵌套子菜单，一般常在Options Menu中嵌套，但是子菜单里不能嵌套子菜单。

Options Menu只能在屏幕最下面显示6个菜单选项，这6个菜单叫作Icon Menu(图标菜单)，图标菜单不能有checkable选项，多于6项的会以More Icon Menu菜单来调出，叫作Expanded Menu（扩展菜单）。Options Menu通过Activity的onCreateOptionsMenu来生成，这个函数只会在Menu第一次生成时调用，想要改变Options Menu只能通过onPrepareOptionsMenu来实现，这个函数会在Menu显示之前被调用。onOptionsItemSeleted用来处理选中的Options Menu菜单项。

选项菜单Options Menu默认样式是在屏幕底部弹出一个菜单，其实现方式有两种：第一种是通过Menu类创建菜单；第二种是通过XML布局文件添加菜单的样式。下面通过一个案例来学习如何使用两种方式创建选项菜单。

☞【实例7-1】创建一个Android工程，工程名为"Chapter07"，在该工程下创建一个Module,命名为"OptionsMenu_Demo"。首先学习在Java代码文件中使用Menu类创建菜单。重载onCreateOptionsMenu(Menu menu)方法，并在此方法中通过Menu类的add方法添加菜单项，该方法的四个参数依次代表组别、ID、显示顺序和菜单的显示文本，最后返回true，如果返回false，菜单则不会显示。

使用onOptionsItemSelected(MenuItem item)方法为菜单项注册事件，其他按需要重载，例如重载onOptionsMenuClosed(Menu menu)可以处理菜单关闭后发生的动作等。打开src/main/java/cn.androidstudy.optionsmenu_demo/MainActivity文件，代码如下所示：

```
package cn.androidstudy.optionsmenu_demo;
import android.support.v7.app.AppCompatActivity;
```

```java
import android.os.Bundle;
import android.view.Menu;
import android.view.MenuItem;
import android.widget.Toast;
import java.lang.reflect.Method;
public class MainActivity extends AppCompatActivity {
    @Override
    protected void onCreate(Bundle savedInstanceState) {
        super.onCreate(savedInstanceState);
        setContentView(R.layout.activity_main);     }
    public boolean onCreateOptionsMenu(Menu menu) {
        /* * * add()方法的四个参数,依次是:
         * 1.组别,如果不分组的话就写Menu.NONE, * *
         * 2. ID,这个很重要,Android根据这个ID来确定不同的菜单 *
         * 3.顺序,那个菜单现在在前面由这个参数的大小决定 * *
         * 4.文本,菜单的显示文本 */
        menu.add(Menu.NONE, Menu.FIRST + 1, 5, "删除").setIcon(
                android.R.drawable.ic_menu_add);
        // setIcon()方法为菜单设置图标,这里使用的是系统自带的图标
        menu.add(Menu.NONE, Menu.FIRST + 2, 2, "保存").setIcon(
                android.R.drawable.ic_menu_edit);
        menu.add(Menu.NONE, Menu.FIRST + 3, 6, "帮助").setIcon(
                android.R.drawable.ic_menu_help);
        menu.add(Menu.NONE, Menu.FIRST + 4, 1, "添加").setIcon(
                android.R.drawable.ic_menu_add);
        menu.add(Menu.NONE, Menu.FIRST + 5, 4, "详细").setIcon(
                android.R.drawable.ic_menu_info_details);
        menu.add(Menu.NONE, Menu.FIRST + 6, 3, "发送").setIcon(
                android.R.drawable.ic_menu_send);
        return true;     }
    public boolean onOptionsItemSelected(MenuItem item) {
        switch (item.getItemId()) {
            case Menu.FIRST + 1:
            Toast.makeText(this, "删除菜单被单击了", Toast.LENGTH_LONG).show();
                break;
            case Menu.FIRST + 2:
             Toast.makeText(this, "保存菜单被单击了", Toast.LENGTH_LONG).show();
                break;
            case Menu.FIRST + 3:
             Toast.makeText(this, "帮助菜单被单击了", Toast.LENGTH_LONG).show();
                break;
            case Menu.FIRST + 4:
```

```
            Toast.makeText(this,"添加菜单被单击了", Toast.LENGTH_LONG).show();
              break;
          case Menu.FIRST + 5:
            Toast.makeText(this,"详细菜单被单击了", Toast.LENGTH_LONG).show();
              break;
          case Menu.FIRST + 6:
            Toast.makeText(this,"发送菜单被单击了", Toast.LENGTH_LONG).show();
              break;            }
        return false;        }
    public void onOptionsMenuClosed(Menu menu) {
        Toast.makeText(this,"选项菜单关闭了", Toast.LENGTH_LONG).show();
    }}
```

运行该项目，效果如图7-1所示。

图7-1　选项菜单示例运行效果图

运行后发现我们给菜单项设置的图标都没有显示出来，这是因为从Android4.0开始已经默认Icon是不显示的，如果想要将菜单项的图标显示出来需要再添加代码，具体操作在第二种使用XML布局文件实现菜单的方法中介绍。

使用XML布局文件的方法需要在该Module的res文件夹下先创建一个menu文件夹，在里面创建一个XML文件命名为optionsmenu.xml。另外需要在strings.xml中新建所需要的字符串，打开res/layout/strings.xml文件，修改代码如下：

```
<resources>
    <string name="app_name">OptionsMenu_Demo</string>
```

```xml
        <string name="add">添加</string>
        <string name="save">保存</string>
        <string name="send">发送</string>
        <string name="detail">详情</string>
        <string name="delete">删除</string>
        <string name="help">帮助</string>
    </resources>
```

打开/res/menu目录下的optionsmenu.xml文件，修改编写代码如下：

```xml
<?xml version="1.0" encoding="utf-8"?>
<menu xmlns:android="http://schemas.android.com/apk/res/android">
    <item
        android:id="@+id/add"
        android:title="@string/add"
        android:icon="@android:drawable/ic_menu_add" />
    <item
        android:id="@+id/sava"
        android:title="@string/save"
        android:icon="@android:drawable/ic_menu_save" />
    <item
        android:id="@+id/send"
        android:title="@string/send"
        android:icon="@android:drawable/ic_menu_send" />
    <item
        android:id="@+id/detail"
        android:title="@string/detail"
        android:icon="@android:drawable/ic_menu_info_details" />
    <item
        android:id="@+id/delete"
        android:title="@string/delete"
        android:icon="@android:drawable/ic_menu_delete" />
    <item
        android:id="@+id/help"
        android:title="@string/help"
        android:icon="@android:drawable/ic_menu_help" />
</menu>
```

创建一个命名为DefaultMenuActivity的Java类，继承自Activity，并在其中重载onCreateOptionsMenu(Menu menu)方法，并使用onOptionsItemSelected(MenuItem item)方法为菜单项注册事件，通过重写调用setIconEnable方法实现显示菜单项的图标。修改编写代码如下：

```java
package cn.androidstudy.optionsmenu_demo;
import android.app.Activity;
import android.os.Bundle;
import android.view.Menu;
import android.view.MenuInflater;
import android.view.MenuItem;
import android.widget.Toast;
import java.lang.reflect.Method;
public class DefaultMenuActivity extends Activity {
    public void onCreate(Bundle savedInstanceState) {
        super.onCreate(savedInstanceState);
        setContentView(R.layout.activity_main);
    }
    private void setIconEnable(Menu menu, boolean enable)
    {   try
            {lass<?> clazz = Class.forName("com.android.internal.view.menu.MenuBuilder");
            Method m = clazz.getDeclaredMethod("setOptionalIconsVisible", boolean.class);
            m.setAccessible(true);
            //MenuBuilder实现Menu接口,创建菜单时,传进来的Menu其实就是MenuBuilder对象(Java的多态特征)
            m.invoke(menu, enable);
        } catch (Exception e)
        {          e.printStackTrace();           }      }
    public boolean onCreateOptionsMenu(Menu menu) {
        MenuInflater inflater = getMenuInflater();
        inflater.inflate(R.menu.optionsmenu, menu);
        setIconEnable(menu,true);
        return true;        }
    public boolean onOptionsItemSelected(MenuItem item) {
        switch (item.getItemId()) {
            case R.id.delete:
                Toast.makeText(this, "删除菜单被单击了", Toast.LENGTH_LONG).show();
                break;
            case R.id.sava:
                Toast.makeText(this, "保存菜单被单击了", Toast.LENGTH_LONG).show();
                break;
            case R.id.help:
                Toast.makeText(this, "帮助菜单被单击了", Toast.LENGTH_LONG).show();
                break;
            case R.id.add:
```

```
                Toast.makeText(this, "添加菜单被单击了", Toast.LENGTH_LONG).show();
                break;
            case R.id.detail:
                Toast.makeText(this, "详细菜单被单击了", Toast.LENGTH_LONG).show();
                break;
            case R.id.send:
                Toast.makeText(this, "发送菜单被单击了", Toast.LENGTH_LONG).show();
                break;            }
        return false;        }
    public void onOptionsMenuClosed(Menu menu) {
        Toast.makeText(this, "选项菜单关闭了", Toast.LENGTH_LONG).show();
}}
```

修改清单配置文件AndroidManifest.xml，为DefaultMenuActivity注册信息，并将其设置为程序运行的初始Activity页面，代码如下：

```xml
<?xml version="1.0" encoding="utf-8"?>
<manifest xmlns:android="http://schemas.android.com/apk/res/android"
    package="cn.androidstudy.optionsmenu_demo">
    <application
        android:allowBackup="true"
        android:icon="@mipmap/ic_launcher"
        android:label="@string/app_name"
        android:supportsRtl="true" android:theme="@style/AppTheme">
        <activity android:name=".MainActivity">
        </activity>
        <activity android:name=".DefaultMenu">
            <intent-filter>
                <action android:name="android.intent.action.MAIN" />
                <category android:name="android.intent.category.LAUNCHER" />
            </intent-filter>
        </activity>
    </application>
</manifest>
```

运行该项目，效果如图7-2所示。

7.1.2 监听菜单事件

除了重写onOptionsItemSelected(MenuItem item)方法处理菜单的单击事件外，可以通过菜单项的setOnMenuItemClickListener方法为不同的菜单项分别绑定监听器。采用这种方式无须为每个菜单项指定ID，而是通过获取所添加的MenuItem对象，然后给对象绑定监听器。以下给出"删除"菜单项的创建、事件注册与响应。由于篇幅有限，其他菜单项的实现与其相同。

图7-2 带图标选项菜单示例运行效果图

```
public boolean onCreateOptionsMenu(Menu menu) {
  MenuItem deleteItem = menu.add(Menu.NONE, Menu.FIRST + 1, 5, "
删除").setIcon(android.R.drawable.ic_menu_delete);
  deleteItem.setOnMenuItemClickListener(new OnMenuItemClickListener() {
  public boolean onMenuItemClick(MenuItem arg0) {
      Toast.makeText(MenuDemo2.this, "删除菜单被单击了",
      Toast.LENGTH_LONG).show();
      return false;        }       });
return true;}
```

说明：

① 该方法实现效果与前一方法完全相同，区别仅在于处理菜单事件的监听方式不同。一般来说通过重载 onOptionsItemSelected(MenuItem item) 方法处理菜单的单击事件更加简洁，因为所有的事件处理代码都控制在该方法内，通过绑定事件监听器使程序具有更清晰的逻辑性，但是代码显得有些臃肿。

② 如果是通过XML布局文件实现的菜单，可以通过MenuItem delete=(MenuItem)findViewById(R.id.delete)语句获取菜单项对象。

③ 如果我们希望所创建的菜单项是单选菜单项或多选菜单项，可以调用菜单项的setCheckable(Boolean chackable)来设置该菜单项是否可以被勾选。通过调用setGroupCheckable设置组里的菜单是否可勾选

④ 可以通过菜单项的setShortcut方法为其设置快捷键。

第7章 菜单和对话框的应用

7.1.3 与菜单项关联的Activity的设置

在应用程序中如果需要单击某个菜单项来启动其他Activity或者Service时，最简单的方法是不需要开发者编写任何事件处理代码，只要调用MenuItem的setIntent(Intent intent)方法即可。该方法实现把菜单项与指定的Intent关联在一起，当用户单击该菜单项时，该Intent所代表的组件将会被启动。

☞【实例7-2】通过菜单项启动另一个Activity，效果如图7-3和图7-4所示。当选择了"跳转到另一个界面"菜单项会启动另一个Activity。

图7-3 菜单项跳转页面示例运行效果图　　图7-4 跳转页面后效果图

在工程"Chapter07"下创建一个Module，命名为"Menu_Start_Activity_Demo"。在初始页面MainActivity中，重载onCreateOptionsMenu方法，代码如下：

```
package cn.androidstudy.menu_start_activity_demo;
import android.content.Intent;
import android.support.v7.app.AppCompatActivity;
import android.os.Bundle;
import android.view.Menu;
import android.view.MenuInflater;
import android.view.MenuItem;
import android.widget.Toast;
public class MainActivity extends AppCompatActivity {
```

```java
@Override
protected void onCreate(Bundle savedInstanceState) {
    super.onCreate(savedInstanceState);
    setContentView(R.layout.activity_main);    }
public boolean onCreateOptionsMenu(Menu menu) {
    MenuItem mi = menu.add("跳转到另一个页面");
    mi.setIntent(new Intent(this, SecondActivity.class));
    return super.onCreateOptionsMenu(menu);}}
```

创建SecondActivity，并给其创建布局文件layout_second_activity.xml，在AndroidManifest.xml中注册，在其中添加：

```xml
<activity android:name=".SecondActivity">
</activity>
```

运行程序出现图7-3界面，单击"跳转到另一个页面"，菜单项就会启动SecondActivity页面。

7.1.4 子菜单Sub Menu

Sub Menu继承了Menu，它代表子菜单，实际就是将功能相同或相似的分组进行多级显示的一种菜单。除了Menu的方法外，还有设置菜单头图标的方法SetHeaderIcon(Drawable icon)、设置菜单头标题的方法 SetHeaderTitle(int titleRes)和使用View来设置菜单头的方法SetHeaderView(View view)等设置属性的方法。

☞【实例7-3】下面通过一个简单的案例学习Sub Menu的使用。在工程"Chapter07"下创建一个Module，命名为"SubMenu_Demo"。创建子菜单同样有两种方法，首先学习第一种方法，使用addSubMenu()方法在Java代码中动态添加子菜单。打开src/main/java/cn.androidstudy.submenu_demo路径下的MainActivity文件，修改并输入部分代码，如下所示：

```java
package cn.androidstudy.submenu_demo;
import android.support.v7.app.AppCompatActivity;
import android.os.Bundle;
import android.view.Menu;
import android.view.MenuInflater;
import android.view.MenuItem;
import android.view.SubMenu;
import android.widget.Toast;
public class MainActivity extends AppCompatActivity {
    @Override
    protected void onCreate(Bundle savedInstanceState) {
        super.onCreate(savedInstanceState);
        setContentView(R.layout.activity_main);    }
    @Override
```

```java
        public boolean onCreateOptionsMenu(Menu menu) {
            // Inflate the menu; this adds items to the action bar if it is present.
            getMenuInflater().inflate(R.menu.main, menu);
            SubMenu file=menu.addSubMenu("文件");
            SubMenu edit=menu.addSubMenu("编辑");
            file.add(1, 1, 1, "新建");
            file.add(1, 2, 1, "打开");
            file.add(1, 3, 1, "保存");
            file.setHeaderTitle("文件操作");
            file.setHeaderIcon(R.mipmap.ic_launcher);
            edit.add(2, 1, 1, "复制");
            edit.add(2, 2, 1, "粘贴");
            edit.add(2, 3, 1, "剪切");
            edit.setHeaderTitle("编辑操作");
            edit.setHeaderIcon(R.mipmap.ic_launcher);
            return true;    }
    @Override
    public boolean onOptionsItemSelected(MenuItem item) {
        // TODO Auto-generated method stub
        if (item.getGroupId()==1) {
            switch (item.getItemId()) {
                case 1:
                    Toast.makeText(MainActivity.this, "单击了新建", Toast.LENGTH_LONG).show();
                    break;
                case 2:
                    Toast.makeText(MainActivity.this, "单击了打开", Toast.LENGTH_LONG).show();
                    break;
                case 3:
                    Toast.makeText(MainActivity.this, "单击了保存", Toast.LENGTH_LONG).show();
                    break;
                default:    break;    }
        }else if (item.getGroupId()==2) {
            switch (item.getItemId()) {
                case 1:
                    Toast.makeText(MainActivity.this, "单击了复制", Toast.LENGTH_LONG).show();
                    break;
                case 2:
```

```
                Toast.makeText(MainActivity.this, "单击了粘贴", Toast.LENGTH_
LONG).show();
                    break;
                case 3:
            Toast.makeText(MainActivity.this, "单击了剪切", Toast.LENGTH_
LONG).show();
                    break;
                default: break;         } }
        return super.onOptionsItemSelected(item);    }}
```

运行该项目,效果如图7-5所示。单击模拟器上的Menu按钮或者按下Page Up键或Ctrl+M组合键,则会显示出定义的子菜单。

图7-5 动态添加SubMenu案例运行效果图

第二种方法是使用XML资源文件添加子菜单,在名为"SubMenu_Demo"的Moudle中找到资源文件夹res,创建menu文件夹,并在其中创建submenu.xml资源文件,修改编辑代码见光盘中。

再打开MainActivity文件,修改并输入部分代码,如下所示:

```
package cn.androidstudy.SubMenu_Demo;
import android.support.v7.app.AppCompatActivity;
import android.os.Bundle;
import android.view.Menu;
import android.view.MenuInflater;
import android.view.MenuItem;
import android.view.SubMenu;
```

```java
import android.widget.Toast;
public class MainActivity extends AppCompatActivity {
    @Override
    protected void onCreate(Bundle savedInstanceState) {
        super.onCreate(savedInstanceState);
        setContentView(R.layout.activity_main);      }
    @Override
    public boolean onCreateOptionsMenu(Menu menu) {
        MenuInflater inflater=getMenuInflater();
        inflater.inflate(R.menu.submenu, menu);
        return true;      }
    @Override
    public boolean onOptionsItemSelected(MenuItem item) {
        // TODO Auto-generated method stub
        switch (item.getItemId()) {
            case R.id.new_file:
                Toast.makeText(this, "单击了新建", Toast.LENGTH_LONG).show();
                break;
            case R.id.open_file:
                Toast.makeText(this, "单击了打开", Toast.LENGTH_LONG).show();
                break;
            case R.id.save_file:
                Toast.makeText(this, "单击了保存", Toast.LENGTH_LONG).show();
                break;
            case R.id.c_edit:
                Toast.makeText(this, "单击了复制", Toast.LENGTH_LONG).show();
                break;
            case R.id.v_edit:
                Toast.makeText(this, "单击了粘贴", Toast.LENGTH_LONG).show();
                break;
            case R.id.x_edit:
                Toast.makeText(this, "单击了剪切", Toast.LENGTH_LONG).show();
                break;
            default:  break;           }
        return super.onOptionsItemSelected(item);     }}
```

运行该项目，效果如图7-6所示。单击模拟器上的Menu按钮或者按下Page Up键或

Ctrl+M组合键,则会显示出定义的子菜单。

图7-6 用XML添加SubMenu案例运行效果图

7.2 上下文菜单Context Menu

由于Android运行的平台一般都为触摸屏,在Android开发中,上下文菜单控件Context Menu也会经常为Android程序开发人员所用。Android用Context Menu来代表上下文菜单,类似于桌面程序的右键弹出式菜单,在Android中不是通过用户单击鼠标右键得到的,通过长时间按住界面上的元素,就会出现事先设计好的上下文菜单。开发上下文菜单的方法与开发选项菜单的方法相似,因为Context Menu也是Menu的子类,所以可用相同的方法为它添加菜单项。其区别在于:开发上下文菜单不是重写onCreateOptionsMenu(Menu menu)方法,而是重写onCreateContextMenu(ContextMenu menu, View source,ContextMenu.ContextMenuIfno menuInfo)方法。该方法在每次启动上下文菜单时都会被调用一次,在该方法中可以通过使用add()方法添加相应的菜单项。

开发上下文菜单的步骤如下。

第一步,重写 onCreateContextMenu()方法。

第二步,调用Activity的registerForContextMenu(View view)为View组件注册上下文菜单。

第三步,重载onContextItemSelected(MenuItem mi)或者绑定事件监听器,对菜单项进行事件响应。

☞【实例7-4】在工程名"Chapter07"下创建一个Module,命名为"ContextMenu_

Demo"。定义上下文菜单，让用户选择颜色，根据用户所选颜色的不同来更改文本框的背景颜色。打开 src/main/java/cn.androidstudy.contextmenu_demo/MainActivity 文件，修改并输入部分代码，如下所示：

```java
package cn.androidstudy.contextmenu_demo;
import android.graphics.Color;
import android.support.v7.app.AppCompatActivity;
import android.os.Bundle;
import android.view.ContextMenu;
import android.view.Menu;
import android.view.MenuInflater;
import android.view.MenuItem;
import android.view.View;
import android.widget.ArrayAdapter;
import android.widget.ListView;
import android.widget.TextView;
import android.widget.Toast;
import java.util.ArrayList;
public class MainActivity extends AppCompatActivity {
    TextView text;
    public void onCreate(Bundle savedInstanceState) {
        super.onCreate(savedInstanceState);
        setContentView(R.layout.activity_main);
        text = (TextView) findViewById(R.id.contextmenu);
        registerForContextMenu(text);// 为文本框注册上下文菜单
    }
    // 每次创建上下文菜单时都会触发该方法
    public void onCreateContextMenu(ContextMenu menu, View v,
                                   ContextMenu.ContextMenuInfo menuInfo) {
        text.setText("这是关于上下文菜单的测试操作！");// 设置提示信息
        menu.add(1, Menu.FIRST + 1, 1, "红色");
        menu.add(1, Menu.FIRST + 2, 2, "绿色");
        menu.add(1, Menu.FIRST + 3, 3, "蓝色");
        menu.add(1, Menu.FIRST + 4, 4, "退出");
        menu.setGroupCheckable(1, true, true);
        menu.setHeaderTitle("请选择：");
        menu.setHeaderIcon(android.R.drawable.ic_dialog_info);    }
    public boolean onContextItemSelected(MenuItem item) {
        switch (item.getItemId()) {
            case Menu.FIRST + 1:
                text.setBackgroundColor(Color.RED);
                break;
```

```
        case Menu.FIRST + 2:
            text.setBackgroundColor(Color.GREEN);
            break;
        case Menu.FIRST + 3:
            text.setBackgroundColor(Color.BLUE);
            break;
        case Menu.FIRST + 4:
            MainActivity.this.finish();
            break;
        default:
            item.setChecked(true);          }
    return true;       }}
```

运行该项目，效果如图7-7所示。单击长按页面中的TextView控件区域，则会显示添加的上下文菜单，并且更改TextView的文本内容，如图7-8所示。单击上下文菜单中的任意选项，能够将选择的颜色设置为TextView的背景颜色，如图7-9所示。

图7-7　运行初始界面　　　图7-8　上下文菜单效果图　　　图7-9　选择上下文菜单项后效果

7.3 Android中的常用对话框

在Android中，对话框是一个非常常见的控件，而且Android中对话框有许多不同的形式以供Android程序开发人员选择使用。Android Studio提供了丰富的对话框支持，这里先介绍常用的4种对话框。

- AlertDialog：功能丰富、应用较广泛。
- ProgressDialog：进度对话框，这个对话框只是对简单进度条的封装。
- DatePickerDialog：日期选择对话框，这个对话框是对DatePicker的包装。
- TimePickerDialog：时间选择对话框，是对TimePicker的包装。

这四种对话框中功能最强、用法最灵活的是AlertDialog，因此应用最为广泛，本节主要介绍AlertDialog的功能和用法。

AlertDialog是一个提示窗口,要求用户做出选择。该对话框中一般会有几个选择按钮、标题信息和提示信息。AlertDialog提供了一些方法来生成四种预定义对话框。

- 带消息、带N个按钮的提示对话框。
- 带列表、带N个按钮的列表对话框。
- 带多个单选列表项、带N个按钮的对话框。
- 带多个多选列表项、带N个按钮的对话框。

AlertDialog的创建方式有两种:一是直接新建一个AlertDialog对象,然后调用AlertDialog对象的show和dismiss方法来控制对话框的显示和隐藏;二是在Activity的onCreateDialog(int id)方法中创建AlertDialog对象并返回,然后调用Activty的showDialog(int id)和dismissDialog(int id)来显示和隐藏对话框。区别在于通过第二种方式创建的对话框会继承Activity的属性,例如获得Activity的menu事件等。

创建AlertDialog的主要步骤如下:

第一步,创建Andorid项目;
第二步,获得AlertDialog的静态内部类Builder对象,由该类创建对话框;
第三步,通过Builder对象设置对话框的标题、按钮以及按钮将要响应的事件;
第四步,调用Builder对象的create()方法创建对话框;
第五步,调用AlertDialog的show()方法显示对话框。

7.3.1 提示对话框

提示对话框是最简单也是最常见的对话框。本节实例中,设定在程序初始页面中添加一个按钮,用来模拟单击触发弹出提示对话框的操作,在对话框中设定标题、图标、提示信息、确认按钮、取消按钮。

【实例7-5】首先在工程"Chapter07"下新建一个Module,命名为"Dialog_Demo",修改其布局文件activity_main.xml。

打开src/main/java/cn.androidstudy.dialog_demo/MainActivity文件,修改并输入部分代码,如下所示:

```
package cn.androidstudy.dialog_demo;
import android.content.DialogInterface;
import android.support.v7.app.AlertDialog;
import android.support.v7.app.AppCompatActivity;
import android.os.Bundle;
import android.view.View;
import android.widget.Toast;
public class MainActivity extends AppCompatActivity {
  @Override
  protected void onCreate(Bundle savedInstanceState) {
    super.onCreate(savedInstanceState);
    setContentView(R.layout.activity_main);
    initClickEvent();    }
//初始化单击事件
```

```java
    private void initClickEvent() {
        findViewById(R.id.btn_promptDialog).setOnClickListener(new View.OnClickListener() {
            @Override
            public void onClick(View arg0) {
                // TODO Auto-generated method stub
                showDialog1();
            }
        });
    }
    //显示提示对话框
    private void showDialog1() {
        AlertDialog.Builder builder=new AlertDialog.Builder(this);
        builder.setTitle("提示对话框");//设置标题
        builder.setIcon(R.mipmap.ic_launcher);//设置图标
        builder.setMessage("确认退出应用程序吗? ");//设置内容
        builder.setPositiveButton("确定",new DialogInterface.OnClickListener() {
            @Override
            public void onClick(DialogInterface arg0, int arg1) {
                // TODO Auto-generated method stub
                Toast.makeText(MainActivity.this,"单击的确认按钮", Toast.LENGTH_LONG).show();
            }
        });
        builder.setNegativeButton("取消", new DialogInterface.OnClickListener() {
            @Override
            public void onClick(DialogInterface arg0, int arg1) {
                // TODO Auto-generated method stub
                Toast.makeText(MainActivity.this,"单击的取消按钮", Toast.LENGTH_LONG).show();
            }
        });
        AlertDialog dialog=builder.create();//获取Dialog
        dialog.show();//显示对话框  }}
```

运行该项目,效果如图7-10所示。单击初始页面中的"提示对话框"按钮,就会弹出提示对话框。单击对话框中的"确认"或者"取消"按钮会弹出相应的提示信息。

7.3.2 单选对话框

在Module的"Dialog_Demo"中,修改activity_main.xml布局文件和MainActivity.java文件中的代码,实现性别选择的单选对话框,部分代码如下所示:

```java
//显示单选对话框
private void showDialog2() {
    AlertDialog.Builder builder=new AlertDialog.Builder(this);
    builder.setTitle("单选对话框");//设置标题
```

第7章 菜单和对话框的应用

图7-10 提示对话框示例运行效果图

```
    builder.setIcon(R.mipmap.ic_launcher);//设置图标
    builder.setSingleChoiceItems(single_list, 0, new DialogInterface.
OnClickListener() {
        @Override
        public void onClick(DialogInterface arg0, int which) {
        // TODO Auto-generated method stub
        str_radio=single_list[which];
        } });
    builder.setPositiveButton("确定",new DialogInterface.OnClickListener() {
        @Override
        public void onClick(DialogInterface arg0, int arg1) {
        // TODO Auto-generated method stub
  Toast.makeText(MainActivity.this,"你选择的性别是"+str_radio,
Toast.LENGTH_LONG).show();
        }  });
    builder.setNegativeButton("取消", new DialogInterface.OnClickListener() {
        @Override
        public void onClick(DialogInterface arg0, int arg1) {
        // TODO Auto-generated method stub
            arg0.dismiss();   }  });
        AlertDialog dialog=builder.create();//获取Dialog
        dialog.show();//显示对话框  }
```

运行该项目，效果如图7-11所示。单击初始页面中的"单选对话框"按钮，就会弹出单选对话框，"男""女"两个选项选择其一。单击对话框中的"确认"按钮会弹出提示信息显示选中项的内容，单击"取消"按钮则会退出并关闭对话框。

图7-11　单选对话框示例运行效果图

7.3.3　多选对话框

在Module的"Dialog_Demo"中，修改activity_main.xml布局文件和MainActivity.java文件中的代码，实现爱好选择的多选对话框，部分代码如下所示：

```
//显示多选对话框
    private void showDialog3() {
        AlertDialog.Builder builder=new AlertDialog.Builder(this);
        builder.setTitle("多选对话框");//设置标题
        builder.setIcon(R.mipmap.ic_launcher);//设置图标
        //设置默认情况下，每个选项是否被勾选的状态
        final boolean b[]=new boolean[]{false,false,false,false,false};
        builder.setMultiChoiceItems(multi_list, b, new DialogInterface.OnMultiChoiceClickListener() {
            @Override
            public void onClick(DialogInterface dialog, int which, boolean isChecked) {
```

```
                    // TODO Auto-generated method stub
                    b[which]=isChecked;  } });
            builder.setPositiveButton("确认", new DialogInterface.
OnClickListener() {
                @Override
                public void onClick(DialogInterface dialog, int which) {
                    // TODO Auto-generated method stub
                    String item="";
                    //取出被选中的内容
                    for (int i = 0; i < b.length; i++) {
                        if (b[i]) { item+=multi_list[i]+","; } }
                    Toast.makeText(MainActivity.this,"你选择的爱好是:"
+item, Toast.LENGTH_LONG).show();
                    dialog.dismiss();//对话框消失  } });
            //隐藏对话框
            builder.setNegativeButton("取消", new DialogInterface.
OnClickListener() {
                @Override
                public void onClick(DialogInterface dialog, int which) {
                    // TODO Auto-generated method stub
                    dialog.dismiss();} });
            AlertDialog dialog=builder.create();//获取Dialog
            dialog.show();//显示对话框}
```

运行该项目，效果如图7-12所示。单击初始页面中的"多选对话框"按钮，就会弹出多选对话框，"足球""篮球""排球""音乐""书法"五个选项可以同时选择多个。单击对话框中的"确认"按钮会弹出提示信息显示选中项的内容，单击"取消"按钮则会退出关闭对话框。

7.3.4 列表对话框

在Module的"Dialog_Demo"中，修改activity_main.xml布局文件和MainActivity.java文件中的代码，实现班级列表项选择的列表对话框，部分代码如下所示：

```
    //显示列表对话框
    private void showDialog4() {
        AlertDialog.Builder builder=new AlertDialog.Builder(this);
        builder.setTitle("列表对话框");//设置标题
        builder.setIcon(R.mipmap.ic_launcher);//设置图标
        builder.setItems(item_list, new DialogInterface.OnClickListener() {
            @Override
            public void onClick(DialogInterface arg0, int which) {
                // TODO Auto-generated method stub
```

图7-12 多选对话框示例运行效果图

```
                Toast.makeText(MainActivity.this,"你单击了"+item_list[which], Toast.LENGTH_LONG).show();     } });
        //隐藏对话框
        builder.setNegativeButton("取消", new DialogInterface.OnClickListener() {
            @Override
            public void onClick(DialogInterface dialog, int which) {
                // TODO Auto-generated method stub
                dialog.dismiss(); } });
        AlertDialog dialog=builder.create();//获取Dialog
        dialog.show();//显示对话框 }
```

运行该项目,效果如图7-13所示。单击初始页面中的"列表对话框"按钮,就会弹出列表对话框。当单击"计科一班""计科二班""软件一班""软件二班""移动一班""移动二班"中任意一个列表项时,会弹出提示信息显示单击列表项的内容。

7.3.5 自定义对话框

前面通过几个示例介绍了几种常见的对话框,但是在实际开发中可能会遇到这些对话框都不满足需求的情况,这时候可以通过自定义对话框的方式来实现想要的效果。

在Module的"Dialog_Demo"中,新建一个XML布局文件,命名为custom_dialog_layout.xml,用来设置自定义对话框的布局。

修改activity_main.xml布局文件和MainActivity.java文件中的代码,实现包含两个输入框、两个按钮、一个图片控件的自定义对话框,部分代码如下所示:

```
//显示自定义对话框
private void showDialog5() {
    LayoutInflater inflater=LayoutInflater.from(this);
    View view=inflater.inflate(R.layout.custom_dialog_layout, null);
    AlertDialog.Builder builder=new AlertDialog.Builder(this);
    builder.setTitle("自定义对话框");//设置标题
    builder.setIcon(R.mipmap.ic_launcher);//设置图标
    builder.setView(view);
    AlertDialog dialog=builder.create();//获取Dialog
    dialog.show();//显示对话框
}
```

运行该项目，效果如图7-14所示。单击初始页面中的"自定义对话框"按钮，就会弹出自定义对话框，对话框页面中包含标题、图标、两个输入框、两个按钮、一个图片控件。单击"取消"按钮会退出并关闭对话框。可以根据需要添加后续代码，实现单击"确认"按钮时提交输入的信息。

图7-13　列表对话框示例运行效果图

图7-14　自定义对话框示例运行效果图

7.4 进度对话框 ProgressDialog

进度对话框 ProgressDialog 的使用非常简单，只要创建 ProgressDialog 实例并将它显示出来就是一个进度对话框。ProgressDialog 类继承自 AlertDialog 类，存放在 android.app 包中。ProgressDialog 有两种形式：一种是圆圈旋转形式；另一种是水平进度条形式，可以通过属性设置来修改其形式。开发者可以通过该类提供的一系列的 set 方法设置对话框里的进度条的风格、进度条的最大值等属性。本节将以实例来展示进度对话框 ProgressDialog 的使用。

☞【实例7-6】在工程"Chapter07"下创建一个 Module，命名为"Progress_Dialog_Demo"。在主界面上放置一个执行按钮，当单击按钮时，弹出一个进度对话框，提示后台程序正在执行，稍等片刻，打开 res/values/strings.xml 文件，添加定义案例中所需的字符串资源，代码如下所示：

```xml
<resources>
    <string name="app_name">Progress_Dialog_Demo</string>
    <string name="execute">执 行</string>
    <string name="str_dialog_title">请稍等片刻</string>
    <string name="str_dialog_body">正在执行...</string>
</resources>
```

修改 res/layout/activity_main.xml 布局文件。

打开 src/main/java/cn.androidstudy.progress_dialog_demo/MainActivity 文件，修改并输入部分代码，如下所示：

```java
package cn.androidstudy.progress_dialog_demo;
import android.app.ProgressDialog;
import android.support.v7.app.AppCompatActivity;
import android.os.Bundle;
import android.view.View;
import android.widget.Button;
public class MainActivity extends AppCompatActivity {
    private Button button=null;
    public ProgressDialog progressDialog=null;
    public void onCreate(Bundle savedInstanceState) {
        super.onCreate(savedInstanceState);
        setContentView(R.layout.activity_main);
        button=(Button)findViewById(R.id.button);
        button.setOnClickListener(new View.OnClickListener(){
            public void onClick(View v) {
                String title =MainActivity.this.getString(R.string.str_dialog_title);
```

```
            String body =MainActivity.this.getString(R.string.str_
dialog_body);
            //显示Progress对话框
            progressDialog=ProgressDialog.show(MainActivity.
this,title,body,true);
            new Thread()      {
                public void run(){
                    try{       //表示后台运行的代码段，以暂停3s代替
                        Thread.sleep(3000);
                    }catch (InterruptedException e){
                        e.printStackTrace();
                    }finally{     //卸载dialog对象
                        progressDialog.dismiss();
                    }              }
            }.start();               }});}}
```

运行该项目，效果如图7-15所示。

图7-15　ProgressDialog案例运行初始界面

7.5 日期对话框和时间对话框

DatePickerDialog与TimePickerDialog的功能和使用方法相对而言比较简单，本节

将对二者进行讲解并使用实例展示其使用方法。在Android应用中，DatePickerDialog与TimePickerDialog分别表示日期对话框和时间对话框，都是以弹出式对话框形式出现的，使用方法基本相同。前者需要实现OnDateSetListener接口中的onDateSet方法，后者需要实现OnTimeSetListener接口中的onTimeSet方法。步骤如下：

① 创建DatePickerDialog或TimePickerDialog对象，通过它们的show方法将它们显示出来。

② 为日期或者时间对话框对象绑定监听者。

☞【实例7-7】在工程"Chapter07"下创建一个Module，命名为"DPD_TPD_Demo"。在主界面上放置两个命令按钮"显示日期"和"显示时间"。当单击"显示日期"命令按钮时，弹出一个日期显示对话框；当单击"显示时间"命令按钮时，弹出一个时间显示对话框。打开res/values/strings.xml文件，添加定义案例中所需的字符串资源，代码如下所示：

```xml
<resources>
    <string name="app_name">DPD_TPD_Demo</string>
    <string name="showdate">显示日期</string>
    <string name="showtime">显示时间</string>
</resources>
```

打开res/layout/activity_main.xml布局文件，修改代码。

打开src/main/java/cn.androidstudy. dpd_tpd_demo /MainActivity文件，修改并输入部分代码，如下所示：

```java
package cn.androidstudy.dpd_tpd_demo;
import android.app.DatePickerDialog;
import android.app.TimePickerDialog;
import android.icu.util.Calendar;
import android.os.Build;
import android.support.annotation.RequiresApi;
import android.support.v7.app.AppCompatActivity;
import android.os.Bundle;
import android.view.View;
import android.widget.Button;
import android.widget.DatePicker;
import android.widget.EditText;
import android.widget.TimePicker;
public class MainActivity extends AppCompatActivity {
public void onCreate(Bundle savedInstanceState) {
        super.onCreate(savedInstanceState);
        setContentView(R.layout.activity_main);
        Button bt1 = (Button) findViewById(R.id.showdate);
        bt1.setOnClickListener(new ClickLis());
        Button bt2 = (Button) findViewById(R.id.showtime);
        bt2.setOnClickListener(new ClickLis());    }
    class ClickLis implements View.OnClickListener {
```

```java
        @RequiresApi(api = Build.VERSION_CODES.N)
        public void onClick(View v) {
            Calendar c = Calendar.getInstance();
            if (v.getId() == R.id.showdate) {
                new DatePickerDialog(MainActivity.this,
                        new DatePickerDialog.OnDateSetListener() {
                            public void onDateSet(DatePicker view,
int year, int monthOfYear, int dayOfMonth) {
                                EditText show = (EditText) findViewById(R.id.show);
                        show.setText("您选择的日期为:" + year + "年" + monthOfYear
                                        + "月" + dayOfMonth+ "日 ");
                }}, c.get(Calendar.YEAR), c.get(Calendar.MONTH),
                        c.get(Calendar.DAY_OF_MONTH)).show();      }
            else if (v.getId() == R.id.showtime) {
                new TimePickerDialog(MainActivity.this,
                        new TimePickerDialog.OnTimeSetListener() {
                            public void onTimeSet(TimePicker
view, int hourOfDay, int minute) {
                                EditText show = (EditText) findViewById(R.id.show);
                    show.setText("您选择的时间为:" + hourOfDay + "点"
                                        + minute + "分");}},
c.get(Calendar.HOUR_OF_DAY),
                        c.get(Calendar.MINUTE),false).show();
            }   }   }
```

运行该项目,效果如图7-16所示。

单击页面中的"显示日期"按钮,会弹出可以选择日期的对话框,如图7-17所示,单击其中的"OK"按钮,会将选择的日期以文本形式显示到EditText控件中,如图7-18所示。同样,单击页面中的"显示时间"按钮,会弹出可以选择时间的对话框,如图7-19所示,单击其中的"OK"按钮,会将选择的时间以文本形式显示到EditText控件中,如图7-20所示。

图7-16 日期、时间对话框示例运行初始界面

图7-17　日期对话框显示界面

图7-18　所选日期显示到页面

图7-19　时间对话框显示界面

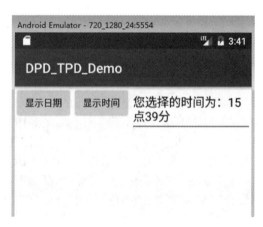

图7-20　所选时间显示到页面

强化训练

菜单（Menu）和对话框（Dialog）是应用程序经常使用到的两种类型控件。本章主要介绍了Android开发中菜单（Menu）和对话框（Dialog）的实现方法，以实例形式较为详细地分析了Android开发实现菜单和对话框的布局及功能的相关技巧，具有一定参考价值。学完本章知识后，不妨通过以下习题来巩固所学的内容吧。

一、填空题

1. Android的SDK提供了三种类型的菜单，分别为_____，_____和_____。

2. Android开发中菜单的创建有两种方式；一种是通过_____；另一种是通过_____。

3. Options Menu只能在屏幕最下面显示___个菜单选项，这___个菜单叫做_____，其不能有checkable选项，多于6项的会以_____菜单来调出，叫作_____。

4. 在应用程序中如果需要单击某个菜单项来启动其他Activity或者Service时，最简单的方法是不需要开发者编写任何事件处理代码，只要调用_____方法即可。该方法实现把菜单项与_____关联在一起，当用户单击该_____时，该Intent所代表的组件将会被启动。

5. _____继承了Menu，它代表了一个_____，实际就是将功能相同或相似的分组进行_____的一种菜单。

6. 除了Menu的方法外还有设置菜单头图标的方法_____、_____和_____等设置属性的方法。

二、选择题

1. 常用的对话框有4种，（   ）是进度对话框。
   A. AlertDialog
   B. TimePickerDialog
   C. ProgressDialog
   D. DatePickerDialog

2. （   ）为创建AlertDialog的主要步骤之一。
   A. 创建Andorid项目
   B. 获得AlertDialog的动态内部类Builder对象
   C. 调用Builder对象的create方法创建进度对话框
   D. 调用AlertDialog的show方法创建对话框

3. ProgressDialog类继承自（   ）类。
   A. Activity
   B. Service
   C. Menu

D. AlertDialog

4. Android用Context Menu来代表上下文菜单，类似于桌面程序的右键弹出式菜单。

A. optionsmenu_demo
B. ContextMenu
C. Menu
D. addSubMenu

三、操作题

1. Activity和Menu都是Android程序经常使用到的组件。练习利用Android Studio开发环境创建一个新的Android应用程序，实现通过菜单项启动另一个Activity。

2. Android提供了丰富的对话框支持，其中功能最强、用法最灵活的就是AlertDialog。练习利用Android Studio开发环境创建一个新的Android应用程序，实现使用AlertDialog创建一个提示窗口，要求用户做出选择。

3. 练习利用Android Studio开发环境创建一个新的Android应用程序，实现以下功能：在主界面上放置一个执行按钮，当单击按钮时创建一个圆圈旋转形式的进度对话框。

第8章 可视化使者之Activity组件

内容导读

活动（Activity）是Android的四大基本组件之一，是开发Android程序的基石，也是Android程序最吸引用户视线以及最易于表现自己个性和作用的地方。Activity作为Android应用程序的表现层，是用于显示可视化的用户界面，并与用户进行交互的地方，是Android应用的门面。在Android应用程序中，一个Activity通常就是一个单独的屏幕。每一个Activity都被实现为一个独立的类，并且从Activity基类中继承而来，Activity类将会显示由视图控件组成的用户接口，并对事件做出响应。所有应用的Activity都继承android.app包中的Activity类，该类是Android提供的基层类，其他的Activity继承该父类后，通过父类的方法来实现各种功能。本章将着重介绍Activity的相关知识并通过实例展示其应用。

学习目标

- 了解什么是Activity；
- 熟悉Activity的生命周期；
- 掌握Activity的创建、配置和使用；
- 掌握Activity的不同启动模式；
- 熟悉Fragment的使用。

8.1 Activity生命周期

对于Android程序开发人员，Activity生命周期极其重要，深入理解并掌握Activity生命周期是极其必要的。图8-1是官方提供的Activity生命周期图，资深Android程序开发人员对该图都再熟悉不过了。当Activity处于Android应用中运行时，它的活动状态由Android以及Activity栈的形式管理。当前活动的Activity位于栈顶。随着不同应用的运行，每个Activity都有可能从活动状态转入非活动状态，也可能从非活动状态转入活动状态。Activity生命周期图中可以看到有三个关键的生命周期循环。

由图8-1可以看到有三个关键的生命周期循环：Activity完整的生命周期、Activity的可视生命周期、Activity的前台生命周期，下面将会逐一进行讲解。一个Activity的生命周期从Activity lanched（加载）开始，到Activity shutdown（销毁）结束，通过调用不同的方法在不同的状态之间进行转换，以达到编程人员对Activity掌控的目的。图8-1中Activity涉及的方法，例如onCreate、onStart、onRestart、onResume等，在表8-1中会详细讲解。

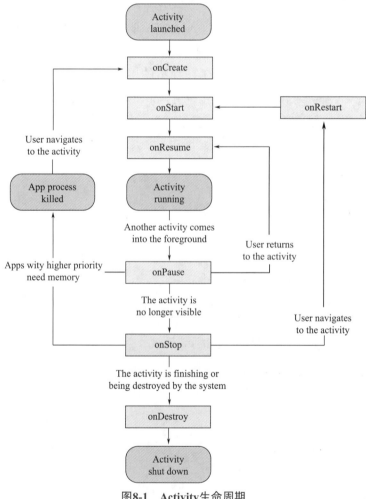

图8-1　Activity生命周期

第8章 可视化使者之Activity组件

表8-1　Activity中方法的调用说明

方法	说明
onCreate	当活动第一次启动的时候，触发该方法，可以在此时完成活动的初始化工作
onStart	该方法的触发表示所属活动将被展现给用户
onResume	当一个活动和用户发生交互的时候，触发该方法
onPause	当一个正在前台运行的活动因为其他的活动需要由前台运行状态而转入后台运行的时候，触发该方法。这时候需要将活动的状态持久化，比如正在编辑的数据库记录等
onStop	当一个活动不再需要展示给用户的时候，触发该方法
onRestart	当处于停止状态的活动需要再次展现给用户的时候，触发该方法
onDestroy	当活动销毁的时候，触发该方法。和onStop方法一样，如果内存紧张，系统会直接结束这个活动而不会触发该方法

（1）一个Activity完整的生命周期

自第一次调用 onCreate(Bundle) 开始，直至调用 onDestroy 为止。Activity 在 onCreate 中设置所有"全局"状态以完成初始化，而在 onDestroy 中释放所有系统资源。比如说，如果 Activity 有一个线程在后台运行用来从网络上下载数据，它会以 onCreate 创建那个线程，而以 onDestroy 销毁那个线程。

（2）一个Activity的可视生命周期

自 onStart 调用开始直到相应的 onStop 调用。在此期间，用户可以在屏幕上看到此 Activity，尽管它也许并不是位于前台或者正在与用户做交互。在这两个方法中，可以管控用来向用户显示这个 Activity 的资源。比如说，可以在 onStart 中注册一个 BroadcastReceiver 来监控会影响到用户 UI 的改变，而在 onStop 中来取消注册，这时用户是无法看到程序显示的内容。onStart 和 onStop 方法可以随着应用程序是否为用户可见而被多次调用。

（3）一个Activity的前台生命周期

自 onResume 调用起至相应的 onPause 调用为止。在此期间，Activity 位于前台最上面并与用户进行交互。Activity 会经常在暂停和恢复之间进行状态转换。比如说当设备转入休眠状态或有新 Activity 启动时，将调用 onPause 方法。当 Activity 获得结果或者接收到新的 intent 的时候会调用 onResume 方法。因此，在这两个方法中的代码应当是轻量级的。

由图 8-1 可以看出，Activity 在整个生命周期中大致会经过如下四个状态。

- 活动状态：当前Activity位于前台，用户可见，可以获得焦点。
- 暂停状态：其他Activity位于前台，该Activity依然可见，只是不能获得焦点。
- 停止状态：该Activity不可见，失去焦点。
- 销毁状态：该Activity结束，或Activity所在的Dalvik进程被结束。

Activity的整个生命周期都定义在下面的接口方法中，所有方法都可以被重载。

```
public class Activity extends ApplicationContext {
    protected void onCreate(Bundle icicle);
    protected void onStart();
    protected void onRestart();
    protected void onResume();
    protected void onFreeze(Bundle outIcicle);
    protected void onPause();
```

```
        protected void onStop();
        protected void onDestroy(); }
```

下面将对这些方法的调用原则进行详细说明。

开发Activity时可以根据需要覆盖指定的方法。其中最常见的就是覆盖onCreate(Bundle saveStatus)方法，之前所有的示例都覆盖了此方法，该方法用于对该Activity执行初始化。除此之外，覆盖onPause方法也很常见，比如用户正在玩一个游戏，此时有电话进来，那么我们需要将当前游戏暂停，并保存该游戏的进行状态，这就可以通过onPause方法来实现。

【实例8-1】下面通过一个示例来演示在一个Activity的生命周期中，这些方法何时被调用。创建名为ActivityLifeCycleDemo的工程，默认的Activity名字为ActivityLifeCycle，其对应的界面非常简单，只有一个退出按钮，单击按钮可以退出程序。代码如下：

扫一扫 看视频

```
package cn.androidstudy.activitylifecycledemo;
import android.support.v10.app.AppCompatActivity;
import android.os.Bundle;
import android.util.Log;
import android.view.View;
import android.view.View.OnClickListener;
import android.widget.Button;
public class ActivityLifeCycle extends AppCompatActivity {
   final String TAG="Hello Activity";
   @Override
   protected void onCreate(Bundle savedInstanceState) {
       super.onCreate(savedInstanceState);
       setContentView(R.layout.activity_life_cycle);
       Log.d(TAG, "----onCreate----");
       Button bt=(Button)findViewById(R.id.button1);
       bt.setOnClickListener(new OnClickListener(){
           public void onClick(View v) {
               ActivityLifeCycle.this.finish();
       } }); }
   protected void onDestroy() {
       super.onDestroy();
       Log.d(TAG, "----onDestroy----");     }
   protected void onPause() {
       super.onPause();
       Log.d(TAG, "----onPause----");       }
   protected void onRestart() {
       super.onRestart();
       Log.d(TAG, "----onRestart----");     }
   protected void onResume() {
       super.onResume();
```

```
        Log.d(TAG, "----onResume----");       }
protected void onStart() {
    super.onStart();
    Log.d(TAG, "----onStart----");        }
protected void onStop() {
    super.onStop();
    Log.d(TAG, "----onStop----");        }}
```

执行该程序,在DDMS的logcat窗口将会看到如图8-2所示信息。

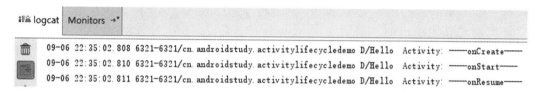

图8-2　启动Activity时调用的回调方法

当程序正在运行时,单击模拟器的 O 按钮,返回桌面,则当前的Activity变得不可见,失去焦点,但并未被销毁,只是处于暂停状态。此时,会看到logcat中的输出信息如图8-3所示。

```
09-06 22:42:26.449 6321-6321/cn.androidstudy.activitylifecycledemo D/Hello  Activity: ----onPause----
09-06 22:42:26.463 6321-6343/cn.androidstudy.activitylifecycledemo D/EGL_emulation: eglMakeCurrent: 0x99cfe2c0: ver 3 1 (tinfo 0xa4a0b620)
09-06 22:42:26.467 6321-6321/cn.androidstudy.activitylifecycledemo D/Hello  Activity: ----onStop----
```

图8-3　暂停Activity时的回调方法

在模拟器列表里找到该应用程序,并再次启动,此时可以在logcat中看到如图8-4所示信息。

```
09-06 22:48:37.258 6321-6321/cn.androidstudy.activitylifecycledemo D/Hello  Activity: ----onRestart----
09-06 22:48:37.259 6321-6321/cn.androidstudy.activitylifecycledemo D/Hello  Activity: ----onStart----
09-06 22:48:37.259 6321-6321/cn.androidstudy.activitylifecycledemo D/Hello  Activity: ----onResume----
```

图8-4　重启Activity时的回调方法

在该程序界面单击退出按钮,该Activity将会结束自己,此时可以在logcat中看到如图8-5所示界面。

```
09-06 22:49:34.700 6321-6321/cn.androidstudy.activitylifecycledemo D/Hello  Activity: ----onStop----
09-06 22:49:34.701 6321-6321/cn.androidstudy.activitylifecycledemo D/Hello  Activity: ----onDestroy----
```

图8-5　结束Activity时的回调方法

通过以上操作,相信读者对于Activity的生命周期状态以及在不同状态之间切换时所回调的方法已经有了清晰的认识。

8.2 Activity 管理栈

当 Activity 在 Android 应用程序中运行时，它的活动状态由 Android 以 Activity 栈的形式进行管理。

一个程序一般由多个 Activity 组成，各活动之间关系很松散，它们之间没有直接的关联。必须有一个 Activity 被指定为主 Activity，它是程序启动时首先显示的界面。每个 Activity 都可以随意启动其他 Activity。每当一个 Activity 被启动，则前一个 Activity 就被停止。一个程序中的所有启动的 Activity 都被放在一个栈中，被停止的 Activity 并没有销毁，而是存于栈中。新启动的 Activity 先被存放于栈中，然后获得输入焦点。在当前活动的 Activity 上点返回键，它被从栈中取出销毁，然后上一个 Activity 被恢复。每个 Activity 的状态是由它在 Activity 栈（是一个后进先出 LIFO，包含所有正在运行 Activity 的队列）中的位置决定的。当一个新的 Activity 启动时，当前活动的 Activity 将会移到 Activity 栈的顶部。如果用户使用后退按钮返回的话，或者前台的 Activity 结束，活动的 Activity 就会被移出栈消亡，而在栈中的上一个活动的 Activity 将会移上来并变为活动状态。

一个应用程序的优先级是受最高优先级的 Activity 影响的。当决定某个应用程序是否要终结并释放资源时，Android 内存管理使用栈来决定基于 Activity 的应用程序的优先级。

8.3 创建、配置和使用 Activity

在使用 Activity 之前需要对其进行的创建、配置，然后才能启动或关闭 Activity。本节将逐一配以实例详细向读者讲解创建、配置和使用 Activity 的方法。

8.3.1 创建 Activity

与开发 Web 应用时建立 Servlet 类类似，建立自己的 Activity 也需要继承 Activity 基类。当然，在不同的应用场景下，有时也要求继承 Activity 的子类。例如，如果应用程序界面只包括列表，则可以让应用程序继承 ListActivity；如果应用程序界面需要实现标签页效果，则可以让应用程序继承 TabActivity。

Activity 类间接或直接地继承了 Context、ContextWrapper、ContextThemeWrapeper 等基类，它的直接子类有 AccountnAuthenticatorActivity、ActivityGroup、Aliasactivity、ExpandableListActivity、FragmentActivity、ListActivity、NativeActivity，间接子类有 LauncherActivity、PreferenceActivity、TabActivity。

当定义了一个 Activity 类，这个 Activity 类何时被实例化、它所包含的方法何时被调用，这些都不是由开发者决定的，而是由 Android 系统决定的。为了响应用户的操作，创建一个 Activity 需要实现一个或多个方法，其中最常见的就是实现 onCreate(Bundle status) 方法。从前面内容可知，该方法将会在 Activity 启动时被回调，该方法调用 Activity 的

setContentView(View view)方法来显示要展现的View。为了管理应用程序界面中的各组件，调用Activity的findViewById(int id)方法来获取程序界面中的组件，按下来去修改各组件的属性和方法即可。这些步骤在之前的章节中经常使用，本节我们将创建一个能处理用户常见动作的Activity。

【实例8-2】创建一个名为ActivityDemo的项目，为其创建一个默认的Activity，名为MainActivity，该Activity能对用户的动作做出反应，并以Toast显示提示信息，所以此Activity需要重写处理事件的相关方法，代码如下：

```java
package cn.androidstudy.activitydemo;
import android.support.v10.app.AppCompatActivity;
import android.os.Bundle;
import android.view.KeyEvent;
import android.view.MotionEvent;
import android.widget.Toast;
public class MainActivity extends AppCompatActivity {
  @Override
    protected void onCreate(Bundle savedInstanceState) {
        super.onCreate(savedInstanceState);
        setContentView(R.layout.activity_main);     }
    public boolean onKeyDown(int keyCode, KeyEvent event) {
        showInfo("按键按下");
        return super.onKeyDown(keyCode, event);     }
    public boolean onTouchEvent(MotionEvent event) {
        float x=event.getX();
        float y=event.getY();
        showInfo("您点击的坐标为:("+x+":"+y+")");
        return super.onTouchEvent(event);     }
    public boolean onKeyUp(int keyCode, KeyEvent event) {
        showInfo("按键弹起！");
        return super.onKeyUp(keyCode, event);     }
    public void showInfo(String info){
        Toast.makeText(MainActivity.this,info, Toast.LENGTH_SHORT).show();   }}
```

执行该程序，当按下按键时，结果如图8-6所示。

当松开按键时，结果如图8-7所示。

当单击屏幕时，结果如图8-8所示。

8.3.2 配置Activity

Android应用要求所有应用程序组件（Activity、Service、ContentProvider、BroadcastReceiver）都必须进行配置。之前我们的项目中都只有一个默认的Activity，只需要编写相关的代码就可以运行，并没有进行配置，这是因为如果一个应用只有一个Activity，系统就默认它是整个程序的入口，并且自动在配置文件中进行配置，如上例ActivityDemo中

图8-6　按键按下时界面　　　图8-7　松开按键时界面　　　图8-8　单击屏幕时界面

的配置文件对MainActivity的默认配置片段如下所示：

```
<activity android:name=".MainActivity">
    <intent-filter>
        <action android:name="android.intent.action.MAIN" />
        <category android:name="android.intent.category.LAUNCHER" />
    </intent-filter>
</activity>
```

其中
```
<intent-filter>
    <action android:name="android.intent.action.MAIN" />
    <category android:name="android.intent.category.LAUNCHER" />
</intent-filter>
```

指明了该Activity是应用程序的入口，即程序一旦启动就会执行该Activity。

但在实际开发中，一个项目不可能只有一个Activity，可能包含多个Activity，那么就需要对每个Activity进行配置，即为<application.../>元素添加<activity.../>子元素。配置Activity时通常指定如下三个属性。

- name：指定该Ativity的实现类。
- icon：指定该Activity对应的图标。
- label：指定该Activity的标签。

除此之外，配置Activity时通常还需要指定一个或多个<intent-filter.../>元素，该元素用于指定该Activity可响应的Intent。关于Intent和IntentFileter的介绍，详见后续章节。

8.3.3 启动关闭Activity

一个应用程序通常包含多个Activity，但是只有一个Activity会作为程序的入口，至于应用中的其他Activity，通常都是由入口Activity启动，或由入口Activity启动的Activity启动。
- Activity启动其他Activity有如下两个方法。
- startActivity(Intent intent)：启动其他Activity。

startActivityForResult(Intent intent,int requestCode)：以指定的请求码（requestCode）启动Activity，而且程序将会等到新启动的Activity的结果［通过重写onActivityResult方法来获取］。

启动Activity时可指定一个requestCode参数，该参数代表了启动Activity的请求码。这个请求码的值由开发者根据业务自行设定，用于标识请求来源。

两个方法都用到了Intent类型的参数，Intent是Android应用中各组件之间通信的重要方式，一个Activity通过Intent来表达自己的"意图"，即想要启动哪个组件；被启动的组件既可以是Activity组件，也可以是Service组件。由一个Activity通过Intent启动另一个Activity的典型方式如图8-9所示。

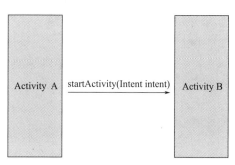

图8-9　两个Activity之间的关系

Android为关闭Activity提供了如下两个方法。
- finish()：结束当前Activity。
- finish(int requestCode)：结束以startActivityForResult(Intent intent,int requestCode)方法启动的Activity。

【实例8-3】下面通过实例演示两个Activity之间的跳转，这两个Activity之间无须传递数据。创建一个名为MultiActivity Demo的项目，默认的Activity名字为FirstActivity，它的界面很简单，只有一个按钮，用来启动第二个Activity，在此不给出布局文件，其代码如下：

```
package cn.androidstudy.multiactivitydemo;
import android.support.v10.app.AppCompatActivity;
import android.os.Bundle;
import android.view.View.OnClickListener;
import android.widget.Button;
import android.content.Intent;
import android.view.View;
public class FirstActivity extends AppCompatActivity {
    @Override
    protected void onCreate(Bundle savedInstanceState) {
```

```java
        super.onCreate(savedInstanceState);
        setContentView(R.layout.activity_first);
        Button bt=(Button)findViewById(R.id.button1);
        bt.setOnClickListener(new OnClickListener(){
            public void onClick(View v) {
                //创建Intent
                Intent intent=new Intent(FirstActivity.this,SecondActivity.class);
                //使用Intent启动另一个Activity
                startActivity(intent);
            } }); }}
```

接着，在该项中再创建一个Activty，名字为SecondActivity，并为该Activty新建一个布局文件，该布局文件也很简单，只有两个按钮，分别用来返回第一个Activity和返回第一个Activity并结束自己，在此不再给出布局文件代码。编写SecondeActivty，其代码如下：

```java
package cn.androidstudy.multiactivitydemo;
import android.content.Intent;
import android.support.v10.app.AppCompatActivity;
import android.os.Bundle;
import android.view.View;
import android.widget.Button;
public class SecondActivity extends AppCompatActivity {
    @Override
    protected void onCreate(Bundle savedInstanceState) {
        super.onCreate(savedInstanceState);
        setContentView(R.layout.activity_second);
        Button previous=(Button)findViewById(R.id.previous);
        Button close=(Button)findViewById(R.id.close);
        previous.setOnClickListener(new View.OnClickListener(){
            public void onClick(View v) {
                Intent intent=new Intent(SecondActivity.this,FirstActivity.class);
                startActivity(intent);
            }    });
        close.setOnClickListener(new View.OnClickListener(){
            public void onClick(View v) {
                Intent intent=new Intent(SecondActivity.this,FirstActivity.class);
                startActivity(intent);
                //结束当前Activity
                finish();      }
        }); }}
```

然后还要在配置文件中对SecondActivity进行配置，才能正常启动，对应的AndroidManifest.xml文件部分内容如下：

```xml
<application
    <application
    android:allowBackup="true"
    android:icon="@mipmap/ic_launcher"
    android:label="@string/app_name"
    android:roundIcon="@mipmap/ic_launcher_round"
    android:supportsRtl="true"
    android:theme="@style/AppTheme">
    <activity android:name=".FirstActivity">
        <intent-filter>
            <action android:name="android.intent.action.MAIN" />
            <category android:name="android.intent.category.LAUNCHER" />
        </intent-filter>
    </activity>
    <activity
        android:name=".SecondActivity"
        android:label="@string/title_activity_second"
        android:theme="@style/AppTheme.NoActionBar">
        <intent-filter>
            <action android:name="android.intent.action.MAIN" />
            <category android:name="android.intent.category.LAUNCHER" />
        </intent-filter>
    </activity>
</application>
```

可见，在此应用中有两个Activity，第一个是默认的入口Activity，程序一启动，自动执行该Activity。运行该程序，效果如图8-10所示。

图8-10　启动Activity示例图

单击按钮，启动第二个Activity，界面如图8-11所示。

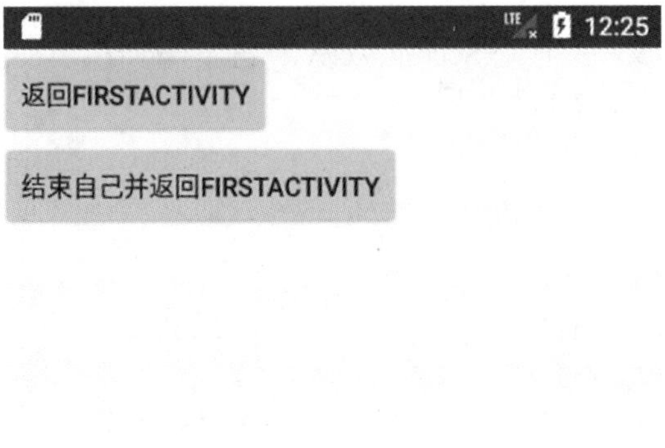

图8-11 第二个Activity启动后的界面

此时，单击两个按钮都会返回到图8-10所示界面，两者的区别在于单击第二个按钮会结束当前Activity，即会调用回调方法onDestroy()。

8.3.4 需要传递参数的Activity启动

前面的例子中，两个Activity进行跳转的过程中不需要传递任何参数，但是有时候当一个Activity启动另一个Activity时，常常需要传递一些数据。这就像Web应用中从一个Servlet跳到另一个Servlet时，Web应用习惯把需要交换的数据放入requestScope、sessionScope中。对于Activity而言，在Activity之间进行数据交换更简单：因为两个Activity之间本来就有一个桥梁——Intent，因此我们只要将需要交换的数据放入Intent即可。

Intent提供了多个重载的方法来"携带"额外的数据，如下所示。

- putExtras(Bundle data)：向Intent中放入需要"携带"的数据。Bundle就是一个简单的数据携带包，该Bundle对象包含了多个方法来存入数据。
- putXxx（String key,Xxx data）：向Bundle放入int、Long等各种类型的数据。
- putSerializable(String key,Serializable data)：向Bundle中放入一个可序列化对象。为了取出Bundle数据携带包中的数据，Bundle提供了如下方法。
- getXxx(String key)：从Bundle取出int、Long等各种类型的数据。
- getSerializable(String key,Serializable data)：从Bundle取出一个可序列化对象。

☞【实例8-4】下面通过一个实例BundleDemo演示两个Activity之间如何通过Bundle交换数据。

本例包含两个Activity，其中LoginActivity用于收集用户的登录信息，单用户单击"登录"按钮时，进入第二个Activity——ResultActivity，此Activity将会获取LoginActivity中的数据，并显示出来。

此界面采用表格布局，表格包含两行，每一行又包含两个组件，第一行用来接收输入的用户名，第二行用来接收输入的密码。

编写LoginActivity的代码如下：

```
package cn.androidstudy.bundledemo;
import android.support.v10.app.AppCompatActivity;
import android.os.Bundle;
import android.content.Intent;
import android.view.View;
```

```java
import android.view.View.OnClickListener;
import android.widget.Button;
import android.widget.EditText;
public class LoginActivity extends AppCompatActivity {
    @Override
    protected void onCreate(Bundle savedInstanceState) {
        super.onCreate(savedInstanceState);
        setContentView(R.layout.activity_login);
        Button bt=(Button)findViewById(R.id.login);
        bt.setOnClickListener(new OnClickListener(){
            public void onClick(View v) {
                EditText name=(EditText)findViewById(R.id.name);
                EditText password=(EditText)findViewById(R.id.password);
                //创建一个Bundle对象
                Bundle data=new Bundle();
                //向Bundle中绑定数据，以键值对的形式
                data.putString("name", name.getText().toString());
                data.putString("password", password.getText().toString());
                Intent intent=new Intent(LoginActivity.this,ResultActivity.class);
                //把Bundle绑定到Intent中
                intent.putExtras(data);
                startActivity(intent);           }        });  }}
```

ResultActivity对应的界面中只有两个TextView，用来显示从LoginActivity中获取的用户名和密码。

编写ResultActivity的代码，内容如下：

```java
package cn.androidstudy.bundledemo;
import android.support.v10.app.AppCompatActivity;
import android.os.Bundle;
import android.content.Intent;
import android.widget.TextView;
public class ResultActivity extends AppCompatActivity {
    @Override
    protected void onCreate(Bundle savedInstanceState) {
        super.onCreate(savedInstanceState);
        setContentView(R.layout.activity_result);
        TextView name=(TextView)findViewById(R.id.nameShow);
        TextView password=(TextView)findViewById(R.id.passwordShow);
        //获取Intent
        Intent intent=getIntent();
        //获取Intent中绑定的Bundle
        Bundle result=intent.getExtras();
```

```
            name.setText("您的用户名为："+result.getString("name"));
            password.setText("您的密码为："+result.getString("password"));
    }}
```

运行此项目，结果如图8-12所示。

图8-12　登录界面

在此界面输入用户名和密码，单击"登录"按钮，则跳转到如图8-13所示界面。

图8-13　登录信息显示界面

8.3.5　启动其他Activity并返回结果

除了能够不带参数以及带参数启动另一个Activity，Activity还提供了一个startActivityForResult(Intent intent,int requestCode)方法来启动其他Activity，并且期望从被启动的Activity中获取指定的结果。这种请求在实际应用中也是比较常见的，例如应用程序的第一个界面需要用户进行选择，但需要选择的列表数据比较复杂，必须启动另一个Activity让用户选择。当用户在第二个Activity选择完成后，程序需要把选择的结果带回给第一个Activity，并在此

Activity中显示。这种情况，也是通过Bundle进行数据交换的。

为了获取被启动的Activity所返回的结果，当前Activity需要重写onActivityResult(int requestCode,int resultCode,Intent intent)。其中，requestCode代表请求码，resultCode代表Activity返回的结果码，这个结果码也是由开发者根据业务自行设定的。

一个Activity中可能包含多个按钮，并调用多个startActivityForResult方法来打开多个不同的Activity处理不同的业务，当这些新Activity关闭后，系统都会调用前面Activity的onActivityResult方法。为了知道该方法是由哪个请求的结果触发的，可利用requestCode请求码；为了知道返回的数据来自于哪个新的Activity，可利用resultCode结果码。

【实例8-5】下面通过实例介绍如何启动Activity并获取被启动的Activity返回的结果。

创建一个名为ForResultDemo的工程，默认的Activity为ForResultActivity，其布局比较简单，界面上只有一个按钮和一个文本框。单击按钮可以进入下一个Activity即SelectActivity，此Activity是ExpandableListActivity的子类，可提供一个可扩展的列表选项，无须布局文件，当选中其中一个选项时，结果返回给ForResultActivity，并把其文本框内容更新为返回的信息。

编写ForResultActivity的代码如下：

```
package cn.androidstudy.forresultdemo;
import android.support.v10.app.AppCompatActivity;
import android.os.Bundle;
import android.content.Intent;
import android.view.View;
import android.view.View.OnClickListener;
import android.widget.Button;
import android.widget.EditText;
public class ForResultActivity extends AppCompatActivity {
    Button bt;
    EditText special;
    @Override
    protected void onCreate(Bundle savedInstanceState) {
        super.onCreate(savedInstanceState);
        setContentView(R.layout.activity_for_result);
        bt=(Button)findViewById(R.id.button1);
        special=(EditText)findViewById(R.id.editText1);
        bt.setOnClickListener(new OnClickListener(){
            public void onClick(View v) {
                Intent intent=new Intent(ForResultActivity.this,SelectActivity.class);
                startActivityForResult(intent, 0);         }
        });    }
    protected void onActivityResult(int requestCode, int resultCode, Intent data) {
        if(requestCode==0&&resultCode==0){
```

```
                Bundle info=data.getExtras();
                String resultSpecial=info.getString("special");
                special.setText(resultSpecial);           }     }}
```

编写 SelectActivity 的代码如下：

```
    package cn.androidstudy.forresultdemo;
    import android.os.Bundle;
    import android.app.ExpandableListActivity;
    import android.content.Intent;
    import android.view.View;
    import android.view.ViewGroup;
    import android.widget.AbsListView;
    import android.widget.BaseExpandableListAdapter;
    import android.widget.ExpandableListAdapter;
    import android.widget.ExpandableListView;
    import android.widget.ExpandableListView.OnChildClickListener;
    import android.widget.ImageView;
    import android.widget.LinearLayout;
    import android.widget.TextView;
    public class SelectActivity extends ExpandableListActivity {
        private String[] colleges=new String[]{"计算机与通信工程学院","
电气信息工程学院","艺术设计学院"};
        private String[][] specials=new String[][]{
            {"计算机","3G","物联网"},
            {"电气工程","自动化","电子信息工程"},
            {"工业设计","艺术设计","动画设计"}        };
        protected void onCreate(Bundle savedInstanceState) {
            super.onCreate(savedInstanceState);
            ExpandableListAdapter adapter=new BaseExpandableListAdapter() {
                public boolean isChildSelectable(int groupPosition, int childPosition) {
                    return true;              }
                public boolean hasStableIds() {
                    return true;              }
                public View getGroupView(int groupPosition, boolean isExpanded,
                                    View convertView, ViewGroup parent) {
                    LinearLayout ll=new LinearLayout(SelectActivity.this);
                    ll.setOrientation(LinearLayout.VERTICAL);
                    ImageView logo=new ImageView(SelectActivity.this);
                    ll.addView(logo);
                    TextView tv=getTextView();
                    tv.setText(getGroup(groupPosition).toString());
```

第8章 可视化使者之Activity组件

```
            ll.addView(tv);
            return ll;                        }
    public long getGroupId(int groupPosition) {
        return groupPosition;                 }
    public int getGroupCount() {
        return colleges.length;               }
    public Object getGroup(int groupPosition) {
        return colleges[groupPosition];       }
    public int getChildrenCount(int groupPosition) {
        return specials[groupPosition].length;         }
    private TextView getTextView(){
        AbsListView.LayoutParams lp=new AbsListView.LayoutParams(
                ViewGroup.LayoutParams.WRAP_CONTENT,64);
        TextView textView=new TextView(SelectActivity.this);
        textView.setLayoutParams(lp);
        textView.setPadding(36, 0, 0, 0);
        textView.setTextSize(20);
        return textView;                      }
    public View getChildView(int groupPosition, int childPosition,
                        boolean isLastChild, View convertView, ViewGroup parent) {
        TextView textView=getTextView();
        textView.setText(getChild(groupPosition,childPosition).toString());
        return textView;                      }
    public long getChildId(int groupPosition, int childPosition) {
        return childPosition;                 }
    public Object getChild(int groupPosition, int childPosition) {
        return specials[groupPosition][childPosition]; } };
    setListAdapter(adapter);
    getExpandableListView().setOnChildClickListener(new OnChildClickListener(){
        public boolean onChildClick(ExpandableListView parent, View v,
                        int groupPosition, int childPosition, long id) {
            Intent intent=getIntent();
            Bundle data=new Bundle();
            data.putString("special", specials[groupPosition][childPosition]);
            intent.putExtras(data);
```

```
                SelectActivity.this.setResult(0, intent);
                SelectActivity.this.finish();
                return false;              }
        });   }}
```

运行此程序,界面如图8-14所示。

图8-14 ForResultActivity界面

单击"请选择您的专业"按钮,则跳转到如图8-15所示界面。
选择计算机与通信工程学院下的物联网专业,则跳转到如图8-16所示界面。

图8-15 专业列表　　　　　　　　　　图8-16 显示返回数据

8.4 启动模式

前文已经讲解了Activity的创建、配置和使用，包括Activity的启动与关闭。由于实际中情况比较复杂，Android为程序开发人员提供了多种启动模式，本节将进一步详细讲解。在Android的多Activity开发中，Activity之间的跳转可能会有多种方式，有时是普通的生成一个新实例，有时希望跳转到原来某个Activity实例，而不是生成大量的重复的Activity。加载模式便是决定以哪种方式启动一个跳转到原来某个Activity实例。

在Activity里，有4种Activity的启动模式，分别为：

- standard：标准模式，一调用startActivity方法就会产生一个新的实例。
- singleTop：如果已经有一个实例位于Activity栈的顶部时，就不产生新的实例，而只是调用Activity中的newInstance方法。如不位于栈顶，会产生一个新的实例。
- singleTask：会在一个新的Task中产生这个实例，以后每次调用都会使用这个，不会去产生新的实例了。
- singleInstance：跟singleTask基本上一样，只有一个区别：在这个模式下的Activity实例所处的task中，只能有这个Activity实例，不能有其他的实例。

这些启动模式可以在功能清单文件AndroidManifest.xml中通过launchMode属性值来设定。相关的代码中也有一些标志可以使用，比如我们想只启用一个实例，则可以使用 Intent.FLAG_ACTIVITY_REORDER_TO_FRONT 标志，这个标志表示：如果这个Activity已经启动，就不产生新的Activity，而只是把这个Activity实例加到栈顶就可以了。例如：

```
Intent intent = new Intent(ReorderFour.this, ReorderTwo.class);
intent.addFlags(Intent.FLAG_ACTIVITY_REORDER_TO_FRONT);
startActivity(intent);
```

Activity的加载模式受启动Activity的Intent对象中设置的Flag和AndroidManifest文件中Activity的元素的特性值交互控制。

下面是影响加载模式的一些特性。

核心的Intent Flag有：
- FLAG_ACTIVITY_NEW_TASK
- FLAG_ACTIVITY_CLEAR_TOP
- FLAG_ACTIVITY_RESET_TASK_IF_NEEDED
- FLAG_ACTIVITY_SINGLE_TOP

核心的特性有：
- taskAffinity
- launchMode
- allowTaskReparenting
- clearTaskOnLaunch
- alwaysRetainTaskState
- finishOnTaskLaunch

上述四种启动模式又可以分为两类，standard和singleTop属于一类，singleTask和

singleInstance 属于另一类。

standard 和 singleTop 属性的 Activity 的实例可以属于任何任务（Task），并且可以位于 Activity 堆栈的任何位置。比较典型的一种情况是，一个任务的代码执行 startActivity，如果传递的 Intent 对象没有包含 FLAG_ACTIVITY_NEW_TASK 属性，指定的 Activity 将被该任务调用，从而装入该任务的 Activity 堆栈中。standard 和 singleTop 的区别在于：standard 模式的 Activity 在被调用时会创建一个新的实例，所有实例处理同一个 Intent 对象；但对于 singleTop 模式的 Activity，如果被调用的任务已经有一个这样的 Activity 在堆栈的顶端，那么不会有新的实例创建，任务会使用当前顶端的 Activity 实例来处理 Intent 对象，换句话说，如果被调用的任务包含一个不在堆栈顶端的 single Top Activity，或者堆栈顶端为 singleTop 的 Activity 的任务不是当前被调用的任务，那么，仍然会有一个新的 Activity 对象被创建。

singleTask 和 singleInstance 模式的 Activity 仅可用于启动任务的情况，这种模式的 Activity 总是处在 Activity 堆栈的最底端，并且一个任务中只能被实例化一次。两者的区别在于：对于 singleInstance 模式的 Activity，任务的 Activity 堆栈中如果有这样的 Activity，那它将是堆栈中唯一的 Activity，当前任务收到的 Intent 都由它处理，由它开启的其他 Activity 将在其他任务中被启动；对于 SingleTask 模式的 Activity，它在堆栈底端，其上方可以有其他 Activity 被创建，但是，如果发给该 Activity 的 Intent 对象到来时该 Activity 不在堆栈顶端，那么该 Intent 对象将被丢弃，但是界面还是会切换到当前的 Activity。

8.5 Fragment的使用

在早期的Android中，Fragment（碎片）是没有的。后期随着Android市场占有率的进一步扩大和Android设备多样化的需求，在Android3.0版本中引入了新API——Fragment（碎片）。Fragment（碎片）与Activity十分相似，程序员甚至可以把Fragment看作Activity中的子模块，它有自己的布局、生命周期，可以单独处理自己的输入等，在Activity运行时可以加载或移除Fragment模块。

8.5.1 Fragment简介

同样作为Android UI的常用手段，Fragment与Activity之间有非常密切的关系：Fragment 是Activity界面中的一部分，一个Activity可以包含多个Fragment，也可以在多个Activity中重用一个Fragment。Fragment不能独立存在，必须嵌入Activity中，而且Fragment的生命周期受其所在的Activity的影响。

Fragment是一种可以嵌入在活动当中的UI片段，它能让程序更加合理和充分地利用大屏幕的空间，因而在平板电脑上应用得非常广泛。虽然Fragment对读者来说应该是个全新的概念，但相信学习起来应该毫不费力，因为它和Activity实在是太像了，同样都能包含布局，同样都有自己的生命周期，甚至可以将Fragment理解成一个迷你型的Activity，虽然这个迷你型的Activity有可能和普通的Activity是一样大的。

下面通过一个例子来说明Fragment的使用。想象正在开发一个新闻应用，其中一个界面使用ListView展示了一组新闻的标题，当单击其中一个标题，就打开另一个界面显示新闻的详细内容。如果是在手机中设计，可以将新闻标题列表放在一个Activity中，将新闻的详细内容放在另一个Activity中，如图8-17所示。

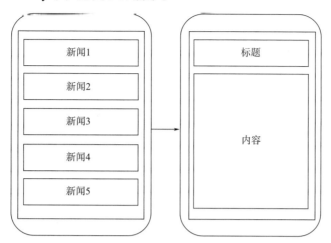

图8-17　手机信息跳转

但在平板电脑上如果也这么设计，那么新闻标题列表将会被拉长至填充满整个平板电脑的屏幕，而由于新闻的标题一般都不会太长，这样将会导致程序app界面上有大量的空白区域，非常不美观。因此一般在设计中，在同一个Activity里引入两个Fragment，将新闻标题列表界面和详细内容界面分别放在两个Fragment中，这样就可以将平板电脑屏幕空间充分地利用起来了，如图8-18所示。

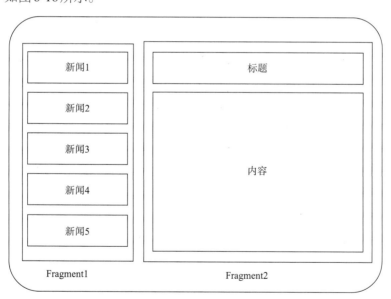

图8-18　平板电脑中引入Fragment

由于平板电脑的屏幕比手机屏幕大，可以容纳较多的UI组件，而且这些UI组件之间存在交互关系。Fragment简化了大屏幕UI的设计，开发者使用Fragment进行分组、模块化管理，就可以直接在运行过程中动态更新Activity的用户界面。

Fragment有如下特征：

① Fragment一般作为Activity的组成部分。Fragment可以调用getActivity()方法获取它所在的Activity，Activity可以调用FragmentManager的findFragmentById或者

findFragmentByTag方法获取Fragment。

② 在Activity运行时，可以调用FragmentManager的remove、replace、add等方法动态删除、替换或添加Fragment。

③ 一个Activity可以包含多个Fragment，也可以在多个Activity中重用一个Fragment。

④ Fragment不能独立存在，必须嵌入Activity中，而且Fragment的生命周期受其所在的Activity的影响。

8.5.2 创建Fragment

在实现Fragment时必须继承Fragment基类或其任意子类（DialogFragment、ListFragment、PreferenceFragment或WebViewFragment等）。

一般创建Fragment有如下三种方法：

• onCreate：系统创建Fragment对象后回调该方法，在实现代码中只初始化想要在Fragment中的必要组件，当Fragment被停止或暂停后可以恢复。

• onCreateView：当Fragment绘制组件时会回调该方法。但该方法必须返回一个View，该View也是该Fragment所显示的View。

• onPause：当用户离开该Fragment时将回调该方法。

☞【实例8-6】创建一个名为FragmentDemo的工程，默认的Activity为MainActivity，其布局界面中有一个Fragment和一个FrameLayout。

添加一个MajorContent类的代码如下：

```java
package cn.androidstudy.fragmentdemo;
import java.util.ArrayList;
import java.util.HashMap;
import java.util.List;
import java.util.Map;
public class MajorContent {
    public static class Major
    {   public Integer id;
        public String title;
        public String desc;
        public Major(Integer id, String title, String desc)
        {   this.id = id;
            this.title = title;
            this.desc = desc;    }
        @Override
        public String toString()
        {    return title;       }    }
    public static List<Major> ITEMS = new ArrayList<Major>();
    public static Map<Integer, Major> ITEM_MAP = new HashMap<Integer, Major>();
    static
    {    addItem(new Major(1, "计算机软件与理论专业"
```

第8章 可视化使者之Activity组件

```
        ,  "计算机软件与理论主要研究软件设计、开发、维护和使用过程
中涉及的软件理论、方法和技术，探讨计算机科学与技术发展的理论基础。"));
        addItem(new Major(2, "计算机网络专业"
        ,  "培养具有一定计算机网络基本理论和开发技术，具备从事程序
设计、Web的软件开发、计算机网络的组建、网络设备配置、网络管理和安全维护能力的网
络高技术应用型人才。"));
        addItem(new Major(3, "嵌入式专业"
        ,  "培养能够完成单片机和嵌入式系统的程序设计和调试，能够利
用主流的嵌入式系统开发平台完成嵌入式产品的开发、测试、维护、技术支持等工作的实用
型人才。"));        }
    private static void addItem(Major major)
    {    ITEMS.add(major);
        ITEM_MAP.put(major.id, major);        }}
```

添加一个基于ListFragment的MajorListFragment，代码如下：

```
package cn.androidstudy.fragmentdemo;

import android.app.Activity;
import android.os.Bundle;
import android.app.ListFragment;
import android.view.View;
import android.widget.ArrayAdapter;
import android.widget.ListView;
import cn.androidstudy.fragmentdemo.MajorContent;
public class MajorListFragment extends ListFragment {
    private Callbacks mCallbacks;
    public interface Callbacks
    {    public void onItemSelected(Integer id);        }
    @Override
    public void onCreate(Bundle savedInstanceState)
    {    super.onCreate(savedInstanceState);
        setListAdapter(new ArrayAdapter<MajorContent.Major>(getActivity(),
android.R.layout.simple_list_item_activated_1,android.R.id.text1,
MajorContent.ITEMS));        }
    @Override
    public void onAttach(Activity activity)
    {    super.onAttach(activity);
        if (!(activity instanceof Callbacks))
            {    throw new IllegalStateException("MajorListFragme
nt所在的Activity必须实现Callbacks接口!");        }
        mCallbacks = (Callbacks)activity;        }
    @Override
```

```java
    public void onDetach()
    {   super.onDetach();
        mCallbacks = null;   }
    @Override
    public void onListItemClick(ListView listView
            , View view, int position, long id)
    {   super.onListItemClick(listView, view, position, id);
        mCallbacks.onItemSelected(MajorContent.ITEMS.get(position).id);   }
    public void setActivateOnItemClick(boolean activateOnItemClick)
                { getListView().setChoiceMode(activateOnItemClick ? ListView.CHOICE_MODE_SINGLE: ListView.CHOICE_MODE_NONE);
    }
}
```

添加一个基于Fragment的MajorDetailsFragment，代码如下：

```java
package cn.androidstudy.fragmentdemo;
import android.os.Bundle;
import android.support.v4.app.Fragment;
import android.view.LayoutInflater;
import android.view.View;
import android.view.ViewGroup;
import android.widget.TextView;
import cn.androidstudy.fragmentdemo.MajorContent;
public class MajorDetailsFragment extends Fragment {
    public static final String ITEM_ID = "item_id";
    MajorContent.Major major;
    @Override
    public void onCreate(Bundle savedInstanceState)
    {   super.onCreate(savedInstanceState);
        if (getArguments().containsKey(ITEM_ID))
        {  major = MajorContent.ITEM_MAP.get(getArguments().getInt(ITEM_ID));
        }  }
    @Override
    public View onCreateView(LayoutInflater inflater, ViewGroup container, Bundle savedInstanceState)
    {   View rootView = inflater.inflate(R.layout.fragment_major,container, false);
        if (major != null)
        {
            ((TextView) rootView.findViewById(R.id.major_title)).setText(major.title);
```

```
                ((TextView) rootView.findViewById(R.id.major_
desc)).setText(major.desc);
        }
        return rootView;  }}
```

主Activity的MainActivity的代码如下：

```
package cn.androidstudy.fragmentdemo;
import android.support.v4.app.Fragment;
import android.support.v7.app.AppCompatActivity;
import android.os.Bundle;
import cn.androidstudy.fragmentdemo.MajorContent;
public class MainActivity extends AppCompatActivity implements
     MajorListFragment.Callbacks
{    @Override
   public void onCreate(Bundle savedInstanceState)
  {    super.onCreate(savedInstanceState);
       setContentView(R.layout.activity_main);  }
   @Override
   public void onItemSelected(Integer id)
  {   Bundle arguments = new Bundle();
       arguments.putInt(MajorDetailsFragment.ITEM_ID, id);
            MajorDetailsFragment fragment = new MajorDetailsFrag
ment();
            fragment.setArguments(arguments);
  getSupportFragmentManager().beginTransaction().replace(R.id.major_
detail_container,fragment).commit();
       }}
```

运行此程序，界面如图8-19所示，单击左边的相关专业，右边就会列出所单击专业的介绍。

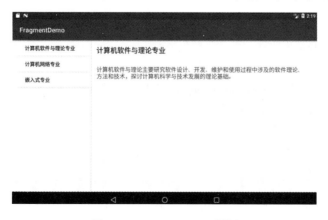

图8-19　FragmentDemo界面

强化训练

本章详细介绍了Android四大基本组件之一的Activity。Activity就相当于Android应用程序的门面，一个Activity通常对应于手机的一屏内容，它负责把组件按照指定的布局呈现给用户，所以普通用户接触最多的就是Activity。学习本章的重点就是在理解Activity生命周期的基础上掌握如何开发、配置Activity。不仅如此，由于Android应用通常包含多个Activity，读者还需要掌握Activity之间的跳转，包括如何利用Bundle在不同的Activity之间传递数据。另外，还介绍了Fragment（碎片）的使用。Fragment是在Android3.0版本引入的新API。可以把Fragment看作Activity中的子模块，它有自己的布局、生命周期，可以单独处理自己的输入等，在Activity运行时可以加载或移除Fragment模块。Fragment能让程序更加合理和充分地利用大屏幕的空间，因而在平板电脑上应用得非常广泛。

学完本章知识后，不妨通过以下习题来巩固所学的内容吧。

一、填空题

1. Activity生命周期有三个关键的生命周期循环：_____，_____，_____。

2. Activity在整个生命周期中大致会经过如下四个状态：_____，_____，_____，_____。

3. 与开发Web应用时建立Servlet类类似，建立自己的Activity也需要继承_____。当然，在不同的应用场景下，有时也要求继承Activity的子类。例如，如果应用程序界面只包括列表，则可以让应用程序继承_____；如果应用程序界面需要实现标签页效果，则可以让应用程序继承_____。

4. Android应用要求所有应用程序组件_____、_____、_____都必须进行配置。

5. 为了取出Bundle数据携带包中的数据，Bundle提供了如下方法：_____，_____。

6. Activity还提供了一个_____方法来启动其他Activity，并且期望从被启动的Activity中获取指定的结果。

二、选择题

1. 在Activity里，有4种Activity的启动模式，其中（　　）会在一个新的Task中产生这个实例，以后每次调用都会使用这个，不会去产生新的实例了。
   A. standard
   B. singleTask
   C. singleInstance
   D. singleTop

2. （　　）和singleTop属性的Activity的实例可以属于任何任务（Task），并且可以位于Activity堆栈的任何位置。
   A. singleTask
   B. singleInstance

C. Intent Flag

D. standard

3. Fragment一般作为Activity的组成部分。Fragment可以调用（　　）方法获取它所在的Activity。

A. getActivity()

B. FragmentManager

C. findFragmentById()

D. findFragmentByTag()

4. 创建Fragment的三种方法中，（　　）方法为当Fragment绘制组件时会回调该方法。但该方法必须返回一个View，该View也是该Fragment所显示的View。

A. onCreate()

B. onPause()

C. onCreateView()

D. Fragment

5. 之前我们的项目中都只有一个默认的（　　），只需要编写相关的代码就可以运行，并没有进行配置，这是因为如果一个应用只有一个（　　），系统就默认为它是整个程序的入口，并且自动在配置文件中进行配置。

A. Activity

B. Service

C. BroadcastReceiver

D. ContentProvider

三、操作题

1. Activity是Android的最基本组件之一，是Android程序的门面。再现Activity生命周期中的各种活动对理解Android程序中Activity非常有用。练习利用Android Studio开发环境创建一个新的Android应用程序项目，演示在一个Activity的生命周期中，Activity中各种方法如何被调用。

2. 作为Android的"门面"，Activity的创建、配置和使用是Android编程必须要掌握的技能之一。练习利用Android Studio开发环境创建一个新的Android应用程序，创建一个Activity，该能对用户的动作做出反应，并以Toast显示提示信息。

3. 一个Activity相当于手机的一屏，用户使用Android程序的过程中绝大部分活动都是在Activity上进行的。练习利用Android Studio开发环境创建一个新的Android应用程序，在该Android应用程序中创建两个Activity，并实现两个Activity之间的跳转。

4. 现在的手机屏幕越来越大，Fragment能让程序更加合理和充分地利用大屏幕的空间。练习利用Android Studio开发环境创建一个新的Android应用程序，在该Android应用程序中实现使用Fragment进行布局。

第9章 信息传递者之Intent机制

内容导读

一个Android应用由多个组件组成,前面讲过的Activity,还有Broadcast Receive、Service、ContentProvider被称为Android四大基本组件。Android组件之间的信息传递这一重要工作,主要是由Intent协助完成,因此Intent被称为"Android信息传递者"。Intent提供了一种通用的信息传递系统,成为Android程序中传输数据的核心对象。Android系统会根据Intent中的描述,找到满足对应的组件,将Intent传递给调用的组件,并完成组件的调用。本章将着重介绍Intent的相关知识并通过实例展示其应用。

学习目标

- 熟悉Intent的概念;
- 掌握Intent Filter的使用;
- 掌握Intent的调用;
- 熟悉常用Intent组件的使用。

9.1 Intent概述

一个Android程序一般会由很多组件组成,而且Android程序作为网络应用程序,其中交互与通信扮演了非常重要的角色。Intent机制就是用来协助Android应用间的交互与通信。

第9章 信息传递者之Intent机制

Intent的中文直译是"意图"的意思,Intent不仅封装了Android应用程序需要启动某个组件的"意图",还是Android程序组件之间消息传递的重要媒介。利用Intent机制不仅可实现不同应用程序之间的交互,还可以实现同一应用程序内不同组件之间的交互,所以Intent是当之无愧的"Android信息传递者",也是Android程序的核心。表9-1示出使用Intent启动不同组件的方法。

表9-1 使用Intent启动不同组件的方法

组件类型	启动方法
Activity	Context.startActivity(Intent intent) Activity.startActivityForResult(Intent intent,int requestCode)
Service	Context.startService(Intent service) Context.bindService(Intent service,ServiceConnection conn, int flags)
BroadcastReceiver	Context.sendBroadcast(Intent intent) Context.sendOrderedBroadcast(Intent intent, String receiverPermission) Context.sendStickyBroadcast(Intent intent)

9.1.1 Intent属性

(1) action

即要执行的动作。使用一个字符串对所将执行的动作进行描述,为了方便引用,Intent类中定义了一些标准的动作,如表9-2所示。

表9-2 action常量

常量	目标组件	动作
ACTION_CALL	Activity	启动一个电话
ACTION_EDIT	Activity	显示数据编辑界面
ACTION_MAIN	Activity	启动项目的初始界面
ACTION_SYNC	Activity	同步服务器和移动设备的数据
ACTION_BATTERY_LOW	BroadcastReceiver	警告电池电量低
ACTION_HEADSET_PLUG	BroadcastReceiver	耳机插入/拔掉设备
ACTION_SCREEN_ON	BroadcastReceiver	屏幕已打开
ACTION_TIMEZONE_CHANGED	BroadcastReceiver	时区的设置已经改变

当然,也可以自定义动作(自定义的动作在使用时,需要加上包名作为前缀,如"com.example.project.SHOW_COLOR"),并可定义相应的Activity来处理我们的自定义动作。

(2) data

即要执行动作所要操作的数据。Android中采用指向数据的一个URI来表示,如在联系人应用中,一个指向某联系人的URI可能为content://contacts/1。对于不同的动作,其URI数据的类型是不同的(可以设置type属性指定特定类型数据),如ACTION_EDIT指定Data为文件URI,打电话为tel:URI,访问网络为http:URI,而由content provider提供的数据则为content:URIs。

(3) type

即显式指定Intent的数据类型（MIME，Multipurpose Internet Mail Extensions，多用途互联网邮件扩展）。一般Intent的数据类型能够根据数据本身进行判定，但是通过设置这个属性，可以强制采用显式指定的类型而不再进行推导。

MIME类型有两种形式：

- 单个记录的格式：vnd.android.cursor.item/vnd.yourcompanyname.content type，如：content://com.example.transportationprovider/trains/122（一条列车信息的URI）的MIME类型是vnd.android.cursor.item/vnd.example.rail；
- 多个记录的格式：vnd.android.cursor.dir/vnd.yourcompanyname.Content type，如：content://com.example.transportationprovider/trains（所有列车信息）的MIME类型是vnd.android.cursor.dir/vnd.example.rail。

(4) category

一个字符串，包含了关于处理该Intent的组件的种类的信息。一个Intent对象可以有任意个category。Intent类定义了许多category常量，如表9-3所示。

表9-3 category常量

常量	作用
CATEGORY_DEFAULT	默认的category
CATEGORY_BROWSABLE	该Activity能被浏览器安全调用
CATEGORY_HOME	设置该Activity随系统启动而运行
CATEGORY_LAUNCHER	Intent的接受者应该在Launcher中作为顶级应用出现
CATEGORY_PREFERENCE	该Activity是参数面板

(5) component

指定Intent的目标组件的类名称。通常Android会根据Intent中包含的其他属性的信息，比如action、data、type、category进行查找，最终找到一个与之匹配的目标组件。但是，如果component这个属性有指定的话，将直接使用它指定的组件，而不再执行上述查找过程。指定了这个属性以后，Intent的其他所有属性都是可选的。

(6) extra（附加信息）

即指其他所有附加信息的集合。使用extra可以为组件提供扩展信息。例如，如果要执行"发送电子邮件"这个动作，可以将电子邮件的标题、正文等保存在extra里，传给电子邮件发送组件。

9.1.2 Intent解析

理解Intent的关键之一是清楚Intent的两种基本用法：一种是显式的Intent，即在构造Intent对象时就指定接收者；另一种是隐式的Intent，即Intent的发送者在构造Intent对象时，并不知道也不关心接收者是谁，有利于降低发送者和接收者之间的耦合。

对于显式Intent，Android不需要去做解析，因为目标组件已经很明确。Android需要解析的是那些隐式Intent，通过解析，将Intent映射给可以处理此Intent的Activity、Service或BroadcastReceiver。

Intent解析机制主要是通过查找已注册在AndroidManifest.xml中的所有<intent-filter>及其中定义的Intent，最终找到匹配的component。在这个解析过程中，Android是通过Intent的action、type、category这三个属性来进行判断的，判断方法如下。

① 如果Intent指明定了action，则目标组件的<intent-filter>的action列表中就必须包含有这个action，否则不能匹配。

② 如果Intent没有提供type，系统将从data中得到数据类型。和action一样，目标组件的数据类型列表中必须包含Intent的数据类型，否则不能匹配。

③ 如果Intent中的数据不是content:类型的URI，而且Intent也没有明确指定它的type，将根据Intent中数据的scheme（比如http:或者mailto:）进行匹配。同上，Intent的scheme必须出现在目标组件的scheme列表中。

④ 如果Intent指定了一个或多个category，这些类别必须全部出现在组建的类别列表中。比如Intent中包含了两个类别：LAUNCHER_CATEGORY 和 ALTERNATIVE_CATEGORY，解析得到的目标组件必须至少包含这两个类别。

9.2 Intent Filter

Filter是"过滤器"的意思，Intent Filter有时直接写作Intent过滤器。活动、服务、广播接收者为了告知系统能够处理哪些隐式Intent，它们可以有一个或多个Intent过滤器。每个过滤器描述组件的一种能力，即乐意接收的一组Intent。实际上，它筛掉不想要的Intent，也仅仅是不想要的隐式Intent。一个显式Intent总是能够传递到它的目标组件，不管它包含什么，不考虑过滤器。但是一个隐式Intent，仅当它能够通过组件的过滤器之一才能传递。

一个组件能够做的每一项工作都有独立的过滤器。例如，记事本中的NoteEditor活动有两个过滤器，一个是启动一个指定的记录，用户可以查看和编辑；另一个是启动一个新的、空的记录，用户能够填充并保存。

一个Intent过滤器是一个IntentFilter类的实例。因为Android系统在启动一个组件之前必须知道它的能力，但是Intent过滤器通常不在Java代码中设置，而是在应用程序的清单文件AndroidManifest.xml中以<intent-filter>元素设置。但有一个例外，广播接收者的过滤器通过调用Context.registerReceiver()动态地注册，直接创建一个IntentFilter对象。

一个过滤器有对应于Intent对象的动作、数据、种类的字段。过滤器要检测隐式Intent的所有这三个字段，其中任何一个失败，Android系统都不会传递Intent给组件。然而，因为一个组件可以有多个Intent过滤器，一个Intent通不过组件的过滤器检测，其他的过滤器可能通过检测。

9.2.1 动作检测

清单文件中的<intent-filter>元素以<action>子元素列出动作，例如：

```
<intent-filter . . . >
```

```
      <action android:name="com.example.project.SHOW_CURRENT" />
      <action android:name="com.example.project.SHOW_RECENT" />
      <action android:name="com.example.project.SHOW_PENDING" />
      . . .
</intent-filter>
```

如例子所展示，虽然一个Intent对象仅是单个动作，但是一个过滤器可以列出不止一个。这个列表不能为空，一个过滤器必须至少包含一个<action>子元素，否则它将阻塞所有的Intent。

要通过检测，Intent对象中指定的动作必须匹配过滤器的动作列表中的一个。如果对象或过滤器没有指定一个动作，结果如下。

① 如果过滤器没有指定动作，没有一个Intent动作匹配，所有的Intent将检测失败，即没有Intent能够通过过滤器。

② 如果Intent对象没有指定动作，将自动通过检查。

9.2.2 种类检测

清单文件中的<intent-filter>元素以<category>子元素列出种类，例如：

```
<intent-filter . . . >
    <category android:name="android.intent.category.DEFAULT" />
    <category android:name="android.intent.category.BROWSABLE" />
    . . .
</intent-filter>
```

注意：前面两个表格列举的动作和种类常量不在清单文件中使用，而是使用全字符串值。例如，例子中所示的"android.intent.category.BROWSABLE"字符串对应于本文前面提到的BROWSABLE常量。类似地，"android.intent.action.EDIT"字符串对应于ACTION_EDIT常量。

对于一个Intent要通过种类检测，Intent对象中的每个种类必须匹配过滤器中的一个，即过滤器能够列出额外的种类。但是Intent对象中的种类都必须能够在过滤器中找到，只有一个种类在过滤器列表中没有，就算种类检测失败。

因此，原则上如果一个Intent对象中没有种类（即种类字段为空），应该总是通过种类测试，而不管过滤器中有什么种类。但是有个例外，Android对待所有传递给Context.startActivity()的隐式Intent至少包含"android.intent.category.DEFAULT"（对应CATEGORY_DEFAULT常量）。因此，活动想要接收隐式Intent必须要在Intent过滤器中包含"android.intent.category.DEFAULT"。

注意："android.intent.action.MAIN"和"android.intent.category.LAUNCHER"设置，它们分别标记活动开始新的任务和带到启动列表界面。它们可以包含"android.intent.category.DEFAULT"种类列表，也可以不包含。

9.2.3 数据检测

清单文件中的<intent-filter>元素以<data>子元素列出数据，例如：

第9章 信息传递者之Intent机制

```
<intent-filter...>
  <data android:mimeType="video/mpeg" android:scheme="http".../>
  <data android:mimeType="audio/mpeg" android:scheme="http".../>
  ...
</intent-filter>
```

每个<data>元素指定一个URI和数据类型（MIME类型）。它有四个属性scheme、host、port、path对应于URI的每个部分：scheme://host:port/path

例如，URI：content://com.example.project:200/folder/subfolder/etc

scheme是content，host是"com.example.project"，port是200，path是"folder/subfolder/etc"。host和port一起构成URI的凭据（authority），如果host没有指定，port也被忽略。这四个属性都是可选的，但它们之间并不都是完全独立的。要让authority有意义，scheme必须也要指定。要让path有意义，scheme和authority也都必须要指定。

当比较Intent对象和过滤器的URI时，仅仅比较过滤器中出现的URI属性。例如，如果一个过滤器仅指定了scheme，所有有此scheme的URI都匹配过滤器；如果一个过滤器指定了scheme和authority，但没有指定path，所有匹配scheme和authority的URI都通过检测，而不管它们的path；如果四个属性都指定了，要都匹配才能算是匹配。然而，过滤器中的path可以包含通配符来要求匹配path中的一部分。

<data>元素的type属性指定数据的MIME类型。Intent对象和过滤器都可以用"*"通配符匹配子类型字段，例如"text/*"，"audio/*"表示任何子类型。

数据检测既要检测URI，也要检测数据类型。规则如下：

① 一个Intent对象既不包含URI，也不包含数据类型：仅当过滤器不指定任何URI和数据类型时，才不能通过检测；否则都能通过。

② 一个Intent对象包含URI，但不包含数据类型：仅当过滤器不指定数据类型，同时它们的URI匹配，才能通过检测。例如，mailto:和tel:都不指定实际数据。

③ 一个Intent对象包含数据类型，但不包含URI：仅当过滤器只包含数据类型且与Intent相同，才通过检测。

④ 一个Intent对象既包含URI，也包含数据类型（或数据类型能够从URI推断出）：数据类型部分，只有与过滤器中之一匹配才算通过；URI部分，它的URI要出现在过滤器中，或者它有content:或file: URI，又或者过滤器没有指定URI。换句话说，如果它的过滤器仅列出了数据类型，组件假定支持content:和file:。

如果一个Intent能够通过不止一个活动或服务的过滤器，用户可能会被问哪个组件被激活。如果没有找到目标，会产生一个异常。

9.2.4 通用情况

前面最后一条规则表明组件能够从文件或内容提供者获取本地数据。因此，它们的过滤器仅列出数据类型且不必明确指出content:和file: scheme的名字。这是一种典型的情况，一个<data>元素像下面这样：

```
<data android:mimeType="image/*" />
```

告诉Android这个组件能够从内容提供者获取image数据并显示它。因为大部分可用数据由内容提供者（content provider）分发，过滤器指定一个数据类型但没有指定URI或许最通用。

另一种通用配置是过滤器指定一个scheme和一个数据类型。例如，一个<data>元素像下面这样：

```
<data android:scheme="http" android:type="video/*" />
```

告诉Android这个组件能够从网络获取视频数据并显示它。考虑当用户点击一个Web页面上的link，浏览器应用程序会做什么？它首先会试图去显示数据（如果link是一个HTML页面，就能显示）。如果它不能显示数据，它将把一个隐式Intent加到scheme和数据类型，去启动一个能够做此工作的活动。如果没有接收者，它将请求下载管理者去下载数据。这将在内容提供者的控制下完成，因此一个潜在的大活动池（它们的过滤器仅有数据类型）能够响应。

大部分应用程序能启动新的活动，而不引用任何特别的数据。活动有指定"android.intent.action.MAIN"的动作的过滤器，能够启动应用程序。如果它们出现在应用程序启动列表中，它们也指定"android.intent.category.LAUNCHER"种类：

```
<intent-filter...>
    <action android:name="code android.intent.action.MAIN" />
    <category android:name="code android.intent.category.LAUNCHER" />
</intent-filter>
```

9.2.5 使用Intent匹配

Intent对照着Intent过滤器匹配，不仅去发现一个目标组件激活，而且去发现设备上的组件的其他信息。例如，Android系统填充应用程序启动列表，最高层屏幕显示用户能够启动的应用程序：是通过查找所有的包含指定了"android.intent.action.MAIN"的动作和"android.intent.category.LAUNCHER"种类的过滤器的活动，然后在启动列表中显示这些活动的图标和标签。类似的，它通过查找有"android.intent.category. HOME"过滤器的活动发掘主菜单。

我们的应用程序也可以类似地使用这种Intent匹配方式。PackageManager有一组query…()方法返回能够接收特定Intent的所有组件，一组resolve…()方法决定最适合的组件响应Intent。例如，queryIntentActivities()返回一组能够给执行指定Intent参数的所有活动，类似的queryIntentServices()返回一组服务。这两个方法都不激活组件，它们仅列出所有能够响应的组件。对应广播接收者也有类似的方法queryBroadcastReceivers()。

9.3 Intent的调用

Android中，通常Intent的调用分为显式调用和隐式调用两种。显式调用中，Intent对象在创建时直接指定接受者；隐式调用中，Intent对象在创建时则不指定具体的接受者。本节将详细向读者讲解这两种Intent的调用方式。

9.3.1 显式调用

显式Intent直接用组件的名称定义目标组件，这种方式很直接。但是由于开发人员往往并不清楚别的应程序的组件名称，因此显式Intent更多用于在应用程序内部传递消息。

```
ComponentName comp = new ComponentName(MainActivity.this,
OtherActivity.class);
Intent intent = new Intent();
intent.setComponent(comp);
startActivity(intent);
```

四行代码用于创建ComponentName对象，并将该对象设置成Intent对象的component属性，这样应用程序即可根据该Intent的"意图"去启动指定组件。简单说，Intent提供了一个构造函数，可以方便地指定要启动的组件。以下代码和前面四行的代码等价。

```
Intent intent = new Intent(MainActivity.this,OtherActivity.
class);
startActivity(intent);
```

如果知道其他应用程序的包名和类名，可以采用如下方式进行调用：

```
Intent intent = new Intent();
intent.setClassName("com.example.test","com.example.test.
OtherActivity");
startActivity(intent);
```

这个方法的前提条件是被调用的类已经安装在运行的模拟器或手机上。

9.3.2 隐式调用

相比显式Intent调用方式，Android还可以利用下面这种隐式Intent调用方式，即Intent发送者不指定特定的组件，而是声明要执行的常规操作，允许其他应用中的组件来处理它。

（1）简单的隐式Intent

【实例9-1】我们先从简单的例子开始。下面的ImplicitIntent Demo程序用来启动Android自带的打电话功能的Dialer程序，程序运行的截图如图9-1所示。
ImplicitIntent程序只包含一个Java源文件MainActivity.java，代码如下所示：

```
package cn.androidstudy.implicitintentdemo;
import android.support.v7.app.AppCompatActivity;
import android.os.Bundle;
import android.content.Intent;
import android.view.View;
import android.view.View.OnClickListener;
import android.widget.Button;
public class MainActivity extends AppCompatActivity {
```

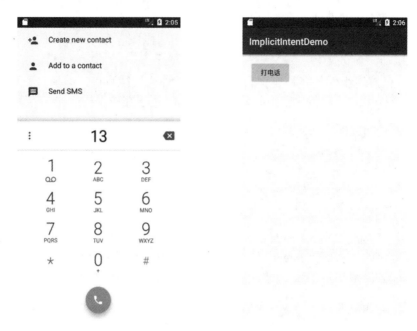

图9-1 ImplicitIntentTest程序运行截图

```
@Override
protected void onCreate(Bundle savedInstanceState) {
    super.onCreate(savedInstanceState);
    setContentView(R.layout.activity_main);
    Button butDial = (Button) findViewById(R.id.btnDial);
    butDial.setOnClickListener(new OnClickListener() {
        //@Override
        public void onClick(View v) {
            Intent intent = new Intent(Intent.ACTION_DIAL);
            startActivity(intent);
        } });  }}
```

该程序在Intent的使用上，与上节中的使用方式有很大的不同，即根本不指定接收者，初始化Intent对象时，只是传入参数，设定action为Intent.ACTION_DIAL：

```
Intent intent = new Intent(Intent.ACTION_DIAL);
startActivity(intent);
```

这里使用的构造函数的原型如下：

```
Intent(String action);
```

有关action的作用，读者可将它理解为描述这个Intent的一种方式。这种使用方式中，Intent的发送者只是指定了action为Intent.ACTION_DIAL，那么如何找到接收者呢？通过下面的例子来进行学习。

（2）增加一个接收者

事实上接收者如果希望能接收某些Intent，需要像上节例子中一样，通过在Android

Manifest.xml中增加Activity的声明,并设置对应的Intent Filter和action,才能被Android的应用程序框架所匹配。为了证明这一点,我们新建一个Activity:MyDialActivity,修改AndroidManifest.xml文件,将MyDialActivity的声明部分改为:

```xml
<activity
    android:name=".MyDialActivity"
    android:label="@string/title_activity_my_dial"
    android:theme="@style/AppTheme.NoActionBar">
    <intent-filter>
        <action android:name="android.intent.action.DIAL" />
        <category android:name="android.intent.category.DEFAULT"/>
    </intent-filter>
</activity>
```

然后再尝试运行ImplicitIntentTest程序,运行的截图如图9-2所示。

图9-2 修改后的运行截图

这个截图中的左图表示可以选择Dialer或者MyDialActivity程序来完成Intent.ACTION_DIAL,也就是说,针对Intent.ACTION_DIAL,Android框架找到了两个符合条件的Activity,因此它将这两个Activity分别列出,供用户选择。当选择"MyDialActivity",点击"JUST ONCE",就打开图9-2的右图。

回过头来看是怎么做到这一点的。我们仅仅在AndroidManifest.xml文件中增加了下面的两行:

```xml
<action android:name="android.intent.action.DIAL" />
<category android:name="android.intent.category.DEFAULT" />
```

这两行修改了原来的Intent Filter,这样这个Activity才能够接收到我们发送的Intent。通过这个改动及其作用,可以进一步理解隐式Intent、Intent Filter及action、category等概念——Intent发送者设定action来说明将要进行的动作,而Intent的接收者在AndroidManifest.xml文件中通过设定Intent Filter来声明自己能接收哪些Intent。

(3) 增加一个数据

修改AndroidManifest.xml文件,将MyDialActivity的声明部分改为:

```xml
<activity
        android:name=".MyDialActivity"
        android:label="@string/title_activity_my_dial"
        android:theme="@style/AppTheme.NoActionBar">
    <intent-filter>
        <action android:name="android.intent.action.DIAL" />
        <category android:name="android.intent.category.DEFAULT"/>
    </intent-filter>
    <intent-filter>
        <action android:name="android.intent.action.MyDIAL" />
        <category android:name="android.intent.category.DEFAULT"/>
        <data android:scheme="wang" />
    </intent-filter>
    <intent-filter>
        <action android:name="android.intent.action.MyDIAL" />
        <category android:name="android.intent.category.DEFAULT"/>
    <data android:scheme="wang" android:host="www.study.com" android:port= "6666" />
    </intent-filter>
</activity>
```

在MainActivity中，增加两个按钮。为第一个按钮的单击事件增加如下代码：

```
Intent intent1 = new Intent("android.intent.action.MyDIAL");
intent1.setData(Uri.parse("wang:123456"));
startActivity(intent1);
```

为第二个按钮的单击事件增加如下代码：

```
Intent intent2 = new Intent("android.intent.action.MyDIAL");
intent2.setData(Uri.parse("wang://www.study.com:6666/123456"));
startActivity(intent2);
```

通过这两个按钮，可以直接打开MyDialActivity，如图9-2的右图所示。

9.3.3 在Intent中传递数据

Intent除了定位目标组件外，另外一个职责就是传递数据信息。Intent之间传递数据一般有两种常用的方法：一种是通过data属性，另一种是通过extra属性。其中，data属性是一种URL，它可以指向HTTP、FTP等网络地址，也可以指向ContentProvider提供的资源。通过调用Intent的setData方法放入数据，使用getData方法取出数据。

例如，需要启动Android内置的浏览器，使用下面的代码可将网址通过data属性传递给它。打开的界面如图9-3所示。

```
Intent intent = new Intent(Intent.ACTION_VIEW);
```

```
intent.setData(Uri.parse("http://www.baidu.com"));
startActivity(intent);
```

图9-3 打开的浏览器界面

由于data属性只能传递数据的URL地址，如果需要传递一下数据对象，则需要使用extra属性。Intent提供了多个重载的方法来"携带"额外的数据，如下：

putExtras(Bundle extras)：向Intent中放入需要传递的数据。

- putExtra(String name, Xxx value)：向Intent中放入Xxx类型的数据。
- putParcelableArrayListExtra(String name, ArrayList<? extends Parcelable> value)：向Intent中放入ArrayList数据。
- putIntegerArrayListExtra(String name, ArrayList<Integer> value)：向Intent中放入ArrayList数据。
- putStringArrayListExtra(String name, ArrayList<String> value)：向Intent中放入ArrayList数据。

Bundle是专门用来在Android的应用组件之间传递数据的一种对象，本质上是一个Map对象，可以将各种基本类型的数据保存在Bundle类中打包传输。

- putXxx(String key, Xxx value)：向Bundle中放入int、long等各种类型的数据。
- putSerializable(String key, Serializable value)：向Bundle中放入一个可系列化的对象。
- putParcelable(String key, Parcelable value)：向Bundle中放入一个可系列化的对象。

为了取出Intent中携带的数据，Intent中提供了如下方法：

- getXxxExtra(String name, Xxx defaultValue)：从Intent中直接取出Xxx类型的数据。

也可以先使用getExtras()方法获取一个Bundle对象，然后使用Bundle的如下方法来获取数据的值。

- getXxx(String key)：从Bundle中取出Xxx类型的数据。
- getXxx(String key, Xxx defaultValue)：从Bundle中取出Xxx类型的数据，如果取不到则使用defaultValue。

两个Activity之间如何使用Bundle交换数据请参考第8章的【实例8-4】。

9.3.4 在Intent中传递复杂对象

Android的Intent之间传递对象有两种方法，一种是"Bundle.putSerializable(Key,Object)"；另一种是"Bundle.putParcelable(Key,Object)"。方法中的Object要满足一定的条件，前者实现了Serializable接口，而后者实现了Parcelable接口。

Serializable和Parcelable这两种接口功能类似，但为什么Android不用内置的Java序列化机制，而偏偏要搞一套新东西呢？这是因为Android设计团队认为Java中的序列化太慢，难以满足Android的进程间通信需求，所以他们构建了Parcelable解决方案。Parcelable要求显式的序列化类的成员，但最终序列化对象的速度将快很多。在Android运行环境中推荐使用Parcelable接口，它不但可以利用Intent传递，还可以在远程方法调用中使用。

实现Parcelable接口需要实现三个方法。

第一，writeToParcel 方法。该方法将类的数据写入外部提供的Parcel中。声明格式为："writeToParcel (Parcel dest, int flags)"。

第二，describeContents方法。返回内容描述信息的资源ID，直接返回0就可以。

第三，静态的Parcelable.Creator<T>接口，本接口有两个方法。

- createFromParcel(Parcel in)：实现从in中创建出类的实例的功能。
- newArray(int size)：创建一个类型为T、长度为size的数组， returnnew T[size]即可。

☞【实例9-2】通过一个示例演示如何使用Serializable和Parcelable接口传递对象。

首先，创建一个新的Android项目——IntentObjectDemo，包含一个Activity——MainActivity。然后创建两个类——SerializableUser和ParcelableUser，分别实现Serializable和Parcelable接口，这两个类的源代码如下：

```
package cn.androidstudy.intentobjectdemo;
import java.io.Serializable;
public class SerializableUser implements Serializable{
   private String userName;
   private String password;
   public SerializableUser(){    }
   public SerializableUser(String userName,String password){
       this.userName = userName;
       this.password = password;    }
   public String getUserName() { return userName;      }
   public void setUserName(String userName) {  this.userName = userName;     }
   public String getPassword() { return password;      }
   public void setPassword(String password) { this.password = password; }}
   package cn.androidstudy.intentobjectdemo;
   import android.os.Parcelable;
   import android.os.Parcel;
   public class ParcelableUser implements Parcelable {
     private String userName;
```

```java
    private String password;
    public ParcelableUser(){    }
    public ParcelableUser(String userName,String password){
        this.userName = userName;
        this.password = password;    }
    public String getUserName() { return userName;    }
    public void setUserName(String userName) { this.userName = userName;    }
    public String getPassword() { return password;    }
    public void setPassword(String password) { this.password = password;    }
     public static final Parcelable.Creator<ParcelableUser> CREATOR = new Creator<ParcelableUser>(){   @Override
        public ParcelableUser createFromParcel(Parcel source) {
            ParcelableUser parcelableUser = new ParcelableUser();
            parcelableUser.userName = source.readString();
            parcelableUser.password = source.readString();
            return parcelableUser;    }
        @Override
        public ParcelableUser[] newArray(int size) {
            return new ParcelableUser[size];    }    };
    @Override
    public int describeContents() {   return 0;    }
    @Override
    public void writeToParcel(Parcel dest, int flags) {
        dest.writeString(userName);
        dest.writeString(password);    }}
```

在MainActivity的布局文件中增加两个按钮。在MainActivity类中增加两个按钮的点击事件处理，分别发送SerializableUser和ParcelableUser对象。修改后的MainActivity.java文件如下：

```java
package cn.androidstudy.intentobjectdemo;
import android.support.v7.app.AppCompatActivity;
import android.os.Bundle;
import android.view.View;
import android.content.Intent;
public class MainActivity extends AppCompatActivity {
    @Override
    protected void onCreate(Bundle savedInstanceState) {
        super.onCreate(savedInstanceState);
        setContentView(R.layout.activity_main);    }
    public void sendData(View view){
```

```java
        switch(view.getId()){
            case R.id.button1:
                SerializableUser sUser = new SerializableUser("John", "123456");
                Intent intent = new Intent(this,ReceiveObjectActivity.class);
                Bundle bundle = new Bundle();
                bundle.putInt("type", 1);
                bundle.putSerializable("serial", sUser);
                intent.putExtras(bundle);
                startActivity(intent);
                break;
            case R.id.button2:
                ParcelableUser pUser = new ParcelableUser("Mike", "123456");
                Intent intent1 = new Intent(this,ReceiveObjectActivity.class);
                Bundle bundle1 = new Bundle();
                bundle1.putInt("type", 2);
                bundle1.putParcelable("parcel", pUser);
                intent1.putExtras(bundle1);
                startActivity(intent1);
                break;
        }}
```

在当前项目中增加一个Activity——ReceiveObjectActivity，用于接收并处理传递过来的对象。修改onCreate方法，修改后的代码如下：

```java
package cn.androidstudy.intentobjectdemo;
import android.os.Bundle;
import android.support.v7.app.AppCompatActivity;
import android.widget.TextView;
public class ReceiveObjectActivity extends AppCompatActivity {
    @Override
    protected void onCreate(Bundle savedInstanceState) {
        super.onCreate(savedInstanceState);
        setContentView(R.layout.activity_receive_object);
        TextView tv = (TextView)findViewById(R.id.textView1);
        Bundle bundle = getIntent().getExtras();
        int type = bundle.getInt("type");
        if(type==1){
            SerializableUser serializableUser = (SerializableUser)getIntent().getSerializableExtra("serial");
            tv.setText(serializableUser.getUserName()+"\n"+serializableUser.getPassword());
        }else{
            ParcelableUser parcelableUser = (ParcelableUser)getIntent().getParcelableExtra("parcel");
```

```
    tv.setText(parcelableUser.getUserName()+"\n"+parcelableUser.
getPassword());
        }}}
```

由于ReceiveObjectActivity的布局文件没有修改，在此就不再给出。运行应用程序，点击第一个按钮，ReceiveObjectActivity接收数据后的界面见图9-4（左图）；点击第二个按钮，接收数据后的界面见图9-4（右图）。

图9-4　使用Intent传递对象

9.3.5　实现Activity之间的协同

如果想在Activity中得到新打开Activity 关闭后返回的数据，需要使用系统提供的startActivityForResult(Intent intent, int requestCode)方法打开新的Activity，新的Activity 关闭后会向前面的Activity传回数据，为了得到传回的数据，必须在前面的Activity中重写onActivityResult(int requestCode, int resultCode, Intent data)方法。

使用startActivityForResult(Intent intent, int requestCode)方法打开新的Activity，我们需要为startActivityForResult()方法传入一个请求码(第二个参数)。请求码的值是根据业务需要由自己设定，用于标识请求来源。例如，一个Activity有两个按钮，点击这两个按钮都会打开同一个Activity，不管是哪个按钮打开新Activity，当这个新Activity关闭后，系统都会调用前面Activity的onActivityResult(int requestCode, int resultCode, Intent data)方法。onActivityResult()方法如果需要知道新Activity是由哪个按钮打开的，并且要做出相应的业务处理，只要使用第一个参数requestCode即可区分开。

新Activity关闭前需要向前面的Activity返回数据，需要使用系统提供的setResult(int resultCode, Intent data)方法。那么这个结果码（resultCode）有什么作用呢？

在一个Activity中，可能会使用startActivityForResult()方法打开多个不同的Activity处理不同的业务，当这些新Activity关闭后，系统都会调用前面Activity的onActivityResult(int requestCode, int resultCode, Intent data)方法。为了知道返回的数据来自于哪个新Activity，只需要它们设置的resultCode不同即可区分。

☞【实例9-3】通过一个示例演示一下如何使用startActivityForResult()方法获取另一个Activity的返回值。

首先，新建一个Android项目——ForResultDemo，该项目包含一个Activity——MainActivity，新建另外一个Activity——OtherActivity。在MainActivity的布局文件增加一个按钮，用来打开OtherActivity。

在MainActivity类中增加按钮的事件处理和onActivityResult()事件处理，修改后的MainActivity.java文件如下：

```
package cn.androidstudy.forresultdemo;
```

```java
import android.support.v7.app.AppCompatActivity;
import android.os.Bundle;
import android.widget.TextView;
import android.content.Intent;
import android.view.View;
public class MainActivity extends AppCompatActivity {
    private TextView tView = null;
    @Override
    protected void onCreate(Bundle savedInstanceState) {
        super.onCreate(savedInstanceState);
        setContentView(R.layout.activity_main);
        tView = (TextView) findViewById(R.id.textView1);     }
    public void forResult(View view){
        Intent intent = new Intent(this,OtherActivity.class);
        intent.putExtra("name", "Tom");
        startActivityForResult(intent, 100);      }
    @Override
    protected void onActivityResult(int requestCode, int resultCode, Intent data) {
        if(requestCode==100 && resultCode == RESULT_OK){
            String string = data.getStringExtra("result");
            tView.setText(string);           }
        super.onActivityResult(requestCode, resultCode, data);   }}
```

在OtherActivity的布局文件中增加一个按钮,用于返回前一个Activity。在OtherActivity类中,为按钮添加事件处理,并由onCreate方法获取传递过来的参数。修改后的OtherActivity.java文件如下:

```java
package cn.androidstudy.forresultdemo;
import android.os.Bundle;
import android.support.v7.app.AppCompatActivity;
import android.view.View;
import android.widget.TextView;
import android.content.Intent;
public class OtherActivity extends AppCompatActivity {
    private TextView tView;
    private String param;
    @Override
    protected void onCreate(Bundle savedInstanceState) {
        super.onCreate(savedInstanceState);
        setContentView(R.layout.activity_other);
        tView = (TextView) findViewById(R.id.textView1);
        param = getIntent().getStringExtra("name");
```

```
            tView.setText(param);    }
    public void fanhui(View view){
        Intent intent = getIntent();
        intent.putExtra("result", "Hi "+param);
        setResult(RESULT_OK, intent);
        finish();      }}
```

运行应用程序,在MainActivity的主界面中点击"OPEN"按钮,打开OtherActivity,可以看到如图9-5(左图)的界面,点击"RETURN"按钮,在图9-5(右图)中可以看到返回值已经获取到了。

图9-5 使用startActivityForResult获取结果

9.4 常用Intent组件的使用

除了前面讲到的内容,本节将常用Intent组件的使用一一列出。用Intent调用系统中经常被用到的组件。

(1)调用拨号程序

```
// 给移动客服10086拨打电话
Uri uri = Uri.parse("tel:10086");
Intent intent = new Intent(Intent.ACTION_DIAL, uri);
startActivity(intent);
```

(2)发送短信或彩信

```
// 给10086发送内容为"Hello"的短信
Uri uri = Uri.parse("smsto:10086");
Intent intent = new Intent(Intent.ACTION_SENDTO, uri);
intent.putExtra("sms_body", "Hello");
startActivity(intent);
// 发送彩信(相当于发送带附件的短信
```

```
    Intent intent = new Intent(Intent.ACTION_SEND);
    intent.putExtra("sms_body", "Hello");
    Uri uri = Uri.parse("content://media/external/images/media/23");
    intent.putExtra(Intent.EXTRA_STREAM, uri);
    intent.setType("image/png");
    startActivity(intent);
```

（3）通过浏览器打开网页

```
    // 打开百度主页
    Uri uri = Uri.parse("http://www.baidu.com");
    Intent intent  = new Intent(Intent.ACTION_VIEW, uri);
    startActivity(intent);
```

（4）发送电子邮件

```
    // 给someone@domain.com发邮件
    Uri uri = Uri.parse("mailto:someone@domain.com");
    Intent intent = new Intent(Intent.ACTION_SENDTO, uri);
    startActivity(intent);
    // 给someone@domain.com发送内容为"Hello"的邮件
    Intent intent = new Intent(Intent.ACTION_SEND);
    intent.putExtra(Intent.EXTRA_EMAIL, "someone@domain.com");
    intent.putExtra(Intent.EXTRA_SUBJECT, "Subject");
    intent.putExtra(Intent.EXTRA_TEXT, "Hello");
    intent.setType("text/plain");
    startActivity(intent);
    // 给多人发邮件
    Intent intent=new Intent(Intent.ACTION_SEND);
    String[] tos = {"1@abc.com", "2@abc.com"};   // 收件人
    String[] ccs = {"3@abc.com", "4@abc.com"};   // 抄送
    String[] bccs = {"5@abc.com", "6@abc.com"};  // 密送
    intent.putExtra(Intent.EXTRA_EMAIL, tos);
    intent.putExtra(Intent.EXTRA_CC, ccs);
    intent.putExtra(Intent.EXTRA_BCC, bccs);
    intent.putExtra(Intent.EXTRA_SUBJECT, "Subject");
    intent.putExtra(Intent.EXTRA_TEXT, "Hello");
    intent.setType("message/rfc822");
    startActivity(intent);
```

（5）显示地图与路径规划

```
    // 打开Google地图中国北京位置（北纬39.9，东经116.3）
```

第9章 信息传递者之Intent机制

```
    Uri uri = Uri.parse("geo:39.9,116.3");
    Intent intent = new Intent(Intent.ACTION_VIEW, uri);
    startActivity(intent);
    // 路径规划:从北京某地(北纬39.9,东经116.3)到上海某地(北纬31.2,东经121.4)
    Uri uri = Uri.parse("http://maps.google.com/maps?f=d&saddr=39.9 116.3&daddr=31.2 121.4");
    Intent intent = new Intent(Intent.ACTION_VIEW, uri);
    startActivity(intent);
```

(6) 播放多媒体

```
    Intent intent = new Intent(Intent.ACTION_VIEW);
    Uri uri = Uri.parse("file:///sdcard/foo.mp3");
    intent.setDataAndType(uri, "audio/mp3");
    startActivity(intent);
    Uri uri = Uri.withAppendedPath(MediaStore.Audio.Media.INTERNAL_CONTENT_URI, "1");
    Intent intent = new Intent(Intent.ACTION_VIEW, uri);
    startActivity(intent);
```

(7) 拍照

```
    // 打开拍照程序
    Intent intent = new Intent(MediaStore.ACTION_IMAGE_CAPTURE);
    startActivityForResult(intent, 0);
    // 取出照片数据
    Bundle extras = intent.getExtras();
    Bitmap bitmap = (Bitmap) extras.get("data");
```

(8) 获取并剪切图片

```
    // 获取并剪切图片
    Intent intent = new Intent(Intent.ACTION_GET_CONTENT);
    intent.setType("image/*");
    intent.putExtra("crop", "true"); // 开启剪切
    intent.putExtra("aspectX", 1); // 剪切的宽高比为1:2
    intent.putExtra("aspectY", 2);
    intent.putExtra("outputX", 20); // 保存图片的宽和高
    intent.putExtra("outputY", 40);
    intent.putExtra("output", Uri.fromFile(new File("/mnt/sdcard/temp"))); // 保存路径
    intent.putExtra("outputFormat", "JPEG");// 返回格式
    startActivityForResult(intent, 0);
    // 剪切特定图片
```

```java
        Intent intent = new Intent("com.android.camera.action.CROP");
        intent.setClassName("com.android.camera", "com.android.camera.CropImage");
        intent.setData(Uri.fromFile(new File("/mnt/sdcard/temp")));
        intent.putExtra("outputX", 1); // 剪切的宽高比为1∶2
        intent.putExtra("outputY", 2);
        intent.putExtra("aspectX", 20); // 保存图片的宽和高
        intent.putExtra("aspectY", 40);
        intent.putExtra("scale", true);
        intent.putExtra("noFaceDetection", true);
        intent.putExtra("output", Uri.parse("file:///mnt/sdcard/temp"));
        startActivityForResult(intent, 0);
```

(9) 打开Google Market

```java
        // 打开Google Market直接进入该程序的详细页面
        Uri uri = Uri.parse("market://details?id=" + "com.demo.app");
        Intent intent = new Intent(Intent.ACTION_VIEW, uri);
        startActivity(intent);
```

(10) 安装和卸载程序

```java
        Uri uri = Uri.fromParts("package", "com.demo.app", null);
        Intent intent = new Intent(Intent.ACTION_DELETE, uri);
        startActivity(intent);
```

(11) 进入无线网络设置界面

```java
        // 进入无线网络设置界面
        Intent intent = new Intent(android.provider.Settings.ACTION_WIRELESS_SETTINGS);
        startActivityForResult(intent, 0);
```

强化训练

本章主要介绍了Android系统中的Intent的功能和用法，Android使用Intent封装了应用程序的启动"意图"，但这种"意图"并未直接与任何程序组件耦合，通过这种方式可很好地提高系统的可扩展性和可维护性。

学习本章需要重点掌握Intent的component、action、category、data、type各属性的功能和用法，并掌握如何在AndroidManifest.xml文件中配置。

学完本章知识后，不妨通过以下习题来巩固所学的内容吧。

第9章 信息传递者之Intent机制

一、填空题

1. 在一个Android应用中，主要由四种组件组成，分别为：____、____、____、____。
2. MIME类型有两种形式：_____，_____。
3. 理解Intent的关键之一是理解清楚Intent的两种基本用法：一种是____，即在构造Intent对象时就指定接收者；另一种是____，即Intent的发送者在构造Intent对象时，并不知道也不关心接收者是谁，有利于降低发送者和接收者之间的耦合。
4. 通常Intent的调用分为____和____两种。
5. Intent除了定位目标组件外，另外一个职责是____。Intent之间传递数据一般有两种常用的方法：一种是____；另一种是____。
6. Android的Intent之间传递对象有两种方法，一种是_____；另一种是_____。

二、选择题

1. Intel的属性中，（　　）指定Intent的目标组件的类名称。
   A. type
   B. data
   C. category
   D. action
2. category常量中，（　　）设置该Activity随系统启动而运行。
   A. CATEGORY_HOME
   B. CATEGORY_LAUNCHER
   C. CATEGORY_BROWSABLE
   D. CATEGORY_PREFERENCE
3. （　　）不是Intel组件的常见使用方式。
   A. 调用拨号程序
   B. 通过浏览器打开网页
   C. 编辑电子邮件
   D. 显示地图与路径规划
4. 在一个Activity中，可能会使用（　　）方法打开多个不同的Activity处理不同的业务。
   A. setResult(int resultCode, Intent data)
   B. onActivityResult(int requestCode, int resultCode, Intent data)
   C. getIntent()
   D. startActivityForResult()
5. 使用Intent启动不同组件的方法，（　　）是启动Activity的方法。
   A. MainActivity.
   B. ACTION_SYNC
   C. Context.startActivity(Intent intent)
   D. OtherActivity
6. action常量中，（　　）是同步服务器和移动设备的数据。
   A. ACTION_BATTERY_LOW
   B. ACTION_SYNC
   C. ACTION_MAIN

D. ACTION_CALL

三、操作题

1. 在Android系统中，电话功能是Android程序中经常使用的功能之一，练习使用Intent在Android应用程序启动Dialer程序的功能。练习利用Android Studio开发环境创建一个新的Android应用程序，实现启动Android自带的打电话功能Dialer程序的功能。

2. 练习利用Android Studio开发环境创建一个新的Android应用程序，实现启动Android内置的浏览器并访问百度的功能。

3. 一个Activity相当于手机的一屏，如何在不同的Activity交换数据是Android应用程序经常使用到的功能之一。练习利用Android Studio开发环境创建一个新的Android应用程序，在该Android应用程序中创建两个Activity，实现两个Activity之间进行数据交换。

第10章

骨干成员之Service组件

内容导读

　　服务（Service）也是Android应用程序中四大基本组件之一。同是四大基本组件，与Activity不同，Service虽然长期在后台运行，却没有界面显示，是不可见的。Android应用程序中，Service主要用于两个目的：后台运行和跨进程访问。通过启动一个Service，可以在不显示界面的前提下在后台运行完成一些持续的、耗时的任务，并不影响用户做其他事情。通过Service可以实现不同进程之间的通信，这也是Service的重要用途之一。本章将着重介绍Service的相关知识并通过实例展示其使用方法。

学习目标

- 熟悉Service的概念；
- 熟悉Service的生命周期；
- 掌握Service的使用方法；
- 熟悉IntentService。

10.1 Service概述

　　Service初期设计思想就是为了实现一个稳定的、生命周期较长的组件。Android中

Service是运行在后台的,为了保障Service的正常运行,Service甚至具有比非活动状态下的Activity更高的优先级,即Service在运行时被Android系统资源管理器回收的可能性会小一些。既然说Service是运行在后台的服务,那么它通常是不可见的,没有界面的。编程人员可以启动一个服务来播放音乐,或者记录智能手机地理信息位置的改变,或者启动一个服务来运行并一直监听某种动作。

Service和其他组件一样,都是运行在主线程中,因此不能用它来做耗时的请求或者动作。编程人员可以在服务中开一个线程,在线程中做耗时动作。

服务一般分为两种。

(1)本地服务

Local Service 用于应用程序内部。在Service可以调用context.startService启动,调用context.stopService结束;在内部可以调用Service.stopSelf或 Service.stopSelfResult来自己停止;无论调用了多少次startService,都只需调用一次stopService来停止。

(2)远程服务

Remote Service 用于Android系统内部的应用程序之间。可以定义接口并把接口暴露出来,以便其他应用进行操作;客户端建立到服务对象的连接,并通过那个连接来调用服务;调用context.bindService方法建立连接并启动,调用 context.unbindService关闭连接,多个客户端可以绑定至同一个服务,如果服务此时还没有加载,bindService会先加载它。提供可被其他应用复用的服务,比如定义一个天气预报服务,提供给其他应用调用即可。

10.2 Service的生命周期

相对于Activity而言,Service生命周期简单很多。不过由于Service可能在用户不知情的情况下在后台运行,Android程序开发人员需要更加关注Service如何创建和销毁。Service生命周期可以从两种启动Service的模式开始讲起,分别是context.startService和context.bindService。

(1)startService的启动模式下的生命周期

当首次使用startService启动一个服务时,系统会实例化一个Service实例,依次调用其onCreate和onStartCommand方法,然后进入运行状态,此后,如果再使用startService启动服务时,不再创建新的服务对象,系统会自动找到刚才创建的Service实例,调用其onStart方法;如果想要停掉一个服务,可使用stopService方法,此时onDestroy方法会被调用,需要注意的是,不管前面使用了多少次startService,只需一次stopService即可停掉服务。

(2)bindService启动模式下的生命周期

在这种模式下,当调用者首次使用bindService绑定一个服务时,系统会实例化一个Service实例,并一次调用其onCreate方法和onBind方法,然后调用者就可以和服务进行交互了,此后,如果再次使用bindService绑定服务,系统不会创建新的Service实例,也不会再调用onBind方法;如果需要解除与这个服务的绑定,可使用unbindService方法,此时onUnbind方法和onDestroy方法会被调用。

两种模式的不同之处在于:startService模式下调用者与服务无必然联系,即使调用者结束了自己的生命周期，只要没有使用stopService方法停止这个服务，服务仍会运行；通常情况下，bindService模式下服务是与调用者"生死与共"的，在绑定结束之后，一旦调用者被销毁，服务也就立即终止。

值得一提的是，在使用startService启动服务时都是习惯重写onStart方法，在Android2.0时系统引进了onStartCommand方法取代onStart方法，为了兼容以前的程序，在onStartCommand方法中其实调用了onStart方法，不过最好是重写onStartCommand方法。

以上两种模式的流程如图10-1所示。在图10-1中左边所示为startService的启动模式下的生命周期（Unbounded）。当首次使用startService启动一个服务时，系统会实例化一个Service，依次调用其onCreate和onStartCommand方法，然后进入运行状态。如果想要停掉服务，可使用stopService方法，此时onDestroy方法会被调用。这种情况下调用者与服务无必然联系，即使调用者结束了自己的生命周期，只要没有使用stopService方法停止这个服务，服务仍会运行。在图10-1中右边所示为bindService启动模式下的生命周期（Bounded），当调用者首次使用bindService绑定一个服务时，系统会实例化一个Service，并调用其onCreate方法和onBind方法。如果需要解除与这个服务的绑定，可使用unbindService方法，此时onUnbind方法和onDestroy方法会被调用。而这种情况下服务是与调用者"生死与共"的，在绑定结束之后，一旦调用者被销毁，服务也就立即终止。

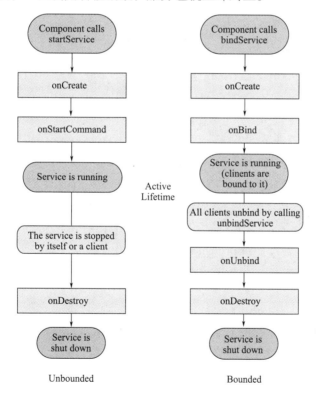

图10-1　Service生命周期

下面将结合实例来演示一下这两种模式的生命周期过程。

10.2.1　startService启动服务

【实例10-1】新建一个名为ServiceTestDemo的项目，然后创建一个MyService服务类，代码如下：

```java
package cn.androidstudy.servicetestdemo;
import android.app.Service;
import android.content.Intent;
import android.os.Binder;
import android.os.IBinder;
import android.util.Log;
public class MyService extends Service {
    private static final String TAG = "MyService";

    @Override
    public void onCreate() {
        super.onCreate();
        Log.i(TAG, "onCreate called.");       }
    @Override
    public int onStartCommand(Intent intent, int flags, int startId) {
        Log.i(TAG, "onStartCommand called.");
        return super.onStartCommand(intent, flags, startId);       }
    @Override
    public void onStart(Intent intent, int startId) {
        super.onStart(intent, startId);
        Log.i(TAG, "onStart called.");      }
    @Override
    public IBinder onBind(Intent intent) {
        Log.i(TAG, "onBind called.");
        return new Binder(){};      }
    @Override
    public boolean onUnbind(Intent intent) {
        Log.i(TAG, "onUnbind called.");
        return super.onUnbind(intent);      }
    @Override
    public void onDestroy() {
        super.onDestroy();
        Log.i(TAG, "onDestroy called.");       }}
```

然后在AndroidManifest.xml中配置服务信息，不然这个服务就无法生效，配置如下：

```xml
<service android:name=".MyService">
    <intent-filter>
        <action android:name="android.intent.action.MyService" />
        <category android:name="android.intent.category.DEFAULT" />
    </intent-filter>
</service>
```

第10章 骨干成员之Service组件

如果服务只是在本应用中使用，可以去掉<intent-filter>属性。

服务搭建完成之后，来关注一下调用者MainActivity，它很简单，只有两个按钮，一个是启动服务，另一个是停止服务，来看一下它们的单击事件：

```
public void startService(View view) {    //启动服务
   Intent intent = new Intent(this, MyService.class);
   startService(intent);    }
public void stopService(View view) {    //停止服务
   Intent intent = new Intent(this, MyService.class);
   stopService(intent);    }
```

接下来先单击一次启动按钮，看看都发生了什么。日志打印结果如图10-2所示。

```
09-08 11:48:20.740 7294-7294/cn.androidstudy.servicetestdemo I/MyService: onCreate called.
09-08 11:48:20.740 7294-7294/cn.androidstudy.servicetestdemo I/MyService: onStartCommand called.
09-08 11:48:20.740 7294-7294/cn.androidstudy.servicetestdemo I/MyService: onStart called.
```

图10-2　LogCat中日志输出

再单击一次，我们会发现结果略有不同，如图10-3所示。

```
09-08 11:50:41.454 7294-7294/cn.androidstudy.servicetestdemo I/MyService: onStartCommand called.
09-08 11:50:41.454 7294-7294/cn.androidstudy.servicetestdemo I/MyService: onStart called.
```

图10-3　第二次启动服务日志输出

看到第二次单击时onCreate方法就不再被调用了，而是直接调用了onStartCommand方法（onStartCommand中又调用了onStart方法）。选择"Settings→Apps&notifacations→App info"打开服务ServiceTestDemo，会发现该服务刚刚启动，可以强制关闭（FORCE STOP），如图10-4所示。

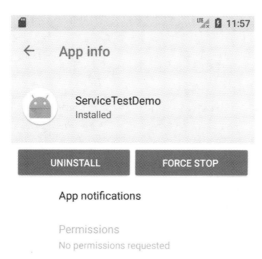

图10-4　查看已运行的服务

然后单击"FORCE STOP"按钮，试图停止服务，会发现onDestroy方法被调用了，此时服务就停止运行了，日志输出如图10-5所示。再次选择"Settings→Apps&notifacations→App info"打开服务ServiceTestDemo，会发现该服务已经停止。

```
09-08 11:52:58.725 7294-7294/cn.androidstudy.servicetestdemo I/MyService: onDestroy called.
```

图10-5　停止服务日志输出

10.2.2　bindSerivce启动服务

bindSerivce的函数原型如下：
bindSerivce(Intent service,ServiceConnection conn,int flags)
参数说明：
- service：通过该参数可以启动指定的Service；
- conn：该参数是一个ServiceConnection对象，这个对象用于监听访问者与Service之间的连接情况，当访问者与Service连接成功时将回调ServiceConnection对象的onServiceConnected(ComponentName name,IBinder service)方法，如果断开将回调onServiceDisconnected(ComponentName name)方法；
- flags：指定绑定时是否自动创建Service。

由于onServiceConnected需要传入一个IBinder接口类型的参数，而前面服务中的onBind方法返回值为null，这样是不行的，要想实现绑定操作，必须返回一个实现了IBinder接口类型的实例，该接口描述了与远程对象进行交互的抽象协议，有了它我们才能与服务进行交互。修改onBind的代码如下：

```java
@Override
public IBinder onBind(Intent intent) {
    Log.i(TAG, "onBind called.");
    return new Binder(){};  }
```

在MainActivity的布局文件中增加两个按钮，增加如下代码，使其可以以bindService的方式启动一个服务，代码如下：

```java
private ServiceConnection conn = new ServiceConnection() {
    @Override
    public void onServiceConnected(ComponentName name, IBinder service) {
        Log.i("MainActivity", "onServiceConnected called.");     }
    @Override
    public void onServiceDisconnected(ComponentName name) {
        Log.i("MainActivity", "onServiceDisconnected called.");    } };
public void bind(View view) {    //绑定服务
    Intent intent = new Intent(this, MyService.class);
    bindService(intent, conn, Context.BIND_AUTO_CREATE); }
public void unbind(View view) {    //解除绑定
    unbindService(conn);   }
```

第10章 骨干成员之Service组件

在使用bindService绑定服务时，我们需要一个ServiceConnection代表与服务的连接，它只有两个方法，onServiceConnected和onServiceDisconnected，前者是在操作者在连接一个服务成功时被调用，而后者是在服务崩溃或被"杀死"导致的连接中断时被调用，如果我们自己解除绑定时则不会被调用。

先单击一下绑定按钮，LogCat的输出如图10-6所示。onServiceConnected方法被调用了，绑定连接已经成功了，onCreate方法和onBind方法被调用，此时服务已进入运行阶段。如果再次单击绑定按钮，onCreate和onBinder不会再次被调用，这个过程中它们仅被调用一次。

```
09-08 12:14:07.308 24521-24521/cn.androidstudy.servicetestdemo I/MyService: onCreate called.
09-08 12:14:07.308 24521-24521/cn.androidstudy.servicetestdemo I/MyService: onBind called.
09-08 12:14:07.315 24521-24521/cn.androidstudy.servicetestdemo I/MainActivity: onServiceConnected called.
```

图10-6　启动绑定服务输出

然后单击解除绑定按钮，LogCat的输出如图10-7所示。可以看到onUnbind方法和onDestroy方法被调用了，此时MyService已被销毁，整个生命周期结束。另一方面，当退出MainActivity时，服务也会随之结束，从这一点看，MyService可以说是"誓死追随"着MainActivity。

```
09-08 12:14:40.882 24521-24521/cn.androidstudy.servicetestdemo I/MyService: onUnbind called.
09-08 12:14:40.882 24521-24521/cn.androidstudy.servicetestdemo I/MyService: onDestroy called.
```

图10-7　解除绑定服务输出

注意：在连接中断状态再去做解除绑定操作会引起一个异常（图10-8），在MainActivity销毁之前没有进行解除绑定也会导致后台出现异常信息，为了确保不会出现此类情况，需要对MainActivity做如下修改。

图10-8　解除绑定引发异常

```
    private boolean binded = false;
    private ServiceConnection conn = new ServiceConnection() {
        @Override
        public void onServiceConnected(ComponentName name, IBinder service) {
            Log.i("MainActivity", "onServiceConnected called.");
            binded = true;       }
        @Override
        public void onServiceDisconnected(ComponentName name) {
            Log.i("MainActivity", "onServiceDisconnected called.");       } };
    public void bind(View view) {   //绑定服务
        Intent intent = new Intent(this, MyService.class);
```

```
        bindService(intent, conn, Context.BIND_AUTO_CREATE);
}
public void unbind(View view) {    //解除绑定
unbindService();}
@Override
protected void onDestroy() {
unbindService();
super.onDestroy();}
private void unbindService() {
if(binded){unbindService(conn);
       binded = false;    }}
```

10.3 Service的使用方法

作为Android系统中四大基本应用程序组件之一，Service如何使用对于Android编程人员非常重要。下面通过具体的实例来讲解Service的使用方法。

10.3.1 编写不需和Activity交互的本地服务

本地服务编写比较简单，在10.2.1【实例10-1】的基础上直接进行修改。这里写了一个计数服务的类，每秒为计数器加1。在服务类的内部，还创建了一个线程，用于实现后台执行上述业务逻辑。只需把MyService类修改如下内容即可。

```
public class MyService extends Service {
    private boolean threadDisable;
    private int count;
    @Override
    public IBinder onBind(Intent intent) {
        return null;         }
    @Override
    public void onCreate() {
        super.onCreate();
        new Thread(new Runnable() {
            @Override
            public void run() {
                while (!threadDisable) {
                    try {
                        Thread.sleep(1000);
                    } catch (InterruptedException e) {            }
```

第10章 骨干成员之Service组件

```
                    count++;
                    Log.v("CountService","Count is " + count);
            }    }    }).start();            }
    @Override
    public void onDestroy() {
        super.onDestroy();
        this.threadDisable= true;
        Log.v("CountService","on destroy");        }
    public int getCount() {        return count;        }}
```

启动服务后,可通过日志查看到后台线程打印的计数内容。

10.3.2 编写本地服务和Activity交互

前面的示例是通过startService和stopService启动关闭服务的,适用于服务和Activity之间没有调用交互的情况。如果之间需要传递参数或者调用方法,需要使用bind和unbind方法。

【实例10-2】创建一个新的项目LocalServiceDemo,新建的Activity命名为LocalServiceDemoActivity。服务类需要增加接口,例如ICountService。另外,服务类需要有一个内部类,这样可以方便访问外部类的封装数据,这个内部类需要继承Binder类并实现ICountService接口。还有,就是要实现Service的onBind方法,不能只传回一个null了。

新建立的接口ICountService源代码如下:

```
package cn.androidstudy.localservicedemo;
public interface ICountService {    public abstract int getCount();}
```

然后建立服务类CountService,源代码如下:

```
package cn.androidstudy.localservicedemo;
import android.app.Service;
import android.content.Intent;
import android.os.Binder;
import android.os.IBinder;
import android.util.Log;
public class CountService extends Service implements ICountService {
    private boolean threadDisable;
    private int count;
    private ServiceBinder serviceBinder = new ServiceBinder();
    public class ServiceBinder extends Binder implements ICountService{
        @Override
        public int getCount() {     return count;        }
    @Override
    public int getCount() {     return count;        }
```

```java
@Override
public IBinder onBind(Intent intent) { return serviceBinder;    }
@Override
public void onCreate() {
    super.onCreate();
    new Thread(new Runnable(){       @Override
        public void run() {    while(!threadDisable){
            try{  Thread.sleep(1000);
            }catch(Exception e){             }
            count++;
            Log.v("CountService", "Count is "+count);
        }   }    }).start();    }
@Override
public void onDestroy() {
    super.onDestroy();
    threadDisable = true;
    Log.v("CountService", "on destroy");     }}
```

修改AndroidManifest.xml文件，注册CountService，源代码如下：

```xml
<?xml version="1.0" encoding="utf-8"?>
<manifest xmlns:android="http://schemas.android.com/apk/res/android"
    package="cn.androidstudy.localservicedemo">
    <application
        android:allowBackup="true"
        android:icon="@mipmap/ic_launcher"
        android:label="@string/app_name"
        android:roundIcon="@mipmap/ic_launcher_round"
        android:supportsRtl="true"
        android:theme="@style/AppTheme">
        <activity android:name=".LocalServiceDemoActivity">
            <intent-filter>
                <action android:name="android.intent.action.MAIN" />
                <category android:name="android.intent.category.LAUNCHER" />
            </intent-filter>
        </activity>
        <service android:name=".CountService">
            <intent-filter >
                <action android:name="android.intent.action.CountService"/>
            </intent-filter>
        </service>
    </application>
```

第10章 骨干成员之Service组件

```
</manifest>
```

Acitity代码需要通过bindService和unbindService启动关闭服务。另外，需要通过ServiceConnection的内部类实现连接Service和Activity。LocalServiceDemoActivity.java的源代码如下：

```java
package cn.androidstudy.localservicedemo;
import android.support.v7.app.AppCompatActivity;
import android.os.Bundle;
import android.os.IBinder;
import android.util.Log;
import android.content.ComponentName;
import android.content.Intent;
import android.content.ServiceConnection;
public class LocalServiceDemoActivity extends AppCompatActivity {
    private ICountService countService;
    private ServiceConnection serviceConnection = new ServiceConnection() {
        @Override
        public void onServiceDisconnected(ComponentName name) {
            countService = null;        }
        @Override
        public void onServiceConnected(ComponentName name, IBinder service) {
            countService = (ICountService)service;
            Log.v("CountService", "on service connected,count is"+countService.getCount());
        }    };
    @Override
    protected void onCreate(Bundle savedInstanceState) {
        super.onCreate(savedInstanceState);
        setContentView(R.layout.activity_local_service_demo);
        bindService(new Intent(this,CountService.class),serviceConnection,BIND_AUTO_CREATE);    }

    @Override
    protected void onDestroy() {
        unbindService(serviceConnection);
        super.onDestroy();        }}
```

运行建立的LocalServiceDemo项目，可以在LogCat看到如图10-9的输出结果，说明服务已经启动了。

```
09-08 13:13:44.124 16495-16495/cn.androidstudy.localservicedemo V/CountService: on service connected,count is0
09-08 13:13:45.032 16495-16537/cn.androidstudy.localservicedemo V/CountService: Count is 1
09-08 13:13:46.034 16495-16537/cn.androidstudy.localservicedemo V/CountService: Count is 2
09-08 13:13:47.036 16495-16537/cn.androidstudy.localservicedemo V/CountService: Count is 3
09-08 13:13:48.037 16495-16537/cn.androidstudy.localservicedemo V/CountService: Count is 4
09-08 13:13:49.039 16495-16537/cn.androidstudy.localservicedemo V/CountService: Count is 5
09-08 13:13:50.041 16495-16537/cn.androidstudy.localservicedemo V/CountService: Count is 6
09-08 13:13:51.044 16495-16537/cn.androidstudy.localservicedemo V/CountService: Count is 7
```

图10-9 本地服务与Activity交互

10.3.3 编写传递基本型数据的远程服务

前述实例可扩展为让其他应用程序复用该服务。这样的服务叫远程（remote）服务，实际上是进程间通信（interprocess communication，IPC）。Java中是不支持跨进程内存共享的，因此要传递对象，需要把对象解析成操作系统能够理解的数据格式，以达到跨界对象访问的目的。在JavaEE中，采用RMI通过序列化传递对象。在Android中，则采用AIDL(Android Interface Definition Language，接口定义语言)方式实现。

AIDL是一种接口定义语言，用于约束两个进程间的通信规则，供编译器生成代码，实现Android设备上的两个进程间通信(IPC)。AIDL的IPC机制和EJB所采用的CORBA很类似，进程之间的通信信息，首先会被转换成AIDL协议消息，然后发送给对方，对方收到AIDL协议消息后再转换成相应的对象。由于进程之间的通信信息需要双向转换，所以Android采用代理类在背后实现了信息的双向转换，代理类由Android编译器生成，对开发人员来说是透明的。

（1）创建AIDL文件

使用AIDL来定义远程服务的接口，而不是简单的Java接口。扩展名为."aidl"而不是."java"。可用前面的ICountService改动成ICountSerivde.aidl，会自动生成相关的Java文件。

☞【实例10-3】为了和10.3.2中的例子区分开，新建一个项目AIDLDemo，然后新建一个ICountSerivde.aidl文件，步骤：File→ New→ AIDL→AIDL File，将【实例10-2】中的ICountService.java中的代码复制过来。ICountSerivde.aidl的源代码如下：

```
package cn.androidstudy.aidldemo;
interface ICountService {    int getCount();   }
```

当完成AIDL文件创建后，IDE会自动在项目的目录中同步生成接口文件。接口文件中生成一个Stub的抽象类，里面包括AIDL定义的方法，还包括一些其他辅助方法。值得关注的是asInterface(IBinder iBinder)，它返回接口类型的实例，对于远程服务调用，远程服务返回给客户端的对象为代理对象，客户端在onServiceConnected(ComponentName name, IBinder service)方法引用该对象时不能直接强转成接口类型的实例，而应该使用asInterface(IBinder iBinder)进行类型转换。

编写AIDL文件时，需要注意下面几点。

- 接口名和AIDL文件名相同。
- 接口和方法前不用加访问权限修饰符public、private、protected等，也不能用final、

static。

- AIDL默认支持的类型包括Java基本类型int、long、boolean等和String、List、Map、CharSequence，使用这些类型时不需要import声明。对于List和Map中的元素类型必须是AIDL支持的类型。如果使用自定义类型作为参数或返回值，自定义类型必须实现Parcelable接口。
- 自定义类型和AIDL生成的其他接口类型在AIDL描述文件中，应该显式import，即便该类和定义的包在同一个包中。
- 在AIDL文件中所有非Java基本类型参数必须加上in、out、inout标记，以指明参数是输入参数、输出参数还是输入输出参数。
- Java原始类型默认的标记为in,不能为其他标记。

（2）建立服务类

定义好AIDL接口之后，接下来就可定义一个服务（Service）类，该Service的onBind方法返回的IBinder对象应该是ICountSerivde.Stub的子类的实例。至于其他部分，则与开发本地Service完全一样。

新建一个CountService类，其源代码如下：

```
package cn.androidstudy.aidldemo;
import android.app.Service;
import android.content.Intent;
import android.os.IBinder;
import android.os.RemoteException;
import android.util.Log;
import cn.androidstudy.aidldemo.ICountService;
public class CountService extends Service{
    private  threadDisable;
    private int count;
    @Override
    public IBinder onBind(Intent intent) {
        return new serviceBinder();        }

    public class serviceBinder extends ICountService.Stub {
        @Override
        public int getCount() throws RemoteException {
            return count;        }        }
    @Override
    public void onCreate() {
        super.onCreate();
        new Thread(new Runnable(){
            @Override
            public void run() {
                while(!threadDisable){
                    try{ Thread.sleep(1000);
                    }catch(Exception e){          }
```

```
                count++;
                Log.v("CountService", "Count is "+count);
            } }   }).start();    }
    @Override
    public void onDestroy() {
        super.onDestroy();
        threadDisable = true;
        Log.v("CountService", "on destroy");    }}
```

（3）注册服务

注册CountService和前面的示例类似，注册后的配置文件AndroidManifest.xml如下：

```xml
<?xml version="1.0" encoding="utf-8"?>
<manifest xmlns:android="http://schemas.android.com/apk/res/android"
    package="cn.androidstudy.aidldemo">
    <application
        android:allowBackup="true"
        android:icon="@mipmap/ic_launcher"
        android:label="@string/app_name"
        android:roundIcon="@mipmap/ic_launcher_round"
        android:supportsRtl="true"
        android:theme="@style/AppTheme">
        <activity android:name=".AIDLDemoActivity">
            <intent-filter>
                <action android:name="android.intent.action.MAIN" />
                <category android:name="android.intent.category.LAUNCHER" />
            </intent-filter>
        </activity>
        <service
            android:name=".CountService">
        </service>
    </application>
</manifest>
```

（4）访问CountService

本例与前面本地服务和Activity交互例子中，在Activity中使用Service服务的方式差别不大，只需要在ServiceConnection中调用远程服务的方法时捕获异常。AIDLDemo项目的结构如图10-10所示。

修改AIDLDemoActivity.java，源代码如下：

```
package cn.androidstudy.aidldemo;
import android.support.v7.app.AppCompatActivity;
import android.os.Bundle;
```

第10章 骨干成员之Service组件

图10-10　AIDLDemo项目结构

```
import android.os.IBinder;
import android.os.RemoteException;
import android.content.ComponentName;
import android.content.Intent;
import android.content.ServiceConnection;
import android.util.Log;
import cn.androidstudy.aidldemo.ICountService;
public class AIDLDemoActivity extends AppCompatActivity {
    private ICountService countService;
    private ServiceConnection serviceConnection = new ServiceConnection() {
        @Override
        public void onServiceDisconnected(ComponentName name) {
            countService = null;       }

        @Override
          public void onServiceConnected(ComponentName name, IBinder service) {
            countService = ICountService.Stub.asInterface(service);
            try {
                Log.v("CountService", "on service connected,count is "+ countService.getCount());
            } catch (RemoteException e) { e.printStackTrace(); } } };
    @Override
    protected void onCreate(Bundle savedInstanceState) {
        super.onCreate(savedInstanceState);
        setContentView(R.layout.activity_aidldemo);
        bindService(new Intent(this, CountService.class),
```

209

```
        serviceConnection, BIND_AUTO_CREATE);    }
    @Override
    protected void onDestroy() {
        unbindService(serviceConnection);
        super.onDestroy();      }}
```

由代码可以看出,在onServiceConnected方法中,countService需要通过ICountService.Stub.asInterface(service)进行赋值,另外需要处理getCount抛出的异常:RemoteException。

运行AIDLDemo项目,就可以看到如图10-11所示的输出。

```
09-10 03:18:48.080 3774-3774/? V/CountService: on service connected, count is 0
09-10 03:18:49.000 3774-3793/cn.androidstudy.aidldemo V/CountService: Count is 1
09-10 03:18:50.002 3774-3793/cn.androidstudy.aidldemo V/CountService: Count is 2
09-10 03:18:51.003 3774-3793/cn.androidstudy.aidldemo V/CountService: Count is 3
09-10 03:18:52.004 3774-3793/cn.androidstudy.aidldemo V/CountService: Count is 4
09-10 03:18:53.005 3774-3793/cn.androidstudy.aidldemo V/CountService: Count is 5
09-10 03:18:54.006 3774-3793/cn.androidstudy.aidldemo V/CountService: Count is 6
```

图10-11 运行AIDLDemo项目

10.3.4 编写传递复杂数据类型的远程服务

远程服务往往不只是传递Java基本数据类型。这时需要注意AIDL的一些限制和规定:
- AIDL支持Java原始数据类型;
- AIDL支持String和CharSequence;
- 如果需要在AIDL中使用其他AIDL接口类型,需要import,即使是在相同包结构下;
- AIDL允许传递实现Parcelable接口的类,需要import;
- AIDL支持集合接口类型List和Map,但是有一些限制,元素必须是基本型或者上述三种情况,不需要import集合接口类,但是需要对元素涉及的类型import;
- 非基本数据类型,也不是String和CharSequence类型的,需要有方向指示,包括in、out和inout,in表示由客户端设置,out表示由服务端设置,inout是两者均可设置。

也就是说,在不同的进程间传递一个类对象,该类必须实现Parcelable接口。Parcelable接口会告诉Android运行时,在封送(marshalling)和解封送(unmarshalling)过程中如何实现序列化和反序列化对象。我们很容易联想到java.io.Serializable接口,用户可能会有疑问,两种接口功能确实类似,但为什么Android不用内置的Java序列化机制,而偏偏要搞一套新东西呢?这是因为Android设计团队认为Java中的序列化太慢,难以满足Android的进程间通信需求,所以他们构建了Parcelable解决方案。Parcelable要求显式地序列化类的成员,但最终序列化对象的速度将快很多。

注意:Android提供了两种机制来将数据传递给另一个进程:第一种是使用Intent将数据束(Bundle)传递给活动;第二种也就是Parcelable传递给服务。这两种机制不可互换,不要混淆。也就是说,Parcelable无法传递给活动,只能用作AIDL定义的一部分。

那么,如何创建这样的类呢?必须满足如下要求:

第10章 骨干成员之Service组件

- 实现Parcelable接口；
- 实现writeToParcel方法，该方法会将对象的属性值写入Parcel；
- 添加一个静态属性Creator，该属性需实现android.os.Parcelable.Creator<T>接口；
- 创建一个声明该类的AIDL文件。

【实例10-4】下面通过一个示例来演示如何在进程间传递复制数据类型。

首先，新建一个Android项目PersonAidlService，然后创建一个Person类，实现Parcelable接口。Person.java的源代码如下：

```java
package cn.androidstudy.personaidlservice;
import android.os.Parcelable;
import android.os.Parcel;
public class Person implements Parcelable {
    private String name;
    private int age;
    public static final Parcelable.Creator<Person> CREATOR =
new Parcelable.Creator<Person>() {
        @Override
        public Person createFromParcel(Parcel source) {
            return new Person(source);        }
        @Override
        public Person[] newArray(int size) {
            return new Person[size];        }    };
    public Person(){   }
    private Person(Parcel source){
        readFromParcel(source);        }
    @Override
    public int describeContents() { return 0;   }
    @Override
    public void writeToParcel(Parcel dest, int flags) {
        dest.writeString(name);
        dest.writeInt(age);        }
    public void readFromParcel(Parcel source){
        name = source.readString();
        age = source.readInt();      }
    public String getName(){  return name;      }
    public void setName(String name){ this.name = name;   }
    public int getAge(){  return age;      }
    public void setAge(int age){  this.age = age;   }}
```

这个Person.aidl文件很简单，就是定义了一个Parcelable类，告诉系统我们需要序列化和反序列化的类型。需要注意的是，readFromParcel方法是从Parcel中读取数据，为了避免出错，应该和writeToParcel方法的写入顺序保持一致。

然后，需要在同一包下建立一个包含复杂类型的Person.aidl文件，代码如下：

211

```
package cn.wang.personaidl;
parcelable Person;
```

接下来，需要创建一个 IGreetService.aidl 文件，以接收类型为 Person 的输入参数，以便客户端可以将 Person 传递给服务。

```
package cn.androidstudy.personaidlservice;
import cn.androidstudy.personaidlservice.Person;
interface IGreetService{
String greet(in Person person);}
```

注意：需要在参数上加入方向指示符 in，代表参数由客户端设置，还需要为 Person 提供一个 import 语句（虽然在同一个包下）。

此时，在插件 ADT 的帮助下，AIDL 编译器会自动编译生成一个 IGreetService.java 文件。

接下来，在 cn.wang.personaidlservice 包中创建 Service 类 PersonService。PersonService.java 的源代码如下：

```
package cn.androidstudy.personaidlservice;
import android.app.Service;
import android.content.Intent;
import android.os.IBinder;
import android.os.RemoteException;
import cn.androidstudy.personaidlservice.Person;
import cn.androidstudy.personaidlservice.IGreetService;
public class PersonService extends Service {
    IGreetService.Stub stub = new IGreetService.Stub() {
        @Override
        public String greet(Person person) throws RemoteException {
            String strRet = "Hello, "+person.getName()+", your age is "+person.getAge();
            return strRet;     }      };
    @Override
    public IBinder onBind(Intent intent) {
        return stub;       }}
```

最后，在 AndroidManifest.xml 中配置该服务，AndroidManifest.xml 的源代码如下：

```
<?xml version="1.0" encoding="utf-8"?>
<manifest xmlns:android="http://schemas.android.com/apk/res/android"
   package="cn.androidstudy.personaidlservice">
   <application    android:allowBackup="true"
      android:icon="@mipmap/ic_launcher"
      android:label="@string/app_name"
      android:roundIcon="@mipmap/ic_launcher_round"
      android:supportsRtl="true"
```

```
            android:theme="@style/AppTheme">
        <activity android:name=".MainActivity">
            <intent-filter>
                <action android:name="android.intent.action.MAIN" />
                <category android:name="android.intent.category.LAUNCHER" />
            </intent-filter>
        </activity>
        <service android:name=".PersonService">
        </service>
    </application>
</manifest>
```

这样，服务端就完成了，服务端的结构如图10-12所示。

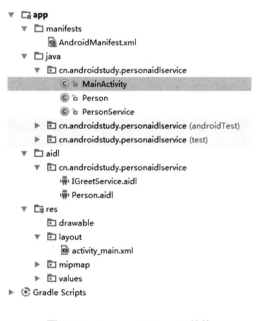

图10-12　PersonAidlService结构

☞【实例10-5】下面创建一个新的Android工程PersonAidlClient，去访问刚创建的PersonService。首先，将PersonAidlService工程里的cn.wang.personaidl包复制到PersonAidlClient中，然后在MainActivity的布局文件中增加三个按钮，分别用来绑定服务、调用greet方法和解除绑定服务。然后，修改MainActivity.java，增加按钮的处理事件。MainActivity.java的源代码如下：

```
package cn.androidstudy.personaidlservice;
import android.support.v7.app.AppCompatActivity;
import android.os.Bundle;
import android.os.IBinder;
import android.os.RemoteException;
import android.view.View;
import android.view.View.OnClickListener;
```

```java
import android.widget.Button;
import android.widget.TextView;
import android.content.ComponentName;
import android.content.Intent;
import android.content.ServiceConnection;
import cn.androidstudy.personaidlservice.Person;
import cn.androidstudy.personaidlservice.IGreetService;
import cn.androidstudy.personaidlservice.PersonService;
public class MainActivity extends AppCompatActivity {
    private Button btnBind,btnHello,btnUnbind;
    private TextView tv;
    private IGreetService greetService;
    private ServiceConnection conn = new ServiceConnection(){
        @Override
        public void onServiceConnected(ComponentName name, IBinder service) {
            greetService = IGreetService.Stub.asInterface(service);        }
        @Override
        public void onServiceDisconnected(ComponentName name) {    }  };
    @Override
    protected void onCreate(Bundle savedInstanceState) {
        super.onCreate(savedInstanceState);
        setContentView(R.layout.activity_main);
        tv = (TextView)findViewById(R.id.textView1);
        btnBind = (Button)findViewById(R.id.btnBind);
        btnBind.setOnClickListener(new OnClickListener() {
            @Override
            public void onClick(View v) {
                Intent intent = new Intent();
                intent.setAction("cn.androidstudy.personaidlservice.PersonService");
                intent.setPackage("cn.androidstudy.personaidlservice");
                bindService(intent, conn, BIND_AUTO_CREATE);
                btnBind.setEnabled(false);
                btnHello.setEnabled(true);
                btnUnbind.setEnabled(true);     }  });
        btnHello = (Button)findViewById(R.id.btnHello);
        btnHello.setOnClickListener(new OnClickListener() {

            @Override
            public void onClick(View v) {
                Person p = new Person();
```

```
                p.setName("Mike");
                p.setAge(30);
                try { String ret = greetService.greet(p);
                    tv.setText(ret);
                } catch (RemoteException e) {
                    e.printStackTrace();    }    }         });
        btnUnbind = (Button)findViewById(R.id.btnUnbind);
        btnUnbind.setOnClickListener(new OnClickListener() {
            @Override
            public void onClick(View v) {
                unbindService(conn);
                btnBind.setEnabled(true);
                btnHello.setEnabled(false);
                btnUnbind.setEnabled(false);    }   });   }}
```

先运行一次 PersonAidlService 工程，然后运行 PersonAidlClient 工程，单击 "BIND" 按钮启动服务，然后单击 "HELLO" 按钮，可以看到 TextView 里已经显示了 greet 返回的字符串（图10-13），说明在进程间传递复杂数据类型已经成功了。

图10-13　进程间传递复杂数据类型

10.4　IntentService

Android 中 IntentService 是基于 Service 的子类，主要用来处理异步请求，其优点是使用简单。不管是何种 Service，它默认都是在应用程序的主线程（亦即 UI 线程）中运行的。所以，如果编程人员的 Service 将要运行非常耗时或者可能被阻塞的操作时，编程人员的应用程序将会被挂起，甚至会出现 ANR 错误。为了避免这一问题，编程人员应该在 Service 中重新启动一个新的线程来进行这些操作。可以通过以下两种方法来解决。

（1）方法一

直接在 Service 的 onStartCommand 方法中重启一个线程来执行，例如：

```
    @Override
    public int onStartCommand(Intent intent, int flags, int startId) {
          MyServiceActivity.updateLog(TAG + " ----> onStartCommand()");
          new Thread(new Runnable() {
                @Override
                public void run() {
                        // 此处进行耗时的操作，这里只是简单地让线程休眠了1s
                        try { Thread.sleep(1000);
                        } catch (Exception e) {    e.printStackTrace();
}}}}).start();
          return START_STICKY;         }
```

（2）方法二

Android SDK中提供了一个现成的Service类来实现该功能，即IntentService，它主要负责以下几个方面。

• 生成一个默认的且与主线程互相独立的工作者线程来执行所有传送至onStartCommand方法的Intent；

• 生成一个工作队列来传送Intent对象给用户的onHandleIntent方法，同一时刻只传送一个Intent对象，这样一来，用户就不必担心多线程的问题；

• 在所有的请求(Intent)都被执行完以后会自动停止服务，所以，用户不需要自己去调用stopSelf方法来停止该服务；

• 提供了一个onBind方法的默认实现，它返回null；

• 提供了一个onStartCommand方法的默认实现，它将Intent先传送至工作队列，然后从工作队列中每次取出一个传送至onHandleIntent方法，在该方法中对Intent做相应的处理。

简单说，IntentService是继承于Service并处理异步请求的一个类，在IntentService内有一个工作线程来处理耗时操作，启动IntentService的方式和启动传统Service一样，同时，当任务执行完后，IntentService会自动停止，而不需要去手动控制。另外，可以启动IntentService多次，而每一个耗时操作会以工作队列的方式在IntentService的onHandleIntent回调方法中执行，并且，每次只会执行一个工作线程，执行完第一个再执行第二个，以此类推。

使用IntentService的好处有：省去了在Service中手动开线程的麻烦；当操作完成时，不用手动停止Service；简单易用。

☞【实例10-6】下面通过一个示例来演示IntentService的使用方法。

首先新建一个Android项目IntentServiceDemo。在项目里增加一个类IntentServiceDemo，该类继承IntentService。继承IntentService时，必须提供一个无参构造函数，且在该构造函数内，需要调用父类的构造函数。在该类中需要实现onHandleIntent，该方法中对Intent做相应的处理，在此模拟了两个耗时的操作，根据不同的参数在LogCat做不同的输出，然后让线程等待2秒。同时，为了便于了解IntentService的生命周期，对其他方法也进行了重写。

IntentServiceDemo.java文件的源代码如下：

```
package cn.androidstudy.intentservicedemo;
import android.app.IntentService;
import android.content.Intent;
import android.os.IBinder;
```

第10章 骨干成员之Service组件

```java
import android.util.Log;
public class IntentServiceDemo extends IntentService {
    private static final String TAG = "IntentServiceDemo";
    public IntentServiceDemo() {   super("IntentServiceDemo");    }
    @Override
    protected void onHandleIntent(Intent intent) {
        //Intent是从Activity发过来的，携带识别参数，根据参数不同执行不同的任务
        String action = intent.getExtras().getString("param");
        long id = Thread.currentThread().getId();
        if (action.equals("oper1")) {
            Log.i(TAG,"Operation1 in thread:"+id);
        }else if (action.equals("oper2")) {
            Log.i(TAG,"Operation2 in thread:"+id);   }
        try {    Thread.sleep(2000);    }
        catch (InterruptedException e) { e.printStackTrace();  }    }
    @Override
    public void setIntentRedelivery(boolean enabled) {
        Log.i(TAG, "setIntentRedelivery");
        super.setIntentRedelivery(enabled);      }
package cn.androidstudy.intentservicedemo;
import android.app.IntentService;
import android.content.Intent;
import android.content.Context;
import android.os.IBinder;
import android.util.Log;
public class IntentServiceDemo extends IntentService {
    private static final String TAG = "IntentServiceDemo";
    public IntentServiceDemo() {
        super("IntentServiceDemo");
    }
    @Override
    protected void onHandleIntent(Intent intent) {
        //Intent是从Activity发过来的，携带识别参数，根据参数不同执行不同的任务
        String action = intent.getExtras().getString("param");
        long id = Thread.currentThread().getId();
        if (action.equals("oper1")) {
            Log.i(TAG,"Operation1 in thread:"+id);
        }else if (action.equals("oper2")) {
            Log.i(TAG,"Operation2 in thread:"+id);          }
        try {    Thread.sleep(2000);
        } catch (InterruptedException e) {
            e.printStackTrace();       }
```

```java
@Override
public void onCreate() {
    long id = Thread.currentThread().getId();
    Log.i(TAG, "onCreate in thread:"+id);
    super.onCreate();       }
@Override
public void onStart(Intent intent, int startId) {
    Log.i(TAG, "onStart");
    super.onStart(intent, startId);        }
@Override
public int onStartCommand(Intent intent, int flags, int startId) {
    Log.i(TAG, "onStartCommand");
    return super.onStartCommand(intent, flags, startId);      }
@Override
public void onDestroy() {
    Log.i(TAG, "onDestroy");
    super.onDestroy();      }
@Override
public IBinder onBind(Intent intent) {
    Log.i(TAG, "onBind");
    return super.onBind(intent);     }}}
```

然后，需要在AndroidManifest.xml文件中注册刚创建的服务，AndroidManifest.xml文件源代码如下：

```xml
<?xml version="1.0" encoding="utf-8"?>
<manifest xmlns:android="http://schemas.android.com/apk/res/android"
    package="cn.androidstudy.intentservicedemo">
    <application    android:allowBackup="true"
        android:icon="@mipmap/ic_launcher"
        android:label="@string/app_name"
        android:roundIcon="@mipmap/ic_launcher_round"
        android:supportsRtl="true"
        android:theme="@style/AppTheme">
        <activity android:name=".MainActivity">
            <intent-filter>
                <action android:name="android.intent.action.MAIN" />
                <category android:name="android.intent.category.LAUNCHER" />
            </intent-filter>
        </activity>
        <service android:name=".IntentServiceDemo">
            <intent-filter>
         <action android:name="cn.androidstudy.intentservicedemo.
```

```
IntentServiceDemo" />
            </intent filter>
        </service>
    </application>
</manifest>
```

最后，修改MainActivity.java文件，在onCreate方法中通过两个Intent对象，携带不同的参数分别启动服务。MainActivity.java文件的源代码如下：

```
package cn.androidstudy.intentservicedemo;
import android.support.v7.app.AppCompatActivity;
import android.os.Bundle;
import android.util.Log;
import android.content.Intent;
import cn.androidstudy.intentservicedemo.IntentServiceDemo;
public class MainActivity extends AppCompatActivity {
    @Override
    protected void onCreate(Bundle savedInstanceState) {
        super.onCreate(savedInstanceState);
        setContentView(R.layout.activity_main);
        Intent intent = new Intent();
        intent.setAction("cn.androidstudy.intentservicedemo.IntentServiceDemo");
        intent.setPackage("cn.androidstudy.intentservicedemo");
        intent.putExtra("param", "oper1");
        startService(intent);
        Intent intent2 = new Intent();
        intent2.setAction("cn.androidstudy.intentservicedemo.IntentServiceDemo");
        intent2.setPackage("cn.androidstudy.intentservicedemo");
        intent2.putExtra("param", "oper2");
        startService(intent2);

        Log.i(".MainActivity", "onCreate in thread:"+Thread.currentThread().getId()); }}
```

运行IntentServiceDemo，在LogCat窗口可以看到如图10-14的输出。

```
09-10 10:03:41.882 26328-26328/cn.androidstudy.intentservicedemo I/.MainActivity: onCreate in thread:2
09-10 10:03:41.905 26328-26328/cn.androidstudy.intentservicedemo I/IntentServiceDemo: onCreate in thread:2
09-10 10:03:41.906 26328-26328/cn.androidstudy.intentservicedemo I/IntentServiceDemo: onStartCommand
09-10 10:03:41.906 26328-26328/cn.androidstudy.intentservicedemo I/IntentServiceDemo: onStart
09-10 10:03:41.906 26328-26348/cn.androidstudy.intentservicedemo I/IntentServiceDemo: Operation1 in thread:366
09-10 10:03:41.907 26328-26328/cn.androidstudy.intentservicedemo I/IntentServiceDemo: onStartCommand
09-10 10:03:41.907 26328-26328/cn.androidstudy.intentservicedemo I/IntentServiceDemo: onStart
09-10 10:03:45.913 26328-26328/cn.androidstudy.intentservicedemo I/IntentServiceDemo: onDestroy
```

图10-14　IntentService调用输出

强化训练

作为Android四大基本组件之一的Service，主要承担着两项职能：长期运行的耗时工作和进程间的交互，因此也对应两种启动模式：startService启动模式和bindSerivce绑定模式。学习Service需要重点掌握如何创建、配置Service组件，以及如何启动、停止Service，同时需要区分清楚两种启动方式的区别。本章最后介绍了Intent Service，Android编程人员需要掌握它的特点，以及它和Service之间的区别，达到能够灵活运用的目的。

学完本章知识后，不妨通过以下习题来巩固所学的内容吧。

一、填空题

1. Android中Service生命周期可以从两种启动Service的模式开始讲起，分别是_____和_____。
2. Android中服务一般分为两种：_____、_____。
3. 定义好AIDL接口之后，接下来就可定义一个_____。
4. 与在Activity中使用服务的差别不大，只需要在_____中调用远程服务的方法时捕获异常。
5. 服务主要用于两个目的：_____和_____。通过启动一个_____，可以在不显示界面的前提下在后台运行指定的任务，这样可以不影响用户做其他事情。
6. 在使用bindService绑定服务时，我们需要一个_____代表与服务的连接，它只有两个方法：_____和_____。
7. 作为Android四大组件之一的Service，主要承担着两项职能：_____和_____，因此也对应两种启动模式：_____和_____。
8. 接口文件中生成一个_____的抽象类，里面包括_____定义的方法，还包括一些其他辅助方法。

二、选择题

1. in、out和inout，in表示由客户端设置，out表示由服务端设置，（　　）是两者均可设置。

    A. out
    B. inout
    C. in
    D. outin

2. （　　）接口会告诉Android运行时，在封送（marshalling）和解封送（unmarshalling）过程中如何实现序列化和反序列化对象。

    A. CharSequence
    B. AIDL
    C. import
    D. Parcelable

3. 不管是何种Service，它默认都是在应用程序的（　　）中运行的。

第10章 骨干成员之Service组件

A. 单线程
B. 多线程
C. 主线程
D. 副线程

4. 下列选项中是编写 AIDL 文件时需要注意的点的是（　　）。

A. 自定义类型和 AIDL 生成的其他接口类型在 AIDL 描述文件中，应该显式 import，即便该类和定义的包是同一个包中
B. Java 原始类型默认的标记为 in，能为其他标记
C. 接口名和 AIDL 文件名相同
D. 接口和方法前不用加访问权限修饰符 public、private、protected 等，能用 final、static

5. 下列是 AIDL 的限制和规定的是（　　）。

A. AIDL 不支持 String 和 CharSequence
B. 如果需要在 AIDL 中使用其他 AIDL 接口类型，需要 inout，即使是在相同包结构下
C. AIDL 允许传递实现 Parcelable 接口的类，不需要 import
D. AIDL 支持 Java 原始数据类型

6. Android SDK 中提供了一个现成的 Service 类来实现该功能，即（　　）。

A. onStartCommand
B. onHandleIntent
C. TextView
D. IntentService

7. 下列不是 Android 中使用 IntentService 的好处的是（　　）。

A. 省去了在 Service 中手动开线程的麻烦
B. 功能强大
C. 当操作完成时，不用手动停止 Service
D. 简单易用

三、操作题

1. Service 是 Android 系统中四大主要应用程序组件之一，熟练使用 Service 是 Android 编程人员的基本任务之一。利用 Android Studio 开发环境创建一个新的 Android 应用程序，练习使用 Service 的启动和停止功能，并在日志中打印结果。

2. 利用 Android Studio 开发环境创建一个新的 Android 应用程序，练习使用 bindService 和 unbindService 启动关闭服务。

3. AIDL 是一种接口定义语言，用于约束两个进程间的通信规则，供编译器生成代码，实现 Android 设备上的两个进程间通信(IPC)。利用 Android Studio 开发环境创建一个新的 Android 应用程序，练习使用 AIDL 实现传递基本类型数据的远程服务功能。

第11章

开发利器之BroadcastReceiver组件

内容导读

广播（Broadcast）是一种广泛运用的在应用程序之间传输信息的机制，广播的优点在于可以一对多，可以异步传输消息，接收者可以自行过滤符合条件的广播。Android中广播接收者（BroadcastReceiver）是事件或消息的监听器，用于响应来自当前应用的其他组件、其他Android应用或Android系统的异步广播消息。异步消息的优点在于消息的接收者与发送者互相独立，发送者只需发送，接收者只需接收，无须等待。Android中广播消息的内容可以是与应用程序密切相关的数据信息，也可以是Android的系统信息，如常用的蓝牙设备搜索与管理、电池电量变化、系统设置变化、网络连接变化或收到短信等。在Android系统中，BroadcastReceiver 是对发送出来的广播进行过滤接收并响应的一类组件。本章将着重介绍BroadcastReceiver的相关知识并通过实例展示其使用方法。

学习目标

- 熟悉BroadcastReceiver的概念；
- 掌握广播消息的使用方法；
- 掌握如何处理系统广播消息；
- 熟悉BroadcastReceiver的生命周期。

第11章 开发利器之BroadcastReceiver组件

11.1 BroadcastReceiver概述

BroadcastReceiver的中文翻译是"广播接收者",顾名思义,它是用来接收来自当前应用的其他组件、其他Android应用或Android系统的异步广播消息的。广播(Broadcast)是一种广泛运用的在应用程序之间传输信息的机制,而广播接收者BroadcastReceiver是对发送出来的广播进行过滤、接收并响应的一类组件。

在Android系统中,广播体现在方方面面,例如当开机完成后系统会产生一条广播,接收到这条广播就能实现开机启动服务的功能;当网络状态改变时系统会产生一条广播,接收到这条广播就能及时地做出提示和保存数据等操作;当电池电量改变时,系统会产生一条广播,接收到这条广播就能在电量低时告知用户及时保存进度。

应用程序可以拥有任意数量的广播接收器以对所有感兴趣的通知信息予以响应。所有的接收器均继承自BroadcastReceiver基类。广播接收器没有用户界面。然而,它们可以启动一个Activity来响应它们收到的信息,或者用NotificationManager来通知用户。通知可以用很多种方式来吸引用户的注意力,主要有闪动背灯、振动、播放声音等,一般来说是在状态栏上放一个持久的图标,用户可以打开它并获取消息。

BroadcastReceiver事件分类:

- 系统广播事件,例如ACTION_BOOT_COMPLETED(系统启动完成后触发)、ACTION_TIME_CHANGED(系统时间改变时触发)、ACTION_BATTERY_LOW(电量低时触发)等。
- 用户自定义的广播事件。

BroadcastReceiver事件的编程流程如下。

(1)注册广播事件

注册方式有两种:一种是静态注册,就是在AndroidManifest.xml文件中定义,注册的广播接收器必须要继承BroadcastReceiver类;另一种是动态注册,是在程序中使用Context.registerReceiver注册,注册的广播接收器相当于一个匿名类。两种方式都需要Intent Filter。

(2)发送广播事件

通过Context.sendBroadcast来发送,由Intent来传递注册时用到的action。

(3)接收广播事件

当发送的广播被接收器监听到后,会调用它的onReceive方法,并将包含消息的Intent对象传给它。onReceive中代码的执行时间不要超过10秒,否则Android会弹出超时窗口。

11.2 广播消息

Android中的广播机制设计得非常出色,很多事情原本需要开发者亲自操作,现在只需等待广播告知自己就可以了,大大减少了开发的工作量和开发周期。例如蓝牙设备的搜索与连接、信息传输或断开等,都可以通过广播的形式进行处理。作为Android应用开发者,需要熟练掌握Android系统提供的一个开发利器,那就是广播接收者BroadcastReceiver。接下来本节将通过实例对BroadcastReceiver逐一地分析和演练,了解和掌握它的各种功能和用法。

11.2.1 自定义BroadcastReceiver

【实例11-1】通过一个实例来演示一下一个自定义的BroadcastReceiver,并让这个BroadcastReceiver能够运行起来。

扫一扫 看视频

首先,创建一个名为MyReceiverTest的工程。要创建自己的BroadcastReceiver类,需要继承android.content.BroadcastReceiver,并实现其onReceive方法。下面创建一个名为MyReceiver广播接收者,代码如下:

```java
package cn.androidstudy.myreceivertest;
import android.content.BroadcastReceiver;
import android.content.Context;
import android.content.Intent;
import android.util.Log;
public class MyReceiver extends BroadcastReceiver {
    @Override
    public void onReceive(Context context, Intent intent) {
        String msg = intent.getStringExtra("msg");
        Log.i("MyReceiver", msg);   }}
```

在创建完BroadcastReceiver之后,还不能够使它进入工作状态,需要为它注册一个指定的广播地址。没有注册广播地址的BroadcastReceiver就像一个缺少选台按钮的收音机,虽然功能具备,但也无法收到电台的信号。下面就来介绍一下如何为BroadcastReceiver注册广播地址。

(1)静态注册

静态注册是在AndroidManifest.xml文件中配置的,在application节点内增加如下代码来为MyReceiver注册一个广播地址。

```xml
<receiver    android:name=".MyReceiver"
    android:enabled="true"
    android:exported="true">
    <intent-filter>
        <action android:name="android.intent.action.MYRECEIVER"/>
        <category android:name="android.intent.category.DEFAULT"/>
    </intent-filter>
</receiver>
```

配置了以上信息之后,只要是android.intent.action.MYRECEIVER这个地址的广播,MyReceiver都能够接收到。

注意:这种方式的注册是常驻型的,也就是说当应用关闭后,如果有广播信息传来,

MyReceiver也会被系统调用而自动运行。

在MainActivity的布局文件中增加一个按钮，该按钮的单击事件如下：

```
public void sendStatic(View view){//启动静态注册Receiver
    Intent intent = new Intent("android.intent.action.MYRECEIVER");
    intent.putExtra("msg", "Hello BroadcastReceiver");
    sendBroadcast(intent);
}
```

启动工程，点击该按钮，即可发送一个广播，MyReceiver接收到广播后，即在LogCat窗口输出接收到的信息，如图11-1所示。

```
09-13 06:53:25.993 3525-3525/cn.androidstudy.myreceivertest I/MyReceiver: Hello BroadcastReceiver
```

图11-1　MyReceiver输出

sendBroadcast方法有如下两个重载：
- sendBroadcast(Intent intent)
- sendBroadcast(Intent intent, String receiverPermission)

前文使用的是第一种，没有对接收权限进行限制。如果使用第二种形式，并且指定接收权限，则只有在AndroidManifest.xml中使用标签<uses-permission>声明了拥有此权限的BroadcastReceiver才有可能接收到发送来的广播。同样，若在注册BroadcastReceiver时指定了可接收的广播权限，则只有在包内的AndroidManifest.xml中用标签<uses-permission>进行声明，拥有此权限的对象发出的广播才能被这个BroadcastReceiver所接收。

如果将前面的sendStatic方法修改为：

public void sendStatic(View view){

Intent intent = new Intent("android.intent.action.MYRECEIVER");

intent.putExtra("msg","Hello BroadcastReceiver");

sendBroadcast(intent,"cn.androidstudy.permission");}

那么，无论我们怎么点击该按钮，MyReceiver都接收不到该广播。若想接收到，需要在AndroidManifest.xml文件中增加如下代码：

```
<permission android:name="cn.androidstudy.permission"></permission>
<uses-permission android:name="cn.androidstudy.permission"/>
```

说明：第一行注册一个permission，第二行使用该permission。

如果要限制广播的发送者必须具有某个权限，那么注册广播接收者时增加android:permission属性，代码如下：

```
<receiver android:name=".MyReceiver" android:permission="cn.androidstudy.test">
    <intent-filter>
        <action android:name="android.intent.action.MYRECEIVER"/>
        <category android:name="android.intent.category.DEFAULT"/>
    </intent-filter>
</receiver>
```

在其他的工程中，给该接收者发送广播，必须在AndroidManifest.xml文件中增加相应的权限，代码如下：

```
<permission android:name="cn.androidstudy.test"></permission>
<uses-permission android:name="cn.androidstudy.test"/>
```

（2）动态注册

动态注册需要在代码中动态地指定广播地址并注册，通常是在Activity或Service中注册一个广播，为了确保Activity或Service启动后已完成注册，可以将注册语句添加到onStart方法中。但是，在Activity或Service中注册了一个BroadcastReceiver，当这个Activity或Service被销毁时如果没有解除注册，系统会报一个异常，提示我们是否忘记解除注册了。所以，需要在onDestroy方法中添加解除注册操作，此处一定要注意，具体代码如下：

```
MyReceiver receiver = new MyReceiver();
@Override
protected void onStart() {
IntentFilter filter = new IntentFilter();
filter.addAction("android.intent.action.RECEIVERTEST");
registerReceiver(receiver, filter);
super.onStart();}
@Override
protected void onDestroy() {
unregisterReceiver(receiver);
super.onDestroy();}
public void sendDyn(View view){
    Intent intent = new Intent("android.intent.action.RECEIVERTEST");
intent.putExtra("msg", "Hi BroadcastReceiver");
    sendBroadcast(intent);}
```

启动工程，点击该按钮，即可发送一个广播，MyReceiver接收到广播后，在LogCat窗口输出接收到的信息，和图11-1类似。

这里的例子只是一个接收者来接收广播，如果有多个接收者都注册了相同的广播地址，又会是什么情况呢？能同时接收到同一条广播吗？相互之间会不会有干扰呢？这就涉及普通广播和有序广播的概念了。

11.2.2 普通广播

普通广播（Normal Broadcast）对于多个接收者来说是完全异步的，通常每个接收者都无须等待即可以接收到广播，接收者相互之间不会有影响。对于这种广播，接收者无法终止广播，即无法阻止其他接收者的接收动作。

为了验证以上论断，在MyReceiverTest工程里新建两个BroadcastReceiver，分别为MyReceiver2和MyReceiver3，代码如下：

```
package cn.androidstudy.myreceivertest;
import android.content.BroadcastReceiver;
import android.content.Context;
```

第11章 开发利器之BroadcastReceiver组件

```java
import android.content.Intent;
import android.util.Log;
public class MyReceiver2 extends BroadcastReceiver {
    @Override
    public void onReceive(Context context, Intent intent) {
        String msg = intent.getStringExtra("msg");
        Log.i("MyReceiver2", msg);      }}
public class MyReceiver3 extends BroadcastReceiver {
@Override
    public void onReceive(Context context, Intent intent) {
        String msg = intent.getStringExtra("msg");
        Log.i("MyReceiver3", msg); }}
```

然后在AndroidManifest.xml文件中注册这两个BroadcastReceiver,代码如下：

```xml
<receiver android:name=".MyReceiver2">
    <intent-filter>
        <action android:name="android.intent.action.MYRECEIVER"/>
        <category android:name="android.intent.category.DEFAULT" />
    </intent-filter>
</receiver>
<receiver android:name=".MyReceiver3">
    <intent-filter>
        <action android:name="android.intent.action.MYRECEIVER"/>
        <category android:name="android.intent.category.DEFAULT" />
    </intent-filter>
</receiver>
```

然后再次点击发送按钮,发送一条广播,在LogCat窗口的输出如图11-2所示。

```
09-13 07:08:38.466 17209-17209/cn.androidstudy.myreceivertest I/MyReceiver: Hello BroadcastReceiver
09-13 07:08:38.471 17209-17209/cn.androidstudy.myreceivertest I/MyReceiver2: Hello BroadcastReceiver
09-13 07:08:38.473 17209-17209/cn.androidstudy.myreceivertest I/MyReceiver3: Hello BroadcastReceiver
```

图11-2 普通广播

看来这三个接收者都接收到这条广播了,修改一下三个接收者,在onReceive方法的最后一行添加以下代码,试图终止广播：

```java
abortBroadcast();
```

再次点击发送按钮,会发现三个接收者仍然都打印了自己的日志,表明接收者并不能终止广播。

11.2.3 有序广播

有序广播（Ordered Broadcast）比较特殊，它每次只发送到优先级较高的接收者那里，然后由优先级高的接收者再传播到优先级低的接收者那里，优先级高的接收者有能力终止这个广播。

【实例11-2】为了演示有序广播的流程，首先创建一个新的工程 MyOrderedReceiverTest，然后创建三个接收者，代码如下：

```java
package cn.androidstudy.myorderedreceivertest;
import android.content.BroadcastReceiver;
import android.content.Context;
import android.content.Intent;
import android.os.Bundle;
import android.util.Log;
public class MyReceiver extends BroadcastReceiver {
    @Override
    public void onReceive(Context context, Intent intent) {
        String msg = intent.getStringExtra("msg");
        Log.i("MyService", msg);
        Bundle bundle = new Bundle();
        bundle.putString("msg", msg + ", MyReceiver");
        setResultExtras(bundle);        }}
public class MyReceiver2 extends BroadcastReceiver {
    @Override
    public void onReceive(Context context, Intent intent) {
        String msg = intent.getStringExtra("msg");
        Log.i("MyService2", msg);
        msg = getResultExtras(true).getString("msg");
        Log.i("MyService2", msg);
        Bundle bundle = new Bundle();
        bundle.putString("msg", msg + ", MyReceiver2");
        setResultExtras(bundle);        }}
public class MyReceiver3 extends BroadcastReceiver {
    @Override
    public void onReceive(Context context, Intent intent) {
        String msg = intent.getStringExtra("msg");
        Log.i("MyService3", msg);
        msg = getResultExtras(true).getString("msg");
        Log.i("MyService3", msg);    }}
```

在 MyReceiver 和 MyReceiver2 中最后都使用了 setResultExtras 方法将一个 Bundle 对象设置为结果集对象，传递到下一个接收者那里，这样一来，优先级低的接收者可以用 getResultExtras 获取到最新的经过处理的信息集合。

第11章 开发利器之BroadcastReceiver组件

接着，需要为三个接收者注册广播地址，注册的方法和前面讲的方法基本相同，但是在这里需要体现广播接收者的优先级，需要在<intent-filter>里增加"android:priority"属性，这个属性的范围是-1000 ～ 1000，数值越大，优先级越高。在AndroidMainfest.xml文件中注册三个广播接收者的代码如下：

```xml
<receiver android:name=".MyReceiver">
    <intent-filter android:priority="1000">
        <action android:name="android.intent.action.MYORDEREDRECEIVER"/>
        <category android:name="android.intent.category.DEFAULT"/>
    </intent-filter>
</receiver>
<receiver android:name=".MyReceiver2">
    <intent-filter android:priority="999">
        <action android:name="android.intent.action.MYORDEREDRECEIVER"/>
        <category android:name="android.intent.category.DEFAULT"/>
    </intent-filter>
</receiver>
<receiver android:name=".MyReceiver3">
    <intent-filter android:priority="998">
        <action android:name="android.intent.action.MYORDEREDRECEIVER"/>
        <category android:name="android.intent.category.DEFAULT" />
    </intent-filter>
</receiver>
```

在MainActivity的布局文件中增加一个按钮，该按钮的单击事件如下：

```java
public void sendStatic(View view){//启动静态注册Receiver
    Intent intent = new Intent("android.intent.action.MYRECEIVER");
    intent.putExtra("msg","Hello BroadcastReceiver");
    sendOrderedBroadcast(intent, null);}
```

启动工程，点击该按钮，即可发送一个广播，MyReceiver接收到广播后，即在LogCat窗口输出接收到的信息，如图11-3所示。

```
09-18 23:58:34.309 6342-6342/cn.androidstudy.myorderedreceivertest I/MyService: Hello BroadcastReceiver
09-18 23:58:34.322 6342-6342/cn.androidstudy.myorderedreceivertest I/MyService2: Hello BroadcastReceiver
09-18 23:58:34.322 6342-6342/cn.androidstudy.myorderedreceivertest I/MyService2: Hello BroadcastReceiver, MyReceiver
09-18 23:58:34.325 6342-6342/cn.androidstudy.myorderedreceivertest I/MyService3: Hello BroadcastReceiver
09-18 23:58:34.325 6342-6342/cn.androidstudy.myorderedreceivertest I/MyService3: Hello BroadcastReceiver, MyReceiver, MyReceiver2
```

图11-3 有序广播

在图11-3中可以看到，MyReceiver2和MyReceiver3分别输出了两次，第一次是使用intent.getStringExtra("msg")获取intent中的数据，而第二次是使用getResultExtras(true).getString("msg")，获取的是优先级高的接收者使用setResultExtras传递下来的数据。

修改一下接收者MyReceiver2，在onReceive方法的最后一行添加以下代码终止广播：

```java
abortBroadcast();
```

229

再次点击发送按钮,在LogCat窗口输出接收到的信息,如图11-4所示。可以发现,MyReceiver3没有接收到广播,说明在有序广播模式下,优先级高的接收者可以中断广播的传播。

```
09-19 00:05:57.621 13071-13071/cn.androidstudy.myorderedreceivertest I/MyService: Hello BroadcastReceiver
09-19 00:05:57.623 13071-13071/cn.androidstudy.myorderedreceivertest I/MyService2: Hello BroadcastReceiver
09-19 00:05:57.623 13071-13071/cn.androidstudy.myorderedreceivertest I/MyService2: Hello BroadcastReceiver, MyReceiver
```

图11-4 广播传输中断

11.3 处理系统广播消息

在广播消息中,有一类特殊的广播消息,它们特殊在只能由Android系统发出,这类广播称为系统广播。很多系统后台服务都会使用广播的形式进行处理。系统广播经常被用来通知一些重要的系统事件,如电池电量的变化、应用软件的安装卸载、SD卡插入拔出、外部电源的插拔等。这些系统广播都被定义为action常量,存放在android.content.Intent中,如表11-1所示。

表11-1 常用的系统广播

action常量名	说明
Intent.ACTION_BATTERY_CHANGED	充电状态,或者电池的电量发生变化
Intent.ACTION_BATTERY_LOW	电池电量低
Intent.ACTION_BATTERY_OKAY	电池电量充足,即从电池电量低变化到饱满时会发出广播
Intent.ACTION_BOOT_COMPLETED	系统启动完成后,这个动作被广播一次
Intent.ACTION_CAMERA_BUTTON	按下照相时的拍照按键(硬件按键)时发出的广播
Intent.ACTION_DATE_CHANGED	设备日期发生改变时会发出此广播
Intent.ACTION_HEADSET_PLUG	在耳机口上插入耳机时发出的广播
Intent.ACTION_PACKAGE_ADDED	成功地安装APK之后发出的广播
Intent.ACTION_PACKAGE_REMOVED	成功地删除某个APK之后发出的广播
Intent.ACTION_POWER_CONNECTED	插上外部电源时发出的广播
Intent.ACTION_POWER_DISCONNECTED	已断开外部电源连接时发出的广播
Intent.ACTION_SCREEN_OFF	屏幕被关闭之后的广播
Intent.ACTION_SCREEN_ON	屏幕被打开之后的广播
Intent.ACTION_SHUTDOWN	关闭系统时发出的广播
Intent.ACTION_TIME_CHANGED	时间被设置时发出的广播

(1)开机启动服务

经常会有这样的应用场合,需要实现开机启动某个服务。要实现这个功能,就可以订阅系统"启动完成"这条广播,接收到这条广播后就可以启动自己的服务了。

【实例11-3】首先创建一个工程BootReveiverTest，接着，新建两个类：BootCompleteReceiver和MyService，其中MyService是开机启动的服务，BootCompleteReceiver在接收到Intent. ACTION_BOOT_COMPLETED广播后，启动MyService，二者的具体实现如下：

```java
package cn.androidstudy.bootreveivertest;
import android.app.Service;
import android.content.Intent;
import android.os.IBinder;
import android.util.Log;
public class MyService extends Service {
    @Override
    public IBinder onBind(Intent intent) {
        return null;    }
    @Override
    public void onCreate() {
        super.onCreate();
        Log.i("MyService", "onCreate called.");    }
    @Override
    public int onStartCommand(Intent intent, int flags, int startId) {
        Log.i("MyService", "onStartCommand called.");
        return super.onStartCommand(intent, flags, startId);    }}
package cn.androidstudy.bootreveivertest;
import android.content.BroadcastReceiver;
import android.content.Context;
import android.content.Intent;
import android.util.Log;
public class BootCompleteReceiver extends BroadcastReceiver {
    @Override
    public void onReceive(Context context, Intent intent) {
        Intent service = new Intent(context, MyService.class);
        context.startService(service);
        Log.i("BootCompleteReceiver", "Boot Complete. Starting MyService...");    }}
```

然后需要在AndroidManifest.xml中注册服务和广播接收者，增加如下代码：

```xml
<!-- 需要开机启动的服务 -->
<service android:name=".MyService"/>
<!-- 开机广播接收者 -->
<receiver android:name=".BootCompleteReceiver">
    <intent-filter>
        <!-- 注册开机广播地址-->
        <action android:name="android.intent.action.BOOT_COMPLETED"/>
        <category android:name="android.intent.category.DEFAULT" />
```

```xml
        </intent-filter>
    </receiver>
```

系统要求必须声明接收开机启动广播的权限，于是再声明使用下面的权限：

```xml
<uses-permission android:name="android.permission.RECEIVE_BOOT_COMPLETED" />
```

将应用运行在模拟器上，然后重启模拟器，在LogCat窗口输出接收到的信息，如图11-5所示。由输出信息可以看出，我们定义的MyService已经启动了，同样也可以通过"Settings→应用→正在运行"查看MyService是否运行。

```
09-19 08:19:23.591 23360-23360/cn.androidstudy.bootreveivertest I/BootCompleteReceiver: Boot Complete. Starting MyService...
09-19 08:19:23.642 23360-23360/cn.androidstudy.bootreveivertest I/MyService: onCreate called.
09-19 08:19:23.642 23360-23360/cn.androidstudy.bootreveivertest I/MyService: onStartCommand called.
```

图11-5　开机启动服务

（2）网络状态变化

在某些场合，例如用户浏览网络信息时，网络突然断开，要及时地提醒用户网络已断开。要实现这个功能，可以接收网络状态改变这样一条广播，当由连接状态变为断开状态时，系统就会发送一条广播，我们接收到之后，再通过网络的状态做出相应的操作。

☞【实例11-4】下面就来实现一下这个功能。首先创建一个工程NetStateReveiver，接着，新建一个类NetReceiver，实现如下：

```java
package cn.androidstudy.netstatereveiver;
import android.content.BroadcastReceiver;
import android.content.Context;
import android.content.Intent;
import android.net.ConnectivityManager;
import android.net.NetworkInfo;
import android.util.Log;
import android.widget.Toast;
public class NetReceiver extends BroadcastReceiver {
    @Override
    public void onReceive(Context context, Intent intent) {
        Log.i("NetStateReceiver", "network state changed.");
        ConnectivityManager mgr = (ConnectivityManager) context.getSystemService(Context.CONNECTIVITY_SERVICE);
        NetworkInfo info = mgr.getActiveNetworkInfo();
        if (info != null && info.isAvailable()) {
        Toast.makeText(context,"Network is available.",Toast.LENGTH_SHORT).show();
        }else{
    Toast.makeText(context,"Network is not available.",Toast.LENGTH_SHORT).show();
        } }}
```

然后需要修改MainActivity，代码如下：

```
package cn.androidstudy.netstatereveiver;
import android.content.Intent;
import android.content.IntentFilter;
import android.support.v7.app.AppCompatActivity;
import android.os.Bundle;
public class MainActivity extends AppCompatActivity {
    private IntentFilter intentFilter;
    private NetReceiver netReceiver;
    @Override
    protected void onCreate(Bundle savedInstanceState) {
        super.onCreate(savedInstanceState);
        setContentView(R.layout.activity_main);
        intentFilter=new IntentFilter();
        intentFilter.addAction("android.net.conn.CONNECTIVITY_CHANGE");
        netReceiver=new NetReceiver();
        registerReceiver(netReceiver,intentFilter);     }
    @Override
    protected void onDestroy() {
        super.onDestroy();
        unregisterReceiver(netReceiver);        }}
```

注意：动态注册的广播接收器一定要注销注册，可以在onDestroy 方法中调用unregisterReceiver来实现。

系统要求必须声明访问网络状态广播的权限，于是再声明使用下面的权限：

```
<uses-permission android:name="android.permission.ACCESS_NETWORK_STATE"/>
```

运行程序，然后按Home键回到主界面（不是按Back键），选择"Settings→Data usage"进入数据使用界面，接着尝试开关Cellularr data按钮来打开网络，将会看到Toast提示网络发生变化，如图11-6所示。

图11-6　打开网络提示

选择Cellularr data按钮来禁用网络,将会看到Toast提示网络发生变化,如图11-7所示。

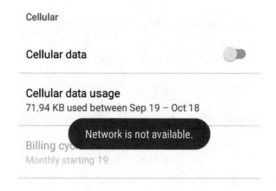

图11-7 禁用网络提示

(3)拦截短信

Android系统在接收到短信时会发送一条广播"android.provider.Telephony.SMS_RECEIVED",这条广播是以有序广播的形式发送的,所以我们可以监听这条信号,在传递给系统的接收程序时,将自定义的广播接收程序的优先级设定大于它,并且取消广播的传播,这样就可以实现拦截短信的功能了。

☞【实例11-5】当运行本例的模拟器收到一条短信时,SMSReceiver拦截到短信后会使用Toast将内容显示出来。需要在AndroidManifest.xml中注册广播接收者SMSReceiver,并在AndroidManifest.xml文件中添加权限。

新建SMSReceiver类继承了BroadcastReceiver,读取接收到的短信内容,然后需要在AndroidManifest.xml中注册广播接收者SMSReceiver,在AndroidManifest.xml中增加如下代码:

```xml
    <receiver android:name="cn.androidstudy.smsreceiverdemo.SMSReceiver">
        <intent-filter android:priority="1000">
            <action android:name="android.provider.Telephony.SMS_RECEIVED" />
            <category android:name="android.intent.category.DEFAULT" />
        </intent-filter>
    </receiver>
```

前面配置中该SMSReceiver的优先级为1000,这样它就可在系统收发短信程序之前被触发。为了使其拥有读取短信的权限,还需要在AndroidManifest.xml文件中添加以下权限:

```xml
    <uses-permission android:name="android.permission.RECEIVE_SMS" />
```

新建SMSReceiver类继承了BroadcastReceiver,其代码如下:

```java
  package cn.androidstudy.smsreceiverdemo;

  import android.content.BroadcastReceiver;
  import android.content.Context;
  import android.content.Intent;
```

```java
import android.os.Bundle;
import android.telephony.SmsMessage;
import android.widget.Toast;

public class SMSReceiver extends BroadcastReceiver {

    // 当接收到短信时被触发
    @Override
    public void onReceive(Context context, Intent intent)
    {
        // 如果是接收到短信
        if (intent.getAction().equals("android.provider.Telephony.SMS_RECEIVED"))
        {
            // 取消广播（这行代码将会让系统收不到短信）
            abortBroadcast();
            StringBuilder sb = new StringBuilder();
            // 接收由SMS传过来的数据
            Bundle bundle = intent.getExtras();
            String format = intent.getStringExtra("format");
            // 判断是否有数据
            if (bundle != null)
            {
                //    通过pdus可以获得接收到的所有短信消息
                Object[] pdus = (Object[]) bundle.get("pdus");
                // 构建短信对象阵列，并依据收到的对象长度来创建阵列的大小
                SmsMessage[] messages = new SmsMessage[pdus.length];
                for (int i = 0; i < pdus.length; i++)
                {
                    messages[i] = SmsMessage.createFromPdu((byte[]) pdus[i]);
                }
                // 将送来的短信合并为自定义信息于StringBuilder当中
                for (SmsMessage message : messages)
                {
                    sb.append("短信来源:");
                    // 获得接收短信的电话号码
                    sb.append(message.getDisplayOriginatingAddress());
                    sb.append("\n------短信内容------\n");
                    // 获得短信的内容
                    sb.append(message.getDisplayMessageBody());
                }
            }
```

```
                Toast.makeText(context, sb.toString(), Toast.LENGTH_LONG).show();
            }
        }
    }
```

将应用运行在模拟器上，使用另一个模拟器给该模拟器发送一条短信，SMSReceiver拦截到短信后会使用Toast将内容显示出来。

11.4 BroadcastReceiver的生命周期

一个广播接收者有一个回调方法void onReceive(Context context, Intent intent)。当一条广播消息到达接收者时，Android调用它的onReceive方法并传递给它包含消息的Intent对象。广播接收者被认为仅当它执行这个方法时是活跃的。当onReceive返回后，它是不活跃的。

有一个活跃的广播接收者的进程是受保护的，不会被"杀死"。但是当占用的内存有别的进程需要时，系统可以在任何时候"杀死"仅有不活跃组件的进程。

如果onReceive方法在10秒内没有执行完毕，Android会认为该程序无响应。所以在BroadcastReceiver里不能做一些比较耗时的操作，否则会弹出"Application No Response"的对话框，即Android的ANR。

这带来一个问题，当一条广播消息的响应是费时的，应该在独立的线程中做这些事，远离用户界面其他组件运行的主线程。如果onReceive衍生线程然后返回，整个进程，包括新的线程，被判定为不活跃的（除非进程中的其他应用程序组件是活跃的），将使它处于被"杀"的危机。解决这个问题的方法是onReceive启动一个服务，及时做这个工作，因此系统知道进程中有活跃的工作在做。

强化训练

广播机制具有很多优点，在Android应用程序中被程序开发人员经常使用。广播接收者（BroadcastReceiver）是实现消息异步处理的组件。学习BroadcastReceiver需要掌握创建和配置BroadcastReceiver组件的方法，还需要掌握在程序中发送Broadcast的方法。注意区分配置BroadcastReceiver组件的两种方法的区别，发送Broadcast的不同方法，以及常用系统广播的接收和处理。

学完本章知识后，不妨通过以下习题来巩固所学的内容吧。

一、填空题

1. Android中_____也就是"广播接收者"的意思，顾名思义，它就是用来接收来自_____中的广播。_____是一种广泛运用的在应用程序之间传输信息的机制。
2. BroadcastReceiver事件分类：_____，_____。
3. BroadcastReceiver事件的编程流程：_____，_____，_____。

4. 为BroadcastReceiver注册广播地址的方法：_____，_____。

5. _____对于多个接收者来说是完全异步的，通常每个接收者都无须等待即可接收到广播，接收者相互之间不会有影响。对于这种广播，接收者无法终止广播，即无法阻止其他接收者的接收动作。

6. _____比较特殊，它每次只发送到优先级较高的接收者那里，然后由优先级高的接收者再传播到优先级低的接收者那里，优先级高的接收者有能力终止这个广播。

7. 在广播消息中，有一类特殊的广播消息，它们特殊在只能由Android系统发出，这类广播称为_____。

8. 这些系统广播都被定义为_____，存放在android.content.Intent中。

9. 一个广播接收者有一个回调方法_____。当一个广播消息到达接收者时，Android调用它的_____方法并传递给它包含消息的Intent对象。广播接收者被认为仅当它执行这个方法时是活跃的。当_____返回后，它是不活跃的。

10. 所有的接收器均继承自_____基类。广播接收器没有_____。然而，它们可以启动一个Activity来响应它们收到的信息，或者用_____来通知用户。

二、选择题

1. 如果onReceive方法在（　　）秒内没有执行完毕，Android会认为该程序无响应。
   A. 10
   B. 8
   C. 7
   D. 5

2. 下列哪一个不会产生广播（　　）。
   A. 开机完成
   B. 网络状态改变
   C. 电池电量改变
   D. 关机完成

3. Android中（　　）是已断开外部电源连接时发出的广播。
   A. Intent.ACTION_HEADSET_PLUG
   B. Intent.ACTION_POWER_DISCONNECTED
   C. Intent.ACTION_SCREEN_OFF
   D. Intent.ACTION_SHUTDOWN

4. 在BroadcastReceiver里不能做一些比较耗时的操作，否则会弹出（　　）的对话框。
   A. Boot Complete Receiver
   B. Net State Reveiver
   C. Application No Response
   D. Hello BroadcastReceiver

5. "android:priority"属性的范围在（　　），数值越大，优先级越高。
   A. −1000~1000
   B. −100~100
   C. −500~500
   D. −1024~1024

6. Android中（　　）方法将一个Bundle对象设置为结果集对象，传递到下一个接收者那里，这样一来，优先级低的接收者可以获取到最新的经过处理的信息集合。

A. getString
B. BroadcastReceiver
C. setResultExtras
D. getStringExtra

三、操作题

1. 广播是一种广泛运用在应用程序之间传输信息的机制，广播消息的内容可以是 Android 系统消息，如电池电量变化、系统设置变化、网络连接变化或收到短信等。利用 Android Studio 开发环境创建一个新的 Android 应用程序，实现自定义 BroadcastReceiver 注册广播地址，并发送广播。

2. 利用 Android Studio 开发环境创建一个新的 Android 应用程序，实现有序广播的流程。

3. 系统广播被用来通知一些重要的系统事件，如电池电量低、应用软件的安装卸载、SD 卡插入拔出、外部电源的插拔等。利用 Android Studio 开发环境创建一个新的 Android 应用程序，通过订阅系统"启动完成"广播，实现开机启动某个服务的功能。

第12章

数据存储精讲

内容导读

在这个以海量信息处理为主要特征的时代,数据是应用程序的核心,也是程序开发人员和用户关注的重点。所有的应用程序都必然涉及大量数据的输入、输出,Android应用也不例外,日常工作经常需要处理大量数据。而且Android应用程序的参数设置、运行状态等重要数据只有保存到外部存储器上,Android系统在关机之后数据才不会丢失。本章重点介绍在Android应用程序中对数据进行存储和读取的方式,并以实例进行详细讲解。

学习目标

- 了解Android系统中数据存储的方式;
- 掌握SharedPreferences进行数据存储与读取的方法;
- 掌握使用PreferenceActivity进行参数设置的方法;
- 掌握使用文件进行数据存储与读取的方法;
- 掌握Android的SD卡上文件存储与读取的方法。

12.1 数据存储概述

Android中提供了多种数据存储的方式。由于存储的这些数据都是其应用程序的各自私有数据,所以如果需要在其他应用程序中使用这些数据,就要使用Android提供的数据存储方式。

Android提供了下列数据存储方式:

• SharedPreferences:它是一个轻量级的键值(key-value)存储机制,只可以存储基本类型数据,主要是针对系统配置信息的保存。

• Files:Android使用的是基于Linux的文件系统,程序开发人员可以建立和访问程序自身的私有文件,也可以访问保存在资源目录中的原始文件和XML文件,还可以在SD卡等外部存储设备中保存文件。

• SQLite:Android提供的标准数据库,支持SQL语句,可以用来存储大量的数据。

• ContentProvider:主要用于在应用程序间的数据共享和交换。

• NetWork:通过网络存储和获得数据。

在此主要介绍前面两种,SQLite和ContentProvider会在后面进行介绍。

12.2 SharedPreferences

SharedPreferences是Android平台上一个轻量级的存储类,是基于XML文件来存储键值对数据。基于XML格式的特点,SharedPreferences主要用于处理一些简单且孤立的数据、文本形式的数据或者需要长久存储的数据。SharedPreferences处理的配置信息一般存储位置在data/data/<包名>/shared_prefs目录下。

12.2.1 使用SharedPreferences

SharedPreferences主要是保存一些常用的配置,例如窗口状态,在Activity中重载窗口状态onSaveInstanceState一般使用SharedPreferences完成,它提供了Android平台常规类型long(长整型)、int(整型)、String(字符串型)的保存。

SharedPreferences是一个接口,程序无法直接创建SharedPreferences实例,只能通过Context提供的getSharedPreferences(String name,int mode)方法来获取SharedPreferences实例,该方法的第一个参数指定XML文件的名字,第二个参数支持如下几个值。

• Context.MODE_PRIVATE:指定该SharedPreferences数据只能被本应用程序读写。

• Context.MODE_WORLD_READABLE:指定该SharedPreferences数据能被其他应用程序读。

• Context.MODE_WORLD_WRITEABLE:指定该SharedPreferences数据能被其他应用程序写。

在Activity中提供了如下方法,可以创建一个SharedPreferences,默认名为当前的Activity的类名。

public SharedPreferences getPreferences(int mode)

也可以使用PreferenceManager中提供的getDefaultSharedPreferences来创建一个SharedPreferences,默认名为"<项目名>_preferences"。

public static SharedPreferences getDefaultSharedPreferences(Context context)

如果查看它们的源代码,就会发现其实还是调用了Context.getSharedPreferences(String

name,int mode)来创建 SharedPreferences。

SharedPreferences 接口主要负责读取应用程序的 Perferences 数据，它提供了如下常用的方法来访问 SharedPreferences 中的 key-value 对。

- boolean contains(String key)：判断 SharedPreferences 是否包含 key 的数据。
- Map<String,?> getAll()：获取 SharedPreferences 中全部的 key-value 对。
- xxx getXxx(String key, xxx defValue)：获取 SharedPreferences 中指定 key 的 value。如果该 key 不存在，返回默认值 defValue。其中，xxx 可以是 boolean、float、int、long、String。
- SharedPreferences.Editor edit()：返回一个 Editor 用于操作 SharedPreferences。
- SharedPreferences 对象本身只能获取数据而不支持存储和修改，存储修改通过 Editor 对象实现。Editor 提供了如下方法向 SharedPreferences 写入数据。
- SharedPreferences.Editor putXxx(String key,xxx value)：向 SharedPreferences 存入指定 key 对应的数据。其中，xxx 可以是 boolean、float、int、long、String。
- SharedPreferences.Editor clear()：清空 SharedPreferences 中所有数据。
- SharedPreferences.Editor remove(String key)：删除 SharedPreferences 中指定 key 对应的数据项。
- boolean commit()：编辑完成后，调用该方法提交修改。

实现 SharedPreferences 存储的步骤如下。

第一步，根据 Context 获取 SharedPreferences 对象。

第二步，利用 edit 方法获取 Editor 对象。

第三步，通过 Editor 对象存储键值对数据。

第四步，通过 commit 方法提交数据。

下面是通过使用 SharedPreferences 存储数据的关键语句代码：

```
SharedPreferences sp = getSharedPreferences("test",Context.MODE_PRIVATE);
//获取它的编辑对象
SharedPreferences.Editor editor =sp.edit();
//写入数据
editor.putString("name","jack");
editor.putBoolean("married",true);
editor.putInt("age", 89);
//保存并提交
editor.commit();
```

生成的 SharedPreferences 文件名为 test.xml，保存在应用程序文件夹下的 shared_prefs 文件夹内，依次打开 "View|Tool Windows|Device File Explorer"，其位置如图 12-1 所示。

若是无法打开查看 data 文件夹的内容，则切换到 SDK 目录 "sdk\platform-tools" 下按住 Shift 键，单击鼠标右键弹出菜单，选择 "在此处打开命令行窗口" 命令，进入命令行窗口后，需要操作的命令（图 12-2）如下。

- 进入 shell 模式： adb shell
- 获得 root 权限： su
- 把文件夹权限全部打开： chmod -R 777 /data

修改成功后，就可以在 "Device 中 File Explorer" 中查看到 test.xml。

可以使用 cat 命令查看文件内容，命令如下：

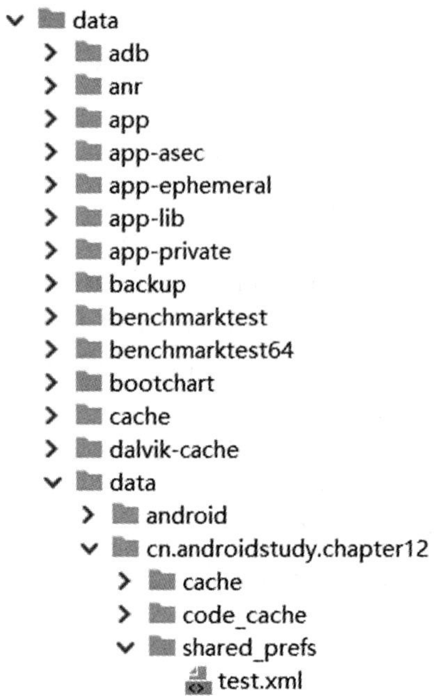

```
v 📁 data
  > 📁 adb
  > 📁 anr
  > 📁 app
  > 📁 app-asec
  > 📁 app-ephemeral
  > 📁 app-lib
  > 📁 app-private
  > 📁 backup
  > 📁 benchmarktest
  > 📁 benchmarktest64
  > 📁 bootchart
  > 📁 cache
  > 📁 dalvik-cache
  v 📁 data
    > 📁 android
    v 📁 cn.androidstudy.chapter12
      > 📁 cache
      > 📁 code_cache
      v 📁 shared_prefs
          📄 test.xml
```

图12-1　SharedPreferences存储文件的位置 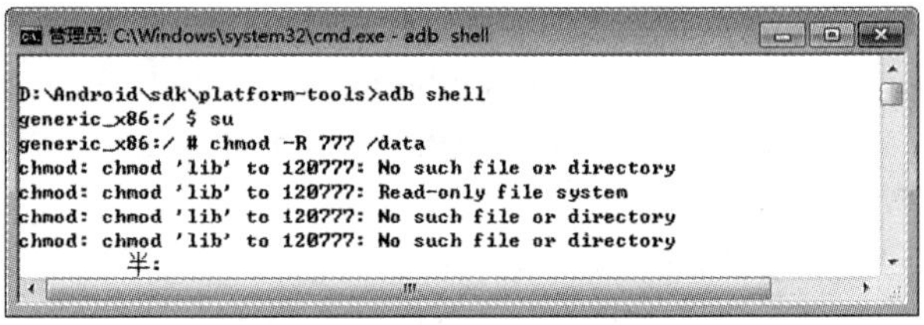　　　图12-2　修改data文件夹的权限

```
cat /data/data/cn.androidstudy.chapter12/shared_prefs/test.xml
```

其内容如下：

```
<?xml version='1.0'encoding='utf-8'standalone='yes'?>
<map>
    <string name="name">jack</string>
    <boolean name="married"value="true"/>
    <int name="age"value="89"/>
</map>
```

如果需要将test.xml文件从模拟器复制到指定文件夹，首先将文件复制到sdcard文件夹，命令如下：

```
cat /data/data/cn.androidstudy.chapter12/shared_prefs/test.xml > /sdcard/test.xml
```

然后运行两次exit直接返回到正常的终端提示符，使用"adb pull"命令将文件放到指定文件夹，如果不指定文件夹，则复制到adb.exe所在文件夹。命令如下：

```
adb pull /sdcard/test.xml D:/xml
```

若将前面程序中的"SharedPreferences sp = getContext().getSharedPreferences("test", Context.MODE_PRIVATE);"用"SharedPreferences sp = getPreferences(MODE_PRIVATE);"或"SharedPreferences sp = PreferenceManager.getDefaultSharedPreferences(this);"替换，请读者自行查看产生的文件及内容。

☞【实例12-1】下面以一个例子来说明如何使用SharedPreferences来存储数据，该例子是在Activity退出时保存界面的基本的信息，当再次运行该程序的时候，会读取上次保存的信息。界面中有一个账号的EditText、密码的EditText、记住密码的CheckBox和显示密码的CheckBox。输入账号和密码，如果选中"记住密码"复选框，下次打开程序时，则会显示账号和密码；如果没有选中"记住密码"复选框，下次打开程序时，则只是显示账号。示例运行的界面如图12-3和图12-4所示。

扫一扫 看视频

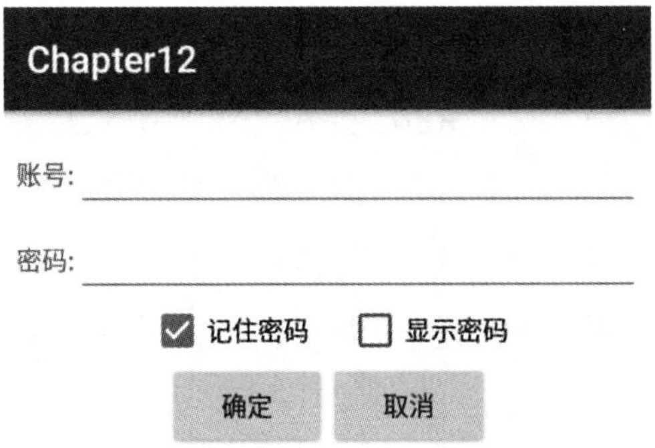

图12-3　首次运行的登录界面

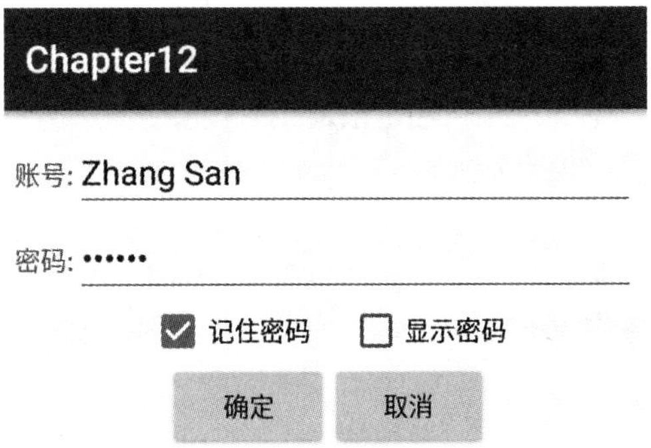

图12-4　再次运行的登录界面

MainActivity中的代码如下：

```java
package cn.androidstudy.chapter12;
import android.content.SharedPreferences;
import android.support.v7.app.AppCompatActivity;
import android.os.Bundle;
import android.text.method.HideReturnsTransformationMethod;
import android.text.method.PasswordTransformationMethod;
import android.view.View;
import android.widget.CheckBox;
import android.widget.CompoundButton;
import android.widget.EditText;
public class MainActivity extends AppCompatActivity {
    private final String PREFERENCES_NAME = "test";
    private EditText username,password;
    private CheckBox cbRemember,cbShow;
    private String userName,passWord;
    private Boolean isRemember = false;
    @Override
    protected void onCreate(Bundle savedInstanceState) {
        super.onCreate(savedInstanceState);
        setContentView(R.layout.activity_main);
        username = (EditText)findViewById(R.id.username);
        password = (EditText)findViewById(R.id.password);
        cbRemember = (CheckBox)findViewById(R.id.ischecked);
        cbRemember.setOnCheckedChangeListener(new CompoundButton.
```

```java
OnCheckedChangeListener() {
            @Override
            public void onCheckedChanged(CompoundButton buttonView, boolean isChecked) {
                isRemember = isChecked;
            }
        });
        cbShow = (CheckBox)findViewById(R.id.isshow);
        cbShow.setOnCheckedChangeListener(new CompoundButton.OnCheckedChangeListener(){
            @Override
            public void onCheckedChanged(CompoundButton compoundButton, boolean b) {
                if(b){
                    //如果选中，显示密码
                    password.setTransformationMethod(HideReturnsTransformationMethod.getInstance());
                }else{
                    //否则隐藏密码
                    password.setTransformationMethod(PasswordTransformationMethod.getInstance());
                }
            }
        });
        //读取userinfo.xml中存储的数据，设置相应的属性
        SharedPreferences preferences = getSharedPreferences(PREFERENCES_NAME, MODE_PRIVATE);
        username.setText(preferences.getString("UserName", null));
        cbRemember.setChecked(preferences.getBoolean("Remember", true));
        if(cbRemember.isChecked()){
            password.setText(preferences.getString("PassWord", null));
        }else{
            password.setText(null);
        }
    }
    //当Activity关闭时，将新的用户信息保存至userinfo.xml
    @Override
    protected void onStop() {
        super.onStop();
        SharedPreferences agPreferences = getSharedPreferences(PREFERENCES_NAME, MODE_PRIVATE);
        SharedPreferences.Editor editor = agPreferences.edit();
```

```
        userName = username.getText().toString();
        passWord = password.getText().toString();
        editor.putString("UserName", userName);
        editor.putString("PassWord", passWord);
        editor.putBoolean("Remember", isRemember);
        editor.commit();
    }
    //单击"确定"按钮执行
    public void login(View view){
        //1.获取用户名和密码
        userName = username.getText().toString();
        passWord = password.getText().toString();
        //2.进行验证

        //3.验证通过,跳转至主界面
    }
    //单击"取消"按钮执行
    public void exit(View view){
        finish();
    }
}
```

12.2.2 PreferenceActivity

在开发应用程序的过程中我们有很大的机会需要用到参数设置功能,那么在 Android 应用中,我们如何实现参数设置界面及参数存储呢?根据刚刚学过的知识很快有一个念头闪过,即 Activity + Preference 组合,前者用于界面构建,后者用于设置数据存放。其实,这是正确的,但是比较烦琐。因为,每个设置选项都要建立与其对应的 Preference。

下面我们来介绍一下 Android 中的一个特殊 Activity——PreferenceActivity。Preference Activity 是 Android 提供的对系统信息和配置进行自动保存的 Activity,它通过 SharedPreferences 方式将信息保存在 XML 文件当中。当然,也可以通过 SharedPreferences 来获取 PreferenceActivity 设置的值。使用 PreferenceActivity 不需要我们对 SharedPreferences 进行操作,系统会自动对 Activity 的各种 View 上的改变进行保存。

PreferenceActivity 与普通的 Activity 不同,它不是使用界面布局文件,而是使用选项设置布局文件。选项设置布局文件以 PreferenceScreen 作为根元素来定义一个参数设置界面布局。

从 Android3.0 以后,Android 不再推荐直接让 PreferenceActivity 加载选项设置布局文件,而是建议将 PreferenceActivity 与 PreferenceFragment 结合使用,其中 PreferenceActivity 只是负责加载选项设置列表的布局文件(图 12-5),单击选项设置列表的某一项,如"个人信息设置",将打开 PreferenceFragment(图 12-6),用来加载具体的设置选项。

【实例 12-2】下面用一个实例来介绍如何使用 PreferenceActivity 和 PreferenceFragment。
首先,在 Chapter12 中新建 Module——PreferenceActivityDemo,在 res 文件夹上单击

图12-5 首次运行的登录界面　　　　图12-6 再次运行的登录界面

鼠标右键，依次选择"New|Android resource directory"命令，打开图12-7所示窗口，设置"Resource type"为"xml"，在"Directory name"编辑框中输入"xml"，点击"OK"，完成文件夹的创建。

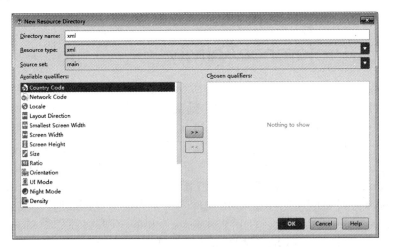

图12-7 创建XML文件夹

接着，在xml文件夹上单击鼠标右键，依次选择"New|XML resource file"命令，依次创建pref_home.xml、pref_my.xml和pref_other.xml文件。

pref_home.xml是选项设置列表的布局文件（图12-5），修改为：

```
<?xml version="1.0" encoding="utf-8"?>
<preference-headers xmlns:android="http://schemas.android.com/apk/res/android">
    <header android:fragment="cn.androidstudy.chapter12.
```

```
MainActivity$Fragment1"
        android:icon="@mipmap/ic_launcher"
        android:title="个人信息设置"
        android:summary="设置个人信息"/>
    <header android:fragment="cn.androidstudy.chapter12.
MainActivity$Fragment2"
        android:icon="@mipmap/ic_launcher_round"
        android:title="其他设置"
        android:summary="设置铃声"/>
    </preference-headers>
```

- 使用preference-headers设置选项列表，可以包含多个header；
- android:fragment：点击时将打开的详细设置；
- android:icon：图标；
- android:title：显示标题（大字体显示）；
- android:summary：副标题（小字体显示）。

pref_my.xml和pref_other.xml为"个人信息设置"和"其他设置"详细设置的布局文件，Android为我们提供两种编辑模式，可视化的结构设计及XML源代码设计。Preference XML文件中的View是有限的，只有下面几个。

① CheckBoxPreference：CheckBox选择项，对应的值的ture或flase。
- android:key：唯一标识；
- android:title：显示标题（大字体显示）；
- android:summary：副标题（小字体显示），分为summaryOn和summaryOff，可以根据CheckBox的取值显示不同的副标题；
- android:defaultValue：默认值（true或false）。

② SwitchPreference：和CheckBoxPreference基本相同，只不过是显示的形式略有不同。

③ EditTextPreference：输入编辑框，值为String类型，会弹出对话框供输入。
- android:key：唯一标识；
- android:title：显示标题（大字体显示）；
- android:summary：副标题（小字体显示）；
- android: dialogTitle：弹出对话框的标题。

④ ListPreference：列表选择，弹出对话框供选择。下拉框内显示的内容和具体的值需要在res/values/array.xml中设置两个array来表示。
- android:key：唯一标识；
- android:title：显示标题（大字体显示）；
- android:dialogTitle：弹出对话框的标题；
- android:entries：列表中显示的值，为一个数组，通过资源文件进行设置；
- androide:entryValues：列表中实际保存的值，与entries对应，为一个数组，通过资源文件进行设置。

⑤ MultiSelectListPreference：和ListPreference基不相同，可以多选。
⑥ PreferenceCategory：用于分组。
- android:title：显示的标题；
- android:key：唯一标识符，SharedPreferences也将通过此key值进行数据保存,也可以通过key值获取保存的信息。

⑦ PreferenceScreen：PreferenceActivity的根元素，设置页面，可嵌套形成二级设置页面，用Title参数设置标题。

⑧ RingtonePreference：系统铃声选择。
- android:title：设置标题；
- android:summary：设置说明；
- android:dialogTitle：设置铃声选择框的标题；
- android:ringtoneType：铃声类型，取值为all、ringtone、alarm、notification；
- android:showDefault：是否在列表中显示"默认铃声"
- android:showSilent：是否在列表中显示"静音"。

下面，可以通过可视化界面进行结构设计或者直接编辑XML源代码，增加相应的组件，修改pref_my.xml文件源代码如下：

```xml
<?xml version="1.0" encoding="utf-8"?>
<PreferenceScreen xmlns:android="http://schemas.android.com/apk/res/android">
    <PreferenceCategory android:title="个人信息设置">
        <EditTextPreference
            android:key="name"
            android:title="用户名"
            android:summary="test EditTextPreference"
            android:dialogTitle="您的用户名为："/>
        <ListPreference
            android:key="gender"
            android:title="性别"
            android:summary="test ListPreference"
            android:dialogTitle="ListPreference"
            android:entries="@array/gender_options"
            android:entryValues="@array/gender_options_values"/>
        <CheckBoxPreference
            android:key="autoSave"
            android:title="自动保存进度"
            android:summaryOn="自动保存：开启"
            android:summaryOff="自动保存：关闭"
            android:defaultValue="true"/>
        <MultiSelectListPreference
            android:dialogTitle="你喜欢"
            android:entries="@array/entries_love"
            android:entryValues="@array/entriesvalue_love"
            android:key="MultiSelect"
            android:summary="你喜欢"
            android:title="爱好" />
    </PreferenceCategory>
```

```xml
</PreferenceScreen>
```

修改 pref_other.xml 文件如下：

```xml
<?xml version="1.0" encoding="utf-8"?>
<PreferenceScreen xmlns:android="http://schemas.android.com/apk/res/android">
    <RingtonePreference
        android:ringtoneType="all"
        android:title="设置铃声"
        android:summary="test RingtonePreference"
        android:showDefault="true"
        android:key="ring_key"
        android:showSilent="true">
    </RingtonePreference>
    <SwitchPreference
        android:key="bluetooth_key"
        android:summaryOn="启用蓝牙"
        android:summaryOff="关闭蓝牙"/>
</PreferenceScreen>
```

由于 ListPreference 和 MultiSelectListPreference 需要用到数组，所以，在 res/values 中新建一个 array.xml 文件，该文件定义了四个数组，源代码如下：

```xml
<?xml version="1.0" encoding="utf-8"?>
<resources>
    <string-array name="gender_options">
        <item>男</item>
        <item>女</item>
        <item>中性</item>
    </string-array>
    <string-array name="gender_options_values">
        <item>0</item>
        <item>1</item>
        <item>2</item>
    </string-array>
    <string-array name="entries_love">
        <item>旅游</item>
        <item>唱歌</item>
        <item>爬山</item>
    </string-array>
    <string-array name="entriesvalue_love">
        <item>1</item>
        <item>2</item>
```

```
        <item>3</item>
    </string-array>
</resources>
```

最后，修改MainActivity.java文件，让MainActivity继承PreferenceActivity类，修改后的MainActivity.java源代码如下：

```
package cn.androidstudy.chapter12;
import android.preference.PreferenceActivity;
import android.preference.PreferenceFragment;
import android.os.Bundle;
import java.util.List;
//1.使MainActivity继承PreferenceActivity
public class MainActivity extends PreferenceActivity {
    @Override
    protected void onCreate(Bundle savedInstanceState) {
        super.onCreate(savedInstanceState);
    }
    //2.重写该方法，负责加载界面布局文件
    @Override
    public void onBuildHeaders(List<Header> target) {
        loadHeadersFromResource(R.xml.pref_home,target);
    }
    //3.兼容4.4之前版本，必须实现，判断是否为有效的Fragment
    @Override
    protected boolean isValidFragment(String fragmentName) {
        return true;
    }
    //4.定义Fragment1,对应pref_home.xml文件中引用的Fragment
    public static class Fragment1 extends PreferenceFragment {
        @Override
        public void onCreate(Bundle savedInstanceState) {
            super.onCreate(savedInstanceState);
            //5.加载布局
            addPreferencesFromResource(R.xml.pref_my);
        }
    }

    public static class Fragment2 extends PreferenceFragment {
        @Override
        public void onCreate(Bundle savedInstanceState) {
            super.onCreate(savedInstanceState);
            addPreferencesFromResource(R.xml.pref_other);
```

```
            }
        ]
    }
```

示例运行的主界面如图12-5所示，进行设置后，会自动保存。在图12-1中就会新增一个文件"cn.androidstudy.chapter12_preferences.xml"，可见文件名的命名规则为：包名_preferences.xml。使用命令查看文件内容如下：

```
<?xml version='1.0'encoding='utf-8'standalone='yes'?>
<map>
    <string name="ring_key">content://media/internal/audio/media/31</string>
    <string name="name">Zhang San</string>
    <set name="MultiSelect">
        <string>1</string>
        <string>3</string>
    </set>
    <string name="gender">0</string>
    <boolean name="bluetooth_key"value="true"/>
    <boolean name="autoSave"value="true"/>
</map>
```

那么在应用程序中怎么读取这些数据呢？只需使用下面语句获取SharedPreferences即可。

```
SharedPreferences sp = PreferenceManager.getDefaultSharedPreferences(this);
```

12.3 文件

数据信息虽然在其他地方进行处理，而文件一般是数据信息在智能手机上最终存放的地方。SharedPreferences用来保存一些配置信息虽然很方便，但只能存储boolean、float、integer、long、String等基本类型的数据，并且保存的数据只能局限在Android应用内部访问。为了在更大的范围内交换复杂内容格式的消息，还是要通过Android文件系统。

Android系统下的文件可以分为两类：一类是共享的文件，如存储在SD卡上的文件，这种文件任何Android应用都可以访问；另外一种是私有文件，即Android应用自己创建的文件。Android的文件读写与JavaSE的文件读写相同，都是使用IO流。但是对于私有文件，只有具有操作权限的用户才能进行操作，故Android提供了一组特有的API来访问私有文件。

FileInputStream openFileInput(String name)
FileOutputStream openFileOutput(String name, int mode)
参数说明：
name：文件名，不能包含路径分隔符；

mode：操作模式，包括：
- Context.MODE_PRIVATE：新内容覆盖原内容；
- Context.MODE_APPEND：新内容追加到原内容后；
- Context.MODE_WORLD_READABLE：允许其他应用程序读取；
- Context.MODE_WORLD_WRITEABLE：允许其他应用程序写入，会覆盖原数据。

可以使用"+"连接这些权限。

Context对象还可以通过调用fileList方法来获得私有文件目录下所有的文件名组成的字符串数组，调用deleteFile(String name)来删除文件名为name的文件。

Activity还提供了getCacheDir和getFilesDir方法：

getCacheDir方法用于获取data/data/<package name>/cache目录（一些临时文件可以放在缓存目录，用完了就删了）；

getFilesDir方法用于获取/data/data/<package name>/files目录。

其他程序获取文件路径的方法如下：
- 绝对路径： data/data/packagename/files/filename；
- context： context.getFilesDir+"/filename";
- 缓存目录： data/data/packagename/Cache或getCacheDir。

12.3.1 应用程序文件读写

☞【实例12-3】下面通过一个示例来演示一下Android下的文件操作。示例运行效果如图12-8所示，然后输入文件名和内容（图12-9），点击"保存"按钮，将会在目录data/data/cn. androidstudy.chapter12/files下创建文件hello。再次运行，输入文件名，点击"读取"，将会显示文件内容。

首先，在Chapter12中新建Module——FileDemo，修改activity_main.xml布局文件。

扫一扫 看视频

图12-8 示例的运行界面

图12-9 文件的读取与保存

其中，MainActivity中的代码如下：

```java
package cn.androidstudy.chapter12;
import android.support.v7.app.AppCompatActivity;
import android.os.Bundle;
import android.view.View;
import android.widget.EditText;
import java.io.FileInputStream;
import java.io.FileNotFoundException;
import java.io.FileOutputStream;
import java.io.IOException;
public class MainActivity extends AppCompatActivity {
    private EditText etTitle,etContent;
    @Override
    protected void onCreate(Bundle savedInstanceState) {
        super.onCreate(savedInstanceState);
        setContentView(R.layout.activity_main);
        //1.获得两个EditText
        etTitle = (EditText) findViewById(R.id.title);
        etContent = (EditText) findViewById(R.id.content);
    }
    //2.保存按钮事件处理
    public void saveFile(View view){
        //3.获得两个EditText中内容
        String title = etTitle.getText().toString();
        String content = etContent.getText().toString();
        FileOutputStream fos = null;
        try {//4.打开并保存
            fos = openFileOutput(title, MODE_PRIVATE);
            fos.write(content.getBytes());
        } catch (FileNotFoundException e) {
            e.printStackTrace();
        } catch (IOException e) {
            e.printStackTrace();
        }finally{
            if(fos!= null){
                try {
                    fos.close();
                } catch (IOException e) {
                    e.printStackTrace();
                }
            }
        }
```

```
    }
    //5.读取按钮事件处理
    public void readFile(View view){
        FileInputStream fis = null;
        //6.读取文件名
        String title = etTitle.getText().toString();
        try {
            fis = openFileInput(title);//打开文件
            byte[] aa = new byte[1024];//存储读取内容
            int len = fis.read(aa);//读取内容,如果大于1KB,请使用循环进行读取
            String string = new String(aa,0,len);//转换为字符串
            etContent.setText(string);//显示
        } catch (FileNotFoundException e) {
            e.printStackTrace();
        } catch (IOException e) {
            e.printStackTrace();
        }finally{
            if(fis != null){
                try {
                    fis.close();
                } catch (IOException e) {
                    e.printStackTrace();
                }
            }
        }
    }
```

12.3.2 操作资源文件

在Android资源文件中,有两个特殊的文件夹——asserts和res/raw,用于存放app所需的特殊文件,且该类文件打包时不会被编码到二进制文件。

assets目录不会被映射到R中,因此,资源无法通过R.id方式获取,必须要通过AssetManager进行操作与获取。

res/raw目录下的资源会被映射到R中,可以通过getResource方法获取资源。

Android的资源文件在编译之前存放在res的特定目录下,系统提供了专用的API。需要注意的是,资源文件只能读取,不能修改。

(1)读取res/raw中的文件

```
public void readRaw(View view){
    String res = "";
```

```java
        try {
            // 在res/raw/hello.txt,
            InputStream in = getResources().openRawResource(R.raw.hello);
            int length = in.available();
            byte[] buffer = new byte[length];
            in.read(buffer);
            //选择合适的编码，如果不调整会乱码
            res = EncodingUtils.getString(buffer, "UTF-8");
            in.close();
        } catch (Exception e) {
            e.printStackTrace();
        }
        // 把得到的内容显示在TextView上
        myTextView.setText(res);
    }
```

（2）读取assets中的文件

```java
    public void readAsset(View view) {
        String fileName = "test.txt";
        String res = "";
        try {
            // assets/test.txt有这样的文件存在
            InputStream in = getResources().getAssets().open(fileName);
            int length = in.available();
            byte[] buffer = new byte[length];
            in.read(buffer);
            res = EncodingUtils.getString(buffer, "UTF-8");
        } catch (Exception e) {
            e.printStackTrace();
        }
        // 把得到的内容显示在TextView上
        myTextView.setText(res);
    }
```

12.3.3 操作SD卡上的文件

使用Activity的openFileOutput方法保存文件，文件存放在手机空间中，一般手机的存储空间不是很大，存放些小文件还行，如果要存放像视频这样的大文件是不可行的。对于像视频这样的大文件，我们可以把它存放在SDCard（SD卡）。SDCard是干什么的？你可以把它看作是移动硬盘或U盘。

在Android Studio中，创建模拟器时，默认的SDCard的容量是100MB（图12-10），也可以使用外部文件，使用DOS命令进行创建，步骤如下：

① 在DOS窗口中进入Android SDK安装路径的tools目录，输入以下命令创建一张容量为2GB的SDCard，文件后缀可以随便取，建议使用.img：

```
mksdcard 2048M D:\AndroidTool\sdcard.img
```

② 然后执行下面的命令将SD卡加载到模拟器中：

```
Emulator -sdcard D:\AndroidTool\sdcard.img -avd myavd
```

启动模拟器后，在Android主界面的选项菜单中选择"设置"，然后选择"存储"，在打开的主界面中查看"USB存储器"的容量即可查看SD卡的加载情况（图12-11）。

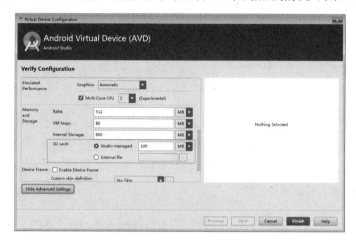

图12-10　模拟器SD卡设置

图12-11　模拟器SD卡查看

在程序中访问SDCard，需要具有访问SDCard的权限。在AndroidManifest.xml中加入访问SDCard的权限如下：

```
<!-- 在SDCard中创建与删除文件权限 -->
<uses-permission android:name="android.permission.MOUNT_UNMOUNT_FILESYSTEMS"/>
<!-- 往SDCard写入数据权限 -->
<uses-permission android:name="android.permission.WRITE_EXTERNAL_STORAGE"/>
```

使用SDCard目录前，需要判断是否有"sdcard：Environment.getExternalStorageState()"方法用于获取SDCard的状态，如果手机装有SDCard，并且可以进行读写，那么方法返回的状态等于Environment.MEDIA_MOUNTED。

Environment类中还提供了以下方法用来获取系统文件夹：

• Environment.getExternalStorageDirectory方法：用于获取SDCard的目录，当然要获取SDCard的目录，也可以这样写：File sdCardDir = new File("/mnt/sdcard")；

• Environment.getDataDirectory方法：用于获取data目录；

• Environment.getDownloadCacheDirectory方法：用于获取cache目录；

• Environment.getRootDirectory方法：用于获取system目录；

• Environment.getExternalStoragePublicDirectory(Environment.DIRECTORY_MUSIC)方法：用于获取mnt/sdcard/Music目录；

• Environment.getExternalStoragePublicDirectory(Environment.DIRECTORY_ALARMS)方法：用于获取mnt/sdcard/Alarms目录；

• Environment.getExternalStoragePublicDirectory(Environment.DIRECTORY_DCIM)方法：用于获取mnt/sdcard/DCIM目录；

• Environment.getExternalStoragePublicDirectory(Environment.DIRECTORY_DOWNLOADS)方法：用于获取mnt/sdcard/Download目录；

• Environment.getExternalStoragePublicDirectory(Environment.DIRECTORY_MOVIES)方法：用于获取mnt/sdcard/Movies目录；

• Environment.getExternalStoragePublicDirectory(Environment.DIRECTORY_NOTIFICATIONS)方法：用于获取mnt/ sdcard /Notifications目录；

• Environment.getExternalStoragePublicDirectory(Environment.DIRECTORY_PICTURES)方法：用于获取mnt/ sdcard /Pictures目录；

• Environment.getExternalStoragePublicDirectory(Environment.DIRECTORY_PODCASTS)方法：用于获取mnt/ sdcard /Podcasts目录；

• Environment.getExternalStoragePublicDirectory(Environment.DIRECTORY_RINGTONES)方法：用于获取mnt/ sdcard /Ringtones目录；

【实例12-4】下面通过一个示例来演示在Android中如何操作SD卡上的文件。示例运行效果如图12-8，然后输入文件名和内容（图12-9），点击"保存"按钮，将会在目录data/data/cn.androidstudy.chapter12/files下创建文件hello。再次运行，输入文件名，点击"读取"，将会显示文件内容。

首先，在Chapter12中新建Module——SDcardDemo，修改activity_main.xml布局文件。

MainActivity的代码如下：

```java
package cn.androidstudy.chapter12;
import android.os.Environment;
import android.support.v7.app.AppCompatActivity;
import android.os.Bundle;
import android.view.View;
import android.widget.EditText;
import android.widget.Toast;
import java.io.BufferedReader;
import java.io.BufferedWriter;
import java.io.File;
import java.io.FileInputStream;
import java.io.FileOutputStream;
import java.io.IOException;
import java.io.InputStream;
import java.io.InputStreamReader;
import java.io.OutputStream;
import java.io.OutputStreamWriter;
public class MainActivity extends AppCompatActivity {
    private EditText etTitle, etContent;
    @Override
    protected void onCreate(Bundle savedInstanceState) {
        super.onCreate(savedInstanceState);
        setContentView(R.layout.activity_main);

        etTitle = (EditText) findViewById(R.id.title);
        etContent = (EditText) findViewById(R.id.content);
    }
    //保存文件
    public void saveFile(View view) {
        String title = etTitle.getText().toString();
        String content = etContent.getText().toString();
        OutputStream out = null;
        BufferedWriter bw = null;
        // 判断是否插入SD卡
        if (Environment.getExternalStorageState().equals(
                Environment.MEDIA_MOUNTED)) {
            String folderName = Environment.getExternalStorageDirectory()
                    .getPath() + "/myTest";
            File folder = new File(folderName);
            if (folder == null || !folder.exists()) {
                // 如果文件夹不存在，则创建
                folder.mkdir();
```

```java
        }
        File saveFile = new File(folderName, title);
        try {
            if (!saveFile.exists()) {
                saveFile.createNewFile();
            }
            out = new FileOutputStream(saveFile);
            bw = new BufferedWriter(new OutputStreamWriter(out, "UTF-8"));
            bw.write(content);
        } catch (IOException e) {
            e.printStackTrace();
        } finally {
            if (bw != null) {
                try {
                    bw.close();
                } catch (IOException e) {
                    e.printStackTrace();
                }
            }
            if(out != null){
                try {
                    out.close();
                } catch (IOException e) {
                    e.printStackTrace();
                }
            }
        }
    }
}
//根据文件名读取文件
public void readFile(View view) {
    String title = etTitle.getText().toString();
    StringBuilder sb = new StringBuilder();
    InputStream in = null;
    BufferedReader br = null;
    // 判断是否插入SD卡
    if (Environment.getExternalStorageState().equals(
            Environment.MEDIA_MOUNTED)) {
        String folderName = Environment.getExternalStorageDirectory()
                .getPath() + "/myTest";
        File folder = new File(folderName);
        if (folder == null || !folder.exists()) {
```

```java
            // 如果文件夹不存在, 则退出
            Toast.makeText(this, folderName + "文件夹不存在", Toast.LENGTH_SHORT)
                    .show();
            return;
        }
        File saveFile = new File(folderName, title);

        try {
            if (!saveFile.exists()) {
                // 如果文件不存在, 则退出
                Toast.makeText(this, folderName + "/" + title + "文件夹不存在",
                        Toast.LENGTH_SHORT).show();
                return;
            }
            in = new FileInputStream(saveFile);
            br = new BufferedReader(new InputStreamReader(in, "UTF-8"));
            String tmp;
            while ((tmp = br.readLine()) != null) {
                sb.append(tmp);
                sb.append("\r\n");
            }
            etContent.setText(sb.toString());
        } catch (IOException e) {
            e.printStackTrace();
        } finally {
            if (br != null) {
                try {
                    br.close();
                } catch (IOException e) {
                    e.printStackTrace();
                }
            }
            if(in != null){
                try {
                    in.close();
                } catch (IOException e) {
                    e.printStackTrace();
                }
            }
        }
```

            }
        }
    }

最后，一定要在AndroidManifest.xml文件中添加SD卡的访问权限语句，如图12-12所示。

```xml
<?xml version="1.0" encoding="utf-8"?>
<manifest xmlns:android="http://schemas.android.com/apk/res/android"
    package="cn.androidstudy.chapter12">

    <!-- 在SDCard中创建与删除文件权限 -->
    <uses-permission android:name="android.permission.MOUNT_UNMOUNT_FILESYSTEMS"/>
    <!-- 往SDCard写入数据权限 -->
    <uses-permission android:name="android.permission.WRITE_EXTERNAL_STORAGE"/>

    <application android:allowBackup="true" android:icon="@mipmap/ic_launcher"
```

图12-12 添加SD卡的访问权限

在模拟器上运行SDcardDemo，并输入文件名和内容，点击"保存"按钮，即可在SD卡上创建一个名为myTest的文件夹，并将文件添加到该文件夹中。如果在Android6.0之前的手机或者模拟器上面运行，文件可以正常地保存和读取，但是如果在Android6.0及之后的版本上面运行，则无法进行文件的保存和读取，为什么呢？

Android6.0 (API 23)之前应用的权限在安装时全部授予，运行时应用不再需要询问用户。在Android6.0或更高版本对权限进行了分类，对某些涉及用户隐私的权限可在运行时根据用户的需要动态授予，这样就不需要在安装时被强迫同意某些权限。

权限分两种：正常权限和危险权限。危险权限在targetSdkVersion ≥ 23就要动态申请权限了。危险权限见表12-1。

表12-1 危险权限

权限组	权限
CALENDAR	READ_CALENDAR WRITE_CALENDAR
CAMERA	CAMERA
CONTACTS	READ_CONTACTS WRITE_CONTACTS GET_ACCOUNTS
LOCATION	ACCESS_FINE_LOCATION ACCESS_COARSE_LOCATION
MICROPHONE	RECORD_AUDIO
SENSORS	BODY_SENSORS
PHONE	READ_PHONE_STATE CALL_PHONE READ_CALL_LOG WRITE_CALL_LOG ADD_VOICEMAIL USE_SIP PROCESS_OUTGOING_CALLS

续表

权限组	权限
SMS	SEND_SMS RECEIVE_SMS READ_SMS RECEIVE_WAP_PUSH RECEIVE_MMS
STORAGE	READ_EXTERNAL_STORAGE WRITE_EXTERNAL_STORAGE

每组权限中，用户只要授权该组下的一个权限，该组中所有权限都可以用。

由表12-1中可以看到，WRITE_EXTERNAL_STORAGE属于危险权限，除了在AndroidManifest.xml文件中添加权限外，还需要在运行时进行动态授权，步骤如下。

① 检查权限。

```
// 检查权限
ContextCompat.checkSelfPermission(Context context, String permission)
// 在Activity中可以使用如下方法检查权限
checkSelfPermission(String permission)
```

返回值（android.content.pm.PackageManager中的常量）：
- 有权限：PackageManager.PERMISSION_GRANTED。
- 无权限：PackageManager.PERMISSION_DENIED。

当应用需要用到危险权限时，在执行权限相关代码前，使用该方法判断是否拥有指定的权限。有权限，则继续执行设计需要权限的代码；无权限，则向用户请求授予权限。

② 请求权限。

```
// 请求权限
ActivityCompat.requestPermissions(Activity activity, String[] permissions, int requestCode)
// 在Activity中可以使用如下方法请求权限
requestPermissions( String[] permissions, int requestCode)
```

当检测到应用没有指定的权限时，调用此方法向用户请求权限。调用此方法将弹出权限请求对话框询问用户"允许"或"拒绝"指定的权限。

③ 处理结果。请求权限的结果返回和接收一个Activity的返回类似，重写FragmentActivity或版本4 Fragment 中的 onRequestPermissionsResult 方法。

按照步骤在MainActivity.java中添加如下方法，用来判断是否已经授权，如未授权，则弹出对话框提示是否授权。

```
//判断是否授权，如未授权，则申请授权
private void grantedAndRequest() {
    if (Build.VERSION.SDK_INT >= Build.VERSION_CODES.M) {//针对Android 6.0及之后版本
```

```java
            int permission=checkSelfPermission(Manifest.permission.WRITE_EXTERNAL_STORAGE);
            //和下面语句等效
            //int permission=checkSelfPermission("android.permission.WRITE_EXTERNAL_STORAGE");
            //如果已授权
            if (permission == PackageManager.PERMISSION_GRANTED) {
                isGranted = true;
            }else{
                //未授权，弹对话框，申请授权
                requestPermissions(new String[]{Manifest.permission.WRITE_EXTERNAL_STORAGE}, 1);
            }
        }else{
            isGranted = true;
        }
    }
```

重写onRequestPermissionsResult方法，获取授权结果代码如下：

```java
/**
 * 处理权限请求结果
 * @param requestCode
 *         请求权限时传入的请求码，用于区别是哪一次请求的
 * @param permissions
 *         所请求的所有权限的数组
 * @param grantResults
 *         权限授予结果，和 permissions 数组参数中的权限一一对应，元素值为两种情况，如下：
 *         授予：PackageManager.PERMISSION_GRANTED
 *         拒绝：PackageManager.PERMISSION_DENIED
 */
@Override
public void onRequestPermissionsResult(int requestCode, @NonNull String[] permissions, @NonNull int[] grantResults) {
    super.onRequestPermissionsResult(requestCode, permissions, grantResults);
    if(requestCode==1){
        if(grantResults[0]==PackageManager.PERMISSION_GRANTED){
            isGranted = true;
        }
    }
}
```

可以在onCreate方法的最后添加如下语句：

```
    grantedAndRequest();
```

这样，在Activity创建后即提醒用户是否授予读写SD卡的权限，如图12-13所示。

注意：由于模拟器不支持中文，对话框内的文字为英文，在真机上面运行就是中文。

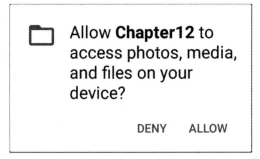

图12-13　是否授予SD卡的访问权限

同理，在saveFile和readFile两个方法的最前面添加如下语句：

```
    if(!isGranted){
        Toast.makeText(this, "未获得权限，请授权！ ", Toast.LENGTH_SHORT).show();
        grantedAndRequest();
        return;
    }
```

在单击保存或者读取时，首先判断是否获得授权，若未授权，重新弹出图12-13让用户授权。

强化训练

数据是应用程序的核心，也是Android程序开发人员和Android用户关注的重点。本章主要介绍了Android的SharedPreferences和文件这两种数据存储方式，要求掌握使用SharedPreferences工具类对简单的配置信息的读写方法，以及PreferenceActivity与PreferenceFragment的使用方法。由于文件操作是从Java中移植过来的，需要重点掌握Android系统为文件操作提供的新方法，以及对SD卡进行操作的方法，需要注意Android 6.0对权限管理的变化，掌握并能熟练地使用它们。

学完本章知识后，不妨通过以下习题来巩固所学的内容吧。

一、填空题

1. Android中应用程序的参数设置、运行状态数据只有保存到_____上，系统在关机之后数据才_____。
2. Android提供的五种数据存储方式为：_____，_____，_____，_____，_____。
3. _____是Android平台上一个轻量级的存储类，是基于XML文件来存储

_____数据，通常用来存储一些简单的配置信息。

4. _____主要是保存一些常用的配置，例如窗口状态，在Activity中重载窗口状态onSaveInstanceState一般使用_____完成，它提供了Android平台常规的_____、_____、_____的保存。

5. Activity + Preference组合，前者用于_____，后者用于_____。

6. PreferenceActivity是Android提供的对系统信息和配置进行自动保存的Activity，它通过_____方式将信息保存在_____当中，当然，也可以通过_____来获取PreferenceActivity设置的值。

7. SharedPreferences用来保存一些配置信息虽然很方便，但只能存储_____、_____、_____、_____、_____等基本类型的数据，并且保存的数据只能局限在Android应用内部访问。为了在更大的范围内交换复杂内容格式的消息，还是要通过_____。

8. Android系统下的文件，可以分为两类：一类是_____；另一类是_____。

9. 在Android资源文件中，有两个特殊的文件夹：_____和_____，用于存放app所需的特殊文件，且该文件打包时不会被编码到二进制文件中。

二、选择题

1. Android中以下（　　）的作用是获取SharedPreferences中全部的key-value对。

A. SharedPreferences.Editor putXxx(String key,xxx value)

B. boolean contains(String key)

C. xxx getXxx(String key, xxx defValue)

D. Map<String,?> getAll()

2. Android中以下（　　）不是实现SharedPreferences存储的步骤。

A. 通过Editor方法提交数据

B. 利用edit方法获取Editor对象

C. 通过Editor对象存储key-value对数据

D. 根据Context获取SharedPreferences对象

3. Android中Preference XML文件中的View是有限的，下面的（　　）不是Preference XML文件中的View。

A. CheckBoxPreference

B. EditTextPreference

C. MainActivity

D. PreferenceScreen

4. Android提供了一组特有的API来访问私有文件，（　　）不属于操作模式mode。

A. Context.MODE_WORLD_WRITEABLE

B. Context.MODE_APPEND

C. Context.MODE_WORLD_READABLE

D. FileInputStream openFileInput

5. Android中WRITE_EXTERNAL_STORAGE属于危险权限，除了在AndroidManifest.xml文件中添加权限外，还需要在运行时进行动态授权，以下（　　）不是步骤。

A. 请求权限

B. 执行权限相关代码

C. 处理结果

D. 检查权限

6. Environment.getExternalStoragePublicDirectory(Environment.DIRECTORY_MUSIC)方法获得的目录是（　　）。
   A. mnt/sdcard/Notifications
   B. mnt/sdcard/DCIM
   C. mnt/sdcard/Music
   D. mnt/sdcard/Download

三、操作题

1. 在项目开发时，可能会经常使用到SharedPreferences。如果每次都像【实例12-1】那样访问，就显得十分烦琐，常用到的方法就是把对SharedPreferences的访问封装为一个类，这样可以提供代码的复用性。请自己完成对SharedPreferences访问的封装，并修改【实例12-1】，调用封装好的类完成登录功能。参考代码如下：

```java
import android.content.Context;
import android.content.SharedPreferences;
import java.util.Map;

public class SPUtils {
    /**
     * 保存在手机里的SP文件名
     */
    public static final String FILE_NAME = "my_sp";

    /**
     * 保存数据
     */
    public static void put(Context context, String key, Object obj) {
        SharedPreferences sp = context.getSharedPreferences(FILE_NAME, context.MODE_PRIVATE);
        SharedPreferences.Editor editor = sp.edit();
        if (obj instanceof Boolean) {
            editor.putBoolean(key, (Boolean) obj);
        } else if (obj instanceof Float) {
            editor.putFloat(key, (Float) obj);
        } else if (obj instanceof Integer) {
            editor.putInt(key, (Integer) obj);
        } else if (obj instanceof Long) {
            editor.putLong(key, (Long) obj);
        } else {
            editor.putString(key, (String) obj);
        }
        editor.commit();
    }
```

```java
/**
 * 获取指定数据
 */
public static Object get(Context context, String key, Object defaultObj) {
    SharedPreferences sp = context.getSharedPreferences(FILE_NAME, context.MODE_PRIVATE);
    if (defaultObj instanceof Boolean) {
        return sp.getBoolean(key, (Boolean) defaultObj);
    } else if (defaultObj instanceof Float) {
        return sp.getFloat(key, (Float) defaultObj);
    } else if (defaultObj instanceof Integer) {
        return sp.getInt(key, (Integer) defaultObj);
    } else if (defaultObj instanceof Long) {
        return sp.getLong(key, (Long) defaultObj);
    } else if (defaultObj instanceof String) {
        return sp.getString(key, (String) defaultObj);
    }
    return null;
}

/**
 * 删除指定数据
 */
public static void remove(Context context, String key) {
    SharedPreferences sp = context.getSharedPreferences(FILE_NAME, context.MODE_PRIVATE);
    SharedPreferences.Editor editor = sp.edit();
    editor.remove(key);
    editor.commit();
}

/**
 * 返回所有键值对
 */
public static Map<String, ?> getAll(Context context) {
    SharedPreferences sp = context.getSharedPreferences(FILE_NAME, context.MODE_PRIVATE);
    Map<String, ?> map = sp.getAll();
    return map;
}
```

```java
/**
 * 删除所有数据
 */
public static void clear(Context context) {
    SharedPreferences sp = context.getSharedPreferences(FILE_NAME, context.MODE_PRIVATE);
    SharedPreferences.Editor editor = sp.edit();
    editor.clear();
    editor.commit();
}

/**
 * 检查key对应的数据是否存在
 */
public static boolean contains(Context context, String key) {
    SharedPreferences sp = context.getSharedPreferences(FILE_NAME, context.MODE_PRIVATE);
    return sp.contains(key);
}
}
```

2. 练习文件操作。

对【实例12-3】的文件操作进行封装。参考代码如下：

```java
public class FileHelper {
  private Context mContext;
  public FileHelper() {
  }

  public FileHelper(Context mContext) {
      super();
      this.mContext = mContext;
  }

  /*
   * 这里定义的是一个文件保存的方法，写入到文件中，所以是输出流
   */
   public void save(String filename, String filecontent) throws Exception {
        //这里我们使用私有模式，创建出来的文件只能被本应用访问，还会覆盖原文件
        FileOutputStream output = mContext.openFileOutput(filename, Context.MODE_PRIVATE);
        output.write(filecontent.getBytes());   //将String字符串以字节
```

流的形式写入到输出流中
```
            output.close();              //关闭输出流
    }

    /*
     * 这里定义的是文件读取的方法
     * */
    public String read(String filename) throws IOException {
        //打开文件输入流
        FileInputStream input = mContext.openFileInput(filename);
        byte[] temp = new byte[1024];
        StringBuilder sb = new StringBuilder("");
        int len = 0;
        //读取文件内容：
        while ((len = input.read(temp)) > 0) {
            sb.append(new String(temp, 0, len));
        }
        //关闭输入流
        input.close();
        return sb.toString();
    }
}
```

3. 利用 Android Studio 开发环境创建一个新的 Android 应用程序，实现在 Android 中操作 SD 上的文件。

第13章

SQLite 数据库精讲

内容导读

前面介绍了如何使用 SharedPreferences 和文件存储来存储数据。但是当需要处理的数据量比较大时,这两种方式显然不适合,这种情况下,Android 系统可以使用 SQLite 数据库来实现结构化数据存储。SQLite 数据库占用资源少、运行效率高、可移植性强,非常适合在资源有限的智能手机或平板电脑上进行数据存储。目前在 Android 系统中,很多程序开发人员采用 SQLite 数据库存储应用程序中的大量数据,并对数据进行管理和维护。本章将针对 SQLite 数据库的相关知识进行讲解分析,并以实例展示其使用方法。

学习目标

- 了解 SQLite 数据库及其特点;
- 熟悉使用 SQL 语句操作 SQLite 数据库的方法;
- 掌握使用 SQLiteDatabase 创建及操作数据库的方法;
- 掌握使用 SQLiteOpenHelper 创建数据库的方法;
- 掌握使用 ListView 显示数据库数据的方法。

13.1 SQLite概述

SQLite 数据库是一个完全开源的轻型嵌入式关系数据库，它在2000年由 D. Richard Hipp发布，最早在iOS智能手机上使用，目前在很多嵌入式产品中应用。SQLite 数据库占用资源非常少，在嵌入式设备中，可能只需要几百"KB"的内存，其应用可以减少应用程序管理数据的开销。SQLite 数据库遵守ACID［即数据库事务正确执行的四要素：原子性（Atomicity）、一致性（Consistency）、隔离性（Isolation）和持久性（Durability）］。SQLite数据库拥有支持Windows、Linux、Unix、Android等多种主流操作系统，可移植性好，容易使用，程度规模非常小，高效而且可靠等众多优点。

目前在Android 系统中集成的是 SQLite3 版本，SQLite不支持静态数据类型，而是使用列关系。这意味着它的数据类型不具有表列属性，而具有数据本身的属性。当某个值插入数据库时，SQLite将检查它的类型。如果该类型与关联的列不匹配，则SQLite 会尝试将该值转换成列类型。如果不能转换，则该值将作为其本身具有的类型存储。SQLite支持 null、integer、real、text 和 blob 数据类型。

例如，可以在 Integer 字段中存放字符串，或者在布尔型字段中存放浮点数，或者在字符型字段中存放日期型值。但是有一种例外，如果用户的主键是 INTEGER，那么只能存储64位整数，当向这种字段中保存除整数以外的数据时，将会产生错误。另外，SQLite在解析CREATE TABLE 语句时，会忽略 CREATE TABLE 语句中跟在字段名后面的数据类型信息。如：

```
CREATE TABLE person (_id integer primary key autoincrement, name varchar(20))
```

SQLite数据库在解析这条语句的时候，会忽略掉跟在 name 字段后面的 varchar（20），也就是说可以往 name 字段填 SQLite 支持的那五种数据类型的任意一种，而且 name字段可以存任意长度的字符，长度 20不起作用。也就是说可以把 SQLite 数据库近似看作是一种无数据类型的数据库，可以把任何类型的资料存放在非 Integer 类型的主键之外的其他字段上去，另外，字段的长度也是没有限度的。

SQLite 的特点如下。

（1）零配置

SQlite3 不用安装、不用配置、不用启动/关闭或者配置数据库实例。当系统崩溃后不用做任何恢复操作，在下次使用数据库的时候自动恢复。

（2）可移植

它是运行在 Windows、Linux 、BSD、Mac OS X 和一些商用 Unix 系统上，例如 Sun 的 Solaris、IBM 的 AIX 。同样，它也可以工作在许多嵌入式操作系统上，比如 Android、QNX、VxWorks、Palm OS、Symbin 和 Windows CE。

（3）紧凑

SQLite 是被设计成轻量级、自包含的。一个头文件、一个 lib 库，用户就可以使用关系数据库了，不用启动任何系统进程。

（4）简单

SQLite 有着简单易用的 API 接口。

（5）可靠

SQLite 的源代码达到 100%分支测试覆盖率。

在 Android 的 sdk 的安装目录下，在 platform-tools 文件夹里面包含了 sqlite3.exe，下面简单了解一下，在 Windows 下如何使用 SQLite。打开 platform-tools 文件夹，按住 Shift 键，单击鼠标右键弹出菜单，选择"在此处打开命令行窗口"命令，打开命令提示符窗口。

```
//创建数据库
D:\Android\sdk\platform-tools>sqlite3 mydb.db
SQLite version 3.16.2 2017-01-06 16:32:41
Enter ".help" for usage hints.
//查看常用命令
sqlite> .help
.auth ON|OFF           Show authorizer callbacks
.backup ?DB? FILE      Backup DB (default "main") to FILE
.bail on|off           Stop after hitting an error.  Default OFF
.binary on|off         Turn binary output on or off.  Default OFF
.changes on|off        Show number of rows changed by SQL
.check GLOB            Fail if output since .testcase does not match
.clone NEWDB           Clone data into NEWDB from the existing database
.databases             List names and files of attached databases
.dbinfo ?DB?           Show status information about the database
.dump ?TABLE? ...      Dump the database in an SQL text format
                         If TABLE specified, only dump tables matching
                         LIKE pattern TABLE.
.echo on|off           Turn command echo on or off
.eqp on|off|full       Enable or disable automatic EXPLAIN QUERY PLAN
.exit                  Exit this program
.explain ?on|off|auto? Turn EXPLAIN output mode on or off or to automatic
.fullschema ?--indent? Show schema and the content of sqlite_stat tables
.headers on|off        Turn display of headers on or off
.help                  Show this message
.import FILE TABLE     Import data from FILE into TABLE
.imposter INDEX TABLE  Create imposter table TABLE on index INDEX
.indexes ?TABLE?       Show names of all indexes
                         If TABLE specified, only show indexes for tables
                         matching LIKE pattern TABLE.
.limit ?LIMIT? ?VAL?   Display or change the value of an SQLITE_LIMIT
.lint OPTIONS          Report potential schema issues. Options:
                         fkey-indexes     Find missing foreign key indexes
.log FILE|off          Turn logging on or off.  FILE can be stderr/stdout
.mode MODE ?TABLE?     Set output mode where MODE is one of:
                         ascii    Columns/rows delimited by 0x1F and 0x1E
```

```
                        csv      Comma-separated values
                        column   Left-aligned columns.  (See .width)
                        html     HTML <table> code
                        insert   SQL insert statements for TABLE
                        line     One value per line
                        list     Values delimited by .separator strings
                        quote    Escape answers as for SQL
                        tabs     Tab-separated values
                        tcl      TCL list elements
.nullvalue STRING       Use STRING in place of NULL values
.once FILENAME          Output for the next SQL command only to FILENAME
.open ?--new? ?FILE?    Close existing database and reopen FILE
                        The --new starts with an empty file
.output ?FILENAME?      Send output to FILENAME or stdout
.print STRING...        Print literal STRING
.prompt MAIN CONTINUE   Replace the standard prompts
.quit                   Exit this program
.read FILENAME          Execute SQL in FILENAME
.restore ?DB? FILE      Restore content of DB (default "main") from FILE
.save FILE              Write in-memory database into FILE
.scanstats on|off       Turn sqlite3_stmt_scanstatus() metrics on or off
.schema ?PATTERN?       Show the CREATE statements matching PATTERN
                        Add --indent for pretty-printing
.separator COL ?ROW?    Change the column separator and optionally the row
                        separator for both the output mode and .import
.shell CMD ARGS...      Run CMD ARGS... in a system shell
.show                   Show the current values for various settings
.stats ?on|off?         Show stats or turn stats on or off
.system CMD ARGS...     Run CMD ARGS... in a system shell
.tables ?TABLE?         List names of tables
                        If TABLE specified, only list tables matching
                        LIKE pattern TABLE.
.testcase NAME          Begin redirecting output to 'testcase-out.txt'
.timeout MS             Try opening locked tables for MS milliseconds
.timer on|off           Turn SQL timer on or off
.trace FILE|off         Output each SQL statement as it is run
.vfsinfo ?AUX?          Information about the top-level VFS
.vfslist                List all available VFSes
.vfsname ?AUX?          Print the name of the VFS stack
.width NUM1 NUM2 ...    Set column widths for "column" mode
                        Negative values right-justify
```

```
//创建表
sqlite> create table person(_id integer primary key autoincrement,name,age);
//查看数据库
sqlite> .databases
main: D:\Android\sdk\platform-tools\mydb.db
//查看表
sqlite> .tables
person
//插入数据
sqlite> insert into person values(null,'zhang',15);
sqlite> insert into person values(null,'li',18);
//查询
sqlite> select * from person;
1|zhang|15
2|li|18
//修改记录
sqlite> update person set age=19 where name='zhang';
//删除记录
sqlite> delete from person where name='li';
//退出
sqlite> .quit
```

使用命令查看操作数据库比较麻烦，还可以使用图形界面的管理工具。现在网络上的SQLite管理工具很多，这里向大家推荐一款好用的工具：Navicat Premium。Navicat Premium是一款可多重连接的数据库管理工具，它可让用户以单一程序同时连接到MySQL、Oracle、PostgreSQL、SQLite及SQLServer数据库，让管理不同类型的数据库更加方便。Navicat Premium结合了其他Navicat成员的功能。有了不同数据库类型的连接能力，Navicat Premium支持在MySQL、Oracle、PostgreSQL、SQLite及SQL Server之间传输数据。它支持大部分MySQL、Oracle、PostgreSQL、SQLite及SQL Server的功能。

打开Navicat Premium，其主界面如图13-1所示。

单击"连接"图标，选择"SQLite"，打开如图13-2所示窗口。

图13-1　Navicat Premium主界面

第13章 SQLite数据库精讲

图13-2 新建连接窗口

单击"确定"按钮，在图13-1所示的窗口左侧会新增一个"mydb"的连接，展开该连接（图13-3），就可以在图形管理工具中进行数据库的操作了。

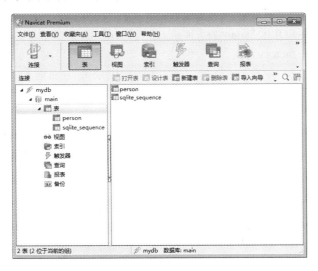

图13-3 增加连接后的主界面

13.2 使用SQLite数据库

SQLite数据库的使用相对来说比较简单，没有服务器进程，通过文件保存数据库，一个文件就是一个数据库,该文件可以跨平台直接使用。SQLite 数据库支持大多数的 SQL92 标准,而且可以在几乎所有的主流操作系统中运行。在 Android 系统中可以比较方便地用 SQLite 数据库来存储应用的数据。Android 提供了创建和使用 SQLite 数据库的 API,可以利用它创建数据库、创建表和执行一些 SQL 语句。

13.2.1　SQLiteDatabase

SQLiteDatabase 代表一个数据库对象,提供了操作数据库的一些方法(表13-1)。

表13-1　SQLiteDatabase 的常用方法

方法名称	作用
openOrCreateDatabase(String path,SQLiteDatabase.CursorFactory factory)	打开或创建数据库
insert(String table,String nullColumnHack,ContentValues values)	插入一条记录
delete(String table,String whereClause,String[] whereArgs)	删除一条记录
query(String table,String[] columns,String selection,String[] selectionArgs,String groupBy,String having,String orderBy,String limit)	查询一条记录
update(String table,ContentValues values,String whereClause,String[] whereArgs)	修改记录
execSQL(String sql)	执行一条 SQL 语句
close()	关闭数据库

使用 SQLiteDatabase 操作数据库的方法如下。

(1)创建数据库

在 Android 中使用 SQLiteDatabase 的静态方法 openOrCreateDatabase(String path, SQLite-Databae.CursorFactory factory)打开或者创建一个数据库。它自动去检测是否存在这个数据库,如果存在则打开,不存在则创建一个数据库。创建成功则返回一个 SQLiteDatabase 对象,否则抛出异常"File Not Found Exception"。

参数说明:

- path:数据库创建的路径;
- factory:游标工厂类,用于产生查询返回的游标,一般设置为 null 就可以了。

下面是创建名为"stu.db"数据库的代码:

```
db=SQLiteDatabase.openOrCreateDatabase("/data/data/
cn.androidstudy.chapter13/databases/stu.db",null);
```

(2)创建表

创建数据表的步骤如下。

① 编写创建表的 SQL 语句。
② 调用 SQLiteDatabase 的 execSQL 方法来执行 SQL 语句。

下面的代码创建了一张用户表,属性列为:id(主键并且自动增加)、sname(用户名)、pass(密码)。

```
private void createTable(SQLiteDatabase db){
    //创建表SQL语句
    String user_table="create table tb_user(_id integer primary
```

```
key autoincrement,sname text,pass text);";
    //执行SQL语句
    db.execSQL(user_table);
}
```

（3）插入数据

插入数据有两种方法。

① SQLiteDatabase 的 insert(String table,String nullColumnHack,ContentValues values) 方法，参数如下：

- table：表名称；
- nullColumnHack：当 values 为空时，将指定列设为 null，插入数据库；
- values：ContentValues 类型，等同于 Java 中的 Map，用来存储键值对（key-value）。

② 编写插入数据的 SQL 语句，直接调用 SQLiteDatabase 的 execSQL 方法来执行。

第一种方法的代码：

```
private void insert(SQLiteDatabase db){
    //实例化常量值
    ContentValues cValue = new ContentValues();
    //添加用户名
    cValue.put("sname","zhangsan");
    //添加密码
    cValue.put("pass","123456");
    //调用insert方法插入数据
    db.insert("tb_user",null,cValue);
}
```

第二种方法的代码：

```
private void insert(SQLiteDatabase db){
    //插入数据SQL语句
    String sql="insert into tb_user(sname,pass) values('zhangsan','123456');";
    //执行SQL语句
    db.execSQL(sql);
}
```

（4）删除数据

删除数据有两种方法。

① 调用 SQLiteDatabase 的 delete(String table,String whereClause,String[] whereArgs) 方法，参数如下：

- table：表名称；
- whereClause：删除条件；
- whereArgs：删除条件值数组。

② 编写删除 SQL 语句，调用 SQLiteDatabase 的 execSQL 方法来执行删除。

第一种方法的代码：

```java
private void delete(SQLiteDatabase db) {
    //删除条件
    String whereClause = "_id=?";
    //删除条件参数
    String[] whereArgs = {String.valueOf(2)};
    //执行删除
    db.delete("tb_user",whereClause,whereArgs);
}
```

第二种方法的代码：

```java
private void delete(SQLiteDatabase db) {
    //删除SQL语句
    String sql = "delete from tb_user where _id = 2;";
    //执行SQL语句
    db.execSQL(sql);
}
```

（5）修改数据

修改数据有两种方法。

① 调用SQLiteDatabase的update(String table,ContentValues values,String whereClause, String[] whereArgs)方法，参数如下：

- table：表名称；
- values：ContentValues类型，要修改的列的键值对（key-value）；
- whereClause：更新条件（where字句）；
- whereArgs：更新条件数组。

② 编写更新的SQL语句，调用SQLiteDatabase的execSQL方法来执行更新。

第一种方法的代码：

```java
private void update(SQLiteDatabase db) {
    //实例化内容值
    ContentValues values = new ContentValues();
    //在values中添加内容
    values.put("pass","password");
    //修改条件
    String whereClause = "_id=?";
    //修改添加参数
    String[] whereArgs={String.valuesOf(1)};
    //修改
    db.update("tb_user",values,whereClause,whereArgs);
}
```

第二种方法的代码：

```java
private void update(SQLiteDatabase db){
```

```
//修改SQL语句
String sql = "update tb_user set pass='password' where _id = 1;";
//执行SQL
db.execSQL(sql);
}
```

(6) 查询数据

在Android中查询数据是通过Cursor类来实现的，当使用SQLiteDatabase.query方法时，会得到一个Cursor对象，Cursor指向的是每一个数据。它提供了很多有关查询的方法，具体方法如下：

public Cursor query(String table,String[] columns,String selection,String[] selectionArgs,String groupBy,String having,String orderBy,String limit);

参数的说明如下：

- table：表名称；
- columns：列名称，数组；
- selection：条件字句，相当于where；
- selectionArgs：条件字句，参数数组；
- groupBy：分组列；
- having：分组条件；
- orderBy：排序列；
- limit：分页查询限制；
- Cursor：返回值，相当于结果集ResultSet。
- Cursor是一个游标接口，提供了遍历查询结果的方法（表13-2），如移动指针方法move、获得列值方法getString等。

表13-2 Cursor游标常用方法

方法名称	方法描述
getCount()	获得行数
getColumnCount()	获得列数
isFirst()	判断是否是第一条记录
isLast()	判断是否是最后一条记录
moveToFirst()	移动到第一条记录
moveToLast()	移动到最后一条记录
move(int offset)	移动到指定记录
moveToNext()	移动到下一条记录
moveToPrevious()	移动到上一条记录
getColumnName(int columnIndex)	从给定的索引返回列名
getColumnIndex(String columnName)	根据列名称获得列索引
getInt(int columnIndex)	获得指定索引的int类型值
getString(int columnIndex)	获得指定列索引的String类型值

下面是用Cursor来查询数据库中的数据，具体代码如下：

```java
    private void query(SQLiteDatabase db) {
        //查询获得游标
        Cursor cursor = db.query ("tb_user",null,null,null,null,null,null);
        //判断游标是否为空
        if(cursor.moveToFirst() {
            //遍历游标
            for(int i=0;i<cursor.getCount();i++){
                cursor.move(i);
                //获得ID
                int id = cursor.getInt(0);
                //获得用户名
                String username=cursor.getString(1);
                //获得密码
                String pass=cursor.getString(2);
                //输出用户信息
                System.out.println(id+":"+username+":"+pass);
            }
        }
    }
```

（7）删除表

编写插入数据的SQL语句，直接调用SQLiteDatabase的execSQL方法来执行。

```java
    private void drop(SQLiteDatabase db){
        //删除表的SQL语句
        String sql ="DROP TABLE tb_user;";
        //执行SQL
        db.execSQL(sql);
    }
```

13.2.2 SQLiteOpenHelper

Android提供了SQLiteOpenHelper帮助用户创建数据库，用户只要继承SQLiteOpenHelper类，就可以轻松地创建数据库。SQLiteOpenHelper类根据开发应用程序的需要，封装了创建和更新数据库使用的逻辑。SQLiteOpenHelper的子类，至少需要实现三个方法：

① 构造函数。调用父类SQLiteOpenHelper的构造函数。这个方法需要四个参数：上下文环境（例如一个Activity）、数据库名字、一个可选的游标工厂（通常是null）、一个代表用户正在使用的数据库模型版本的整数。

② onCreate方法。它需要一个SQLiteDatabase对象作为参数，根据需要对这个对象填充表和初始化数据。

③ onUpgrade方法。它需要三个参数：一个SQLiteDatabase对象、一个旧的版本号和一个新的版本号。这样用户就清楚如何把一个数据库从旧的模型转变到新的模型。

☞【实例13-1】下面示例代码展示了如何继承 SQLiteOpenHelper 创建数据库。

```java
package cn.androidstudy.chapter13;
import android.content.Context;
import android.database.sqlite.SQLiteDatabase;
import android.database.sqlite.SQLiteOpenHelper;
import android.util.Log;
public class DatabaseHelper extends SQLiteOpenHelper {
    public DatabaseHelper(Context context) {
        super(context, "mydb.db", null, 1);
    }
    //第一次获取数据库时,自动执行,主要用来创建数据表
    @Override
    public void onCreate(SQLiteDatabase sqLiteDatabase) {
        String createTable = "create table person"+
                "(_id integer primary key,name text,age integer);";
        sqLiteDatabase.execSQL(createTable);
    }
    @Override
    public void onUpgrade(SQLiteDatabase sqLiteDatabase, int oldVersion, int newVersion) {
        //提示版本升级
        Log.i("Database update......", "Update database from "+oldVersion+" to "
                + newVersion);
        //删除旧的表
        sqLiteDatabase.execSQL("drop table if it exists person");
        //创建新表
        onCreate(sqLiteDatabase);
    }
}
```

使用 SQLiteOpenHelper 访问数据库,需要调用 getWritableDatabase 或者 getReadableDatabase 来获取一个可写或者只读的数据库实例。如果数据库不存在,辅助类就会执行它的 onCreate 方法;如果数据库已经创建,则返回建好的数据库。也就是说,onCreate 方法会在第一次创建数据库的时候自动运行。

☞【实例13-2】下面通过一个例子来演示在 Android 中如何操作数据库,系统运行后,插入三条记录(图13-4),然后修改第二条记录(图13-5)。

MainActivity 的布局文件如下:

```xml
<?xml version="1.0" encoding="utf-8"?>
<android.support.constraint.ConstraintLayout xmlns:android="http://schemas.android.com/apk/res/android"
    xmlns:app="http://schemas.android.com/apk/res-auto"
```

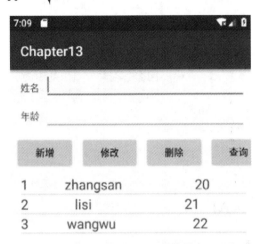

图13-4 新增三条记录后

图13-5 修改第二条记录后

```xml
    xmlns:tools="http://schemas.android.com/tools"
    android:layout_width="match_parent"
    android:layout_height="match_parent"
    tools:context="cn.androidstudy.chapter13.MainActivity">
    <TextView
        android:id="@+id/textView2"
        android:layout_width="wrap_content"
        android:layout_height="wrap_content"
        android:text="姓名"
        app:layout_constraintTop_toTopOf="parent"
        android:layout_marginTop="16dp"
        android:layout_marginLeft="16dp"
        app:layout_constraintLeft_toLeftOf="parent" />
    <EditText
        android:id="@+id/et_name"
        android:layout_width="0dp"
        android:layout_height="43dp"
        android:layout_marginLeft="8dp"
        android:layout_marginRight="8dp"
        android:layout_marginTop="0dp"
        android:ems="10"
        android:inputType="textPersonName"
        app:layout_constraintHorizontal_bias="0.0"
        app:layout_constraintLeft_toRightOf="@+id/textView2"
        app:layout_constraintRight_toRightOf="parent"
        app:layout_constraintTop_toTopOf="parent" />
    <TextView
        android:id="@+id/textView"
```

```xml
        android:layout_width="wrap_content"
        android:layout_height="wrap_content"
        android:text="年龄"
        android:layout_marginLeft="16dp"
        app:layout_constraintLeft_toLeftOf="parent"
        android:layout_marginTop="22dp"
        app:layout_constraintTop_toBottomOf="@+id/textView2" />
    <EditText
        android:id="@+id/et_age"
        android:layout_width="0dp"
        android:layout_height="44dp"
        android:ems="10"
        android:inputType="number"
        app:layout_constraintLeft_toRightOf="@+id/textView"
        android:layout_marginLeft="8dp"
        android:layout_marginRight="8dp"
        app:layout_constraintRight_toRightOf="parent"
        android:layout_marginTop="0dp"
        app:layout_constraintTop_toBottomOf="@+id/et_name"
        app:layout_constraintHorizontal_bias="0.111" />
    <Button
        android:id="@+id/button"
        android:layout_width="wrap_content"
        android:layout_height="wrap_content"
        android:layout_marginLeft="8dp"
        android:text="新增"
        app:layout_constraintLeft_toLeftOf="parent"
        android:layout_marginTop="8dp"
        app:layout_constraintTop_toBottomOf="@+id/et_age"
        android:onClick="insert" />
    <Button
        android:id="@+id/button2"
        android:layout_width="wrap_content"
        android:layout_height="wrap_content"
        android:layout_marginLeft="9dp"
        android:text="修改"
        app:layout_constraintLeft_toRightOf="@+id/button"
        android:layout_marginTop="8dp"
        app:layout_constraintTop_toBottomOf="@+id/et_age"
        android:onClick="update"/>
    <Button
        android:id="@+id/button3"
```

```xml
        android:layout_width="wrap_content"
        android:layout_height="wrap_content"
        android:layout_marginLeft="8dp"
        android:text="删除"
        app:layout_constraintLeft_toRightOf="@+id/button2"
        android:layout_marginTop="8dp"
        app:layout_constraintTop_toBottomOf="@+id/et_age"
        android:onClick="delete"/>

    <Button
        android:id="@+id/button4"
        android:layout_width="wrap_content"
        android:layout_height="wrap_content"
        android:layout_marginLeft="8dp"
        android:text="查询"
        app:layout_constraintLeft_toRightOf="@+id/button3"
        android:layout_marginTop="8dp"
        app:layout_constraintTop_toBottomOf="@+id/et_age"
        android:onClick="select"/>
    <ListView
        android:id="@+id/list"
        android:layout_width="0dp"
        android:layout_height="0dp"
        android:layout_marginBottom="16dp"
        android:layout_marginLeft="16dp"
        android:layout_marginRight="8dp"
        android:layout_marginTop="8dp"
        app:layout_constraintBottom_toBottomOf="parent"
        app:layout_constraintHorizontal_bias="0.0"
        app:layout_constraintLeft_toLeftOf="parent"
        app:layout_constraintRight_toRightOf="parent"
        app:layout_constraintTop_toBottomOf="@+id/button3"
        app:layout_constraintVertical_bias="0.0" />
</android.support.constraint.ConstraintLayout>
```

为了方便数据库的访问，定义DataBaseHelper类，该类继承了SQLiteOpenHelper。因该类在前文中已经给出，在此不再重复。

MainActivity的代码如下：

```java
package cn.androidstudy.chapter13;
import android.content.ContentValues;
import android.database.Cursor;
import android.database.sqlite.SQLiteDatabase;
```

```java
import android.support.v7.app.AppCompatActivity;
import android.os.Bundle;
import android.view.View;
import android.widget.EditText;
import android.widget.ListView;
import android.widget.SimpleCursorAdapter;
public class MainActivity extends AppCompatActivity {
    private EditText et_name;
    private EditText et_age;
    private ListView listView;
    DatabaseHelper helper = null;
    SQLiteDatabase db = null;
    @Override
    protected void onCreate(Bundle savedInstanceState) {
        super.onCreate(savedInstanceState);
        setContentView(R.layout.activity_main);
        helper = new DatabaseHelper(this);
        et_name = (EditText)findViewById(R.id.et_name);
        et_age = (EditText)findViewById(R.id.et_age);
        listView = (ListView)findViewById(R.id.list);
    }
    public void insert(View view){
        db = helper.getWritableDatabase();
        String name = et_name.getText().toString();
        String age = et_age.getText().toString();
        ContentValues values = new ContentValues();
        values.put("name",name);
        values.put("age",age);
        db.insert("person",null,values);
    }
    //根据姓名修改年龄
    public void update(View view){
        db = helper.getWritableDatabase();
        String name = et_name.getText().toString();
        String age = et_age.getText().toString();
        ContentValues values = new ContentValues();
        values.put("age",age);
        db.update("person",values,"name=?",new String[]{name});
    }
    public void delete(View view){
        db = helper.getWritableDatabase();
        String name = et_name.getText().toString();
```

```java
        String age = et_age.getText().toString();
        db.delete("person","name=? and age=?",new String[]{name,age});
    }
    public void select(View view){
        db = helper.getReadableDatabase();
        //查询所有数据
        Cursor cursor = db.query("person",null,null,null,null,null,null);
        SimpleCursorAdapter sca = new SimpleCursorAdapter(this,
                R.layout.list_item,
                cursor,
                new String[]{"_id","name","age"},
                new int[]{R.id.textView1,R.id.textView2,R.id.textView3},
                SimpleCursorAdapter.FLAG_REGISTER_CONTENT_OBSERVER);
        listView.setAdapter(sca);
    }
}
```

在select方法中，查询所有数据到Cursor中，使用SimpleCursorAdapter 将Cursor中数据显示到ListView，SimpleCursorAdapter的构造函数有六个参数，说明如下：
- context：当前上下文；
- layout：ListView的每一行的布局，在此使用自定义的布局：list_item.xml，布局文件如下：

```xml
<?xml version="1.0" encoding="utf-8"?>
<LinearLayout xmlns:android="http://schemas.android.com/apk/res/android"
    android:layout_width="match_parent"
    android:layout_height="match_parent"
    android:orientation="horizontal" >
    <TextView
        android:id="@+id/textView1"
        android:layout_width="wrap_content"
        android:layout_height="wrap_content"
        android:text="TextView"
        android:textSize="20sp"
        android:layout_weight="1" />
    <TextView
        android:id="@+id/textView2"
        android:layout_width="wrap_content"
        android:layout_height="wrap_content"
        android:text="TextView"
        android:textSize="20sp"
        android:layout_weight="2"  />
    <TextView
        android:id="@+id/textView3"
```

```xml
            android:layout_width="wrap_content"
            android:layout_height="wrap_content"
            android:text="TextView"
            android:textSize="20sp"
            android:layout_weight="1"   />
</LinearLayout>
```

- cursor：查询的数据；
- from[]：字符串数组，cursor中要显示的列名；
- to[]：和from[]一一对应，将"列名"的值显示到对应的控件中；
- flags：标识当数据改变调用onContentChanged的时候，是否通知ContentProvider数据有改变，如果无须监听ContentProvider的改变，则可以传0。

数据库的增、删、改也可以通过execSQL方法实现，请读者自行验证，在此不再一一说明。

强化训练

Android经常会使用SQLite数据库来实现结构化数据存储。本章主要介绍了Android中内置的SQLite数据库，系统提供了对数据进行管理和维护的辅助类，要求掌握使用SQLiteOpenHelper创建数据库和数据表，使用SQLiteDatabase对数据进行增、删、改、查等基本操作，掌握游标的使用及ListView显示数据的方法。

学完本章知识后，不妨通过以下习题来巩固所学的内容吧。

一、填空题

1. 当需要存储大量数据时，Android系统中提供了_____，它可以存储应用程序中的大量数据，并对数据进行管理和维护。
2. 目前在Android系统中集成的是SQLite3版本，SQLite不支持_____，而是使用_____。这意味着它的数据类型不具有表列属性，而具有数据本身的属性。
3. SQLite支持_____、_____、_____、_____和_____数据类型。
4. Android中SQLiteOpenHelper的子类，至少需要实现三个方法：_____，_____，_____。
5. SQLite的特点：_____，_____，_____，_____，_____。
6. Android中使用SQLiteDatabase操作数据库的常用方法有：_____，_____，_____，_____，_____，_____。
7. 编写删除SQL语句，调用SQLiteDatabase的_____方法来执行删除。

二、选择题

1. Navicat Premium支持在MySQL、Oracle、(　　)、SQLite及SQLServer之间传输数据。
   A. PostgreSQL
   B. STRING
   C. FILE
   D. CMD

2. （　　）是一个可多重连接的数据库管理工具，它可让用户以单一程序同时连接到MySQL、Oracle、PostgreSQL、SQLite 及 SQL Server 数据库，让管理不同类型的数据库更加方便。

A. Palm OS
B. VxWorks
C. Navicat Premium
D. Symbin

3. （　　）是插入数据的方法。

A. 调用 SQLiteDatabase 的 update(String table,ContentValues values,String whereClause, String[] whereArgs) 方法
B. SQLiteDatabase 的 insert(String table,String nullColumnHack,ContentValues values) 方法
C. 调用 SQLiteDatabase 的 delete(String table,String whereClause,String[] whereArgs) 方法
D. 编写删除 SQL 语句，调用 SQLiteDatabase 的 execSQL 方法来执行删除。

4. Cursor 游标常用方法中，（　　）根据列名称获得列索引。

A. getString(int columnIndex)
B. getColumnCount()
C. moveToFirst()
D. getColumnIndex(String columnName)

5. onCreate 方法需要一个（　　）对象作为参数，根据需要对这个对象填充表和初始化数据。

A. Activity
B. execSQL
C. SQLiteDatabase
D. null

6. 为了方便数据库的访问，定义 DataBaseHelper 类，该类继承了（　　）。

A. SQLiteOpenHelper
B. execSQL
C. SQLiteDatabase
D. null

7. SimpleCursorAdapter 的构造函数有六个参数，（　　）为查询的数据。

A. from[]
B. to[]
C. flags
D. cursor

三、操作题

1. SQLite 数据库是 Android 系统中经常使用的轻型数据库。使用 SQLiteOpenHelper 创建数据库（my.db）和数据表（课程表：Course），其中课程表需要包含如下字段：课程编号、课程名、学分。

2. SQLite 数据库是 Android 系统中经常使用的轻型数据库。学习并练习 SQLite 数据库的基本操作，实现课程的增、删、改、查操作。

3. Android 提供了 SQLiteOpenHelper 帮助编程人员创建数据库，学习并实现继承 SQLiteOpenHelper 创建数据库。

第14章

数据共享机制精讲

内容导读

内容提供者ContentProvider也被称为Android四大基本组件之一，是开发Android程序的基石，其重要性不言而喻。ContentProvider主要用于在不同应用程序之间实现数据共享，它拥有一套完整的机制，在保证被访问数据安全性的同时允许一个程序访问另一个程序中的数据。系统的多个应用程序之间，虽然有时候需要进行数据的共享与交换，但Android系统采用的Linux内核，因此继承了Linux的严格权限管理机制。在前文讲过的Intent只适合用于传递数据量小的场合，对于大的数据量交互明显不适合。为了解决这个难题，Android提供了ContenProvider机制。

学习目标

- 了解ContentProvider共享数据的方法；
- 掌握使用ContentResolver操作数据的方法；
- 掌握自定义ContentProvider的方法；
- 掌握调用系统ContentProvider的方法。

14.1 ContentProvider概述

不同Android应用程序之间是相互独立的，系统没有提供不同应用都能访问的数据公共存储区域。需要在多个应用程序间共享数据时，共享数据可以存放到内容提供者ContentProvider中，所有的ContentProvider都通过通用的方法来实现数据的增、删、查、改等功能，好处是实现统一的数据访问方式。ContentProvider在Android中的作用是对外共享数据，也就是说用户可以通过ContentProvider把应用中的数据共享给其他应用访问，其他应用可以通过ContentProvider对用户应用中的数据进行增、删、改、查。关于数据共享，前文我们学习过文件操作模式，知道通过指定文件的操作模式为Context.MODE_WORLD_READABLE或Context.MODE_WORLD_WRITEABLE同样可以对外共享数据。那么，这里为何要使用ContentProvider对外共享数据呢？如果采用文件操作模式对外共享数据，数据的访问方式会因数据存储的方式而不同，导致数据的访问方式无法统一。如采用XML文件对外共享数据，需要进行XML解析才能读取数据；采用SharedPreferences共享数据，需要使用SharedPreferences API读取数据。使用ContentProvider对外共享数据的好处是统一了数据的访问方式。

ContentProvider类实现了一组标准的方法接口，从而能够让其他的应用程序保存或读取此ContentProvider的各种类型数据。在程序内可以通过实现ContentProvider的抽象接口将自己的数据显示出来，而外界根本不用看到这个显示的数据在应用程序中是如何存储的，以及究竟是用文件存储还是用数据库存储的，外界正是通过这个统一的接口来实现数据的增、删、改、查。

当应用需要通过ContentProvider对外共享数据时，第一步需要继承ContentProvider并重写下面的方法：

```
public class PersonProvider extends ContentProvider{
    public boolean onCreate()
    public Uri insert(Uri uri, ContentValues values)
    public int delete(Uri uri, String selection, String[] selectionArgs)
    public int update(Uri uri, ContentValues values, String selection, String[] selectionArgs)
    public Cursor query(Uri uri, String[] projection, String selection, String[] selectionArgs, String sortOrder)
    public String getType(Uri uri)
}
```

第二步需要在AndroidManifest.xml使用\<provider\>对该ContentProvider进行配置。为了能让其他应用找到该ContentProvider，ContentProvider采用了authorities（主机名/域名）对它进行唯一标识，用户可以把ContentProvider看作是一个网站，authorities就是它的域名。

```
<manifest>
    <application
            android:icon="@drawable/icon"
            android:label="@string/app_name">
        <provider android:name=".PersonProvider"
                android:authorities="cn.androidstudy.providers.personprovider"/>
    </application>
```

```
</manifest>
```

当外部应用需要对ContentProvider中的数据进行添加、删除、修改和查询操作时，可以使用ContentResolver类来完成，要获取ContentResolver对象，可以使用Activity提供的getContentResolver方法。

ContentResolver类提供了与ContentProvider类相同签名的四个方法：

• public Uri insert(Uri uri, ContentValues values)：该方法用于往ContentProvider中添加数据。

• public int delete(Uri uri, String selection, String[] selectionArgs)：该方法用于从ContentProvider中删除数据。

• public int update(Uri uri, ContentValues values, String selection, String[] selectionArgs)：该方法用于更新ContentProvider中的数据。

• public Cursor query(Uri uri, String[] projection, String selection, String[] selectionArgs, String sortOrder)：该方法用于从ContentProvider中获取数据。

很显然，使用ContentResolver类访问ContentProvider，Uri起到了关键作用，因为它决定了去访问哪个ContentProvider。那么，什么是Uri呢？

Uri代表了要操作的数据，Uri主要包含了两部分信息：

第一部分，需要操作的ContentProvider；

第二部分，对ContentProvider中的什么数据进行操作。

一个Uri由以下几部分组成。

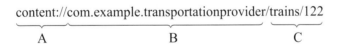

A：ContentProvider（内容提供者）的scheme已经由Android所规定，scheme为"content://"。

B：主机名（或叫Authority）用于唯一标识这个ContentProvider，外部调用者可以根据这个标识找到它。

C：路径（path）可以用来表示要操作的数据，路径的构建应根据业务而定，具体如下：

• 要操作person表中id为10的记录，可以构建这样的路径：/person/10；
• 要操作person表中id为10的记录的name字段：/person/10/name；
• 要操作person表中的所有记录，可以构建这样的路径：/person；
• 要操作xxx表中的记录，可以构建这样的路径：/xxx。

当然要操作的数据不一定来自数据库，也可以是文件、XML文件或网络等其他存储方式，具体如下：要操作XML文件中person节点下的name节点，可以构建这样的路径：/person/name。

如果要把一个字符串转换成Uri，可以使用Uri类中的parse方法，具体如下：

```
Uri uri = Uri.parse("content://cn.androidstudy.provider.personprovider/person");
```

在几乎所有的ContentProvider操作中都会用到Uri，因此一般来讲，如果是自己开发的ContentProvider，最好将Uri定义为常量，这样在简化开发的同时也提高了代码的可维护性。

因为Uri代表了要操作的数据，所以经常需要解析Uri，并从Uri中获取数据。Android系统提供了两个用于操作Uri的工具类，分别为UriMatcher和ContentUris。掌握它们的使用，会便于我们的开发工作。

（1）UriMatcher

用于匹配Uri，它的用法如下：

首先，把需要匹配Uri路径全部给注册上，具体如下：

```
//常量UriMatcher.NO_MATCH表示不匹配任何路径的返回码
UriMatcher sMatcher = new UriMatcher(UriMatcher.NO_MATCH);
//如果match方法匹配content://cn.androidstudy.provider.personprovider/person路径，返回匹配码为1
sMatcher.addURI("cn.androidstudy.provider.personprovider", "person", 1);
//如果match方法匹配content://cn.androidstudy.provider.personprovider/person/11路径，返回匹配码为2
sMatcher.addURI("cn.androidstudy.provider.personprovider", "person/#", 2);//#号为通配符
```

注册完需要匹配的Uri后，就可以使用uriMatcher.match(uri)方法对输入的Uri进行匹配，如果匹配就返回匹配码，匹配码是调用addURI方法传入的第三个参数，如"sMatcher.match(Uri.parse("content://cn.androidstudy.provider.personprovider/person/10"))；"返回的匹配码为2。

（2）ContentUris

用于获取Uri路径后面的ID部分，它有两个比较实用的方法：

① withAppendedId(uri, id)方法用于为路径加上ID部分。

```
Uri uri =Uri.parse("content://cn.androidstudy.provider.personprovider/person");
Uri resultUri =ContentUris.withAppendedId(uri, 10);
//生成后的Uri为content://cn.androidstudy.provider.personprovider/person/10
```

② parseId(uri)方法用于从路径中获取ID部分。

```
Uri uri =Uri.parse("content://cn.androidstudy.provider.personprovider/person/10");
long personid = ContentUris.parseId(uri);//获取的结果为10
```

14.2 自定义ContentProvider

在Android中还支持程序开发人员创建自己的Content Provider，以便将自己程序的数据对其他应用程序共享，下面将通过实例来进行讲解。

第14章 数据共享机制精讲

☞【实例14-1】下面通过一个示例来向读者展示如何自定义Content Provider。

（1）数据存储

数据存储常用的方式是文件和数据库，为了操作方便，采用SQLite数据库对数据进行存储。为了简单起见，将13.2.2节创建的DataBaseHelper类复制到当前项目的包中，仍然使用mydb.db数据库和person表。

（2）继承ContentProvider

新建一个类PersonProvider，继承ContentProvider，在Android Studio中可以在左侧的Android视图中，点击鼠标右键，依次选择"New|Other|Content Provider"命令，打开图14-1所示对话框。输入类名和主机名后，系统自动完成Content Provider的创建和注册（可以省略第三步）。

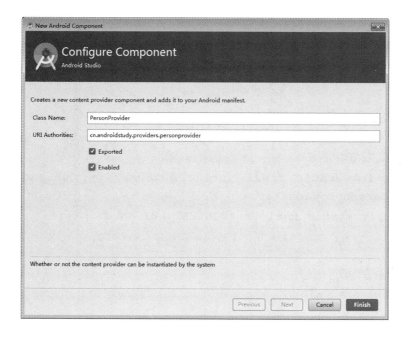

图14-1　新建Content Provider

需要实现该类的六个方法，这些方法的作用如下。

• public boolean onCreate()：该方法在ContentProvider创建后就会被调用，Android开机后，ContentProvider在其他应用第一次访问它时才会被创建。

• public Uri insert(Uri uri, ContentValues values)：该方法用于供外部应用往ContentProvider添加数据。

• public int delete(Uri uri, String selection, String[] selectionArgs)：该方法用于供外部应用从ContentProvider删除数据。

• public int update(Uri uri, ContentValues values, String selection, String[] selectionArgs)：该方法用于供外部应用更新ContentProvider中的数据。

• public Cursor query(Uri uri, String[] projection, String selection, String[] selectionArgs, String sortOrder)：该方法用于供外部应用从ContentProvider中获取数据。

• public String getType(Uri uri)：该方法用于返回当前Uri所代表数据的MIME类型。

如果操作的数据属于集合类型，那么MIME类型字符串应该以"vnd.android.cursor.dir/"

开头,例如要得到所有person记录的Uri为"content://cn.androidstudy.provider.personprovider/person",那么返回的MIME类型字符串应该为"vnd.android.cursor.dir/person"。

如果要操作的数据属于非集合类型数据,那么MIME类型字符串应该以"vnd.android.cursor.item/"开头,例如得到id为10的person记录,Uri为"content://cn.androidstudy.provider.personprovider/person/10",那么返回的MIME类型字符串为"vnd.android.cursor.item/person"。

PersonProvider类文件代码如下:

```java
package cn.androidstudy.chapter14;
import android.content.ContentProvider;
import android.content.ContentUris;
import android.content.ContentValues;
import android.content.UriMatcher;
import android.database.Cursor;
import android.database.sqlite.SQLiteDatabase;
import android.net.Uri;

public class PersonProvider extends ContentProvider {
    private DatabaseHelper helper = null;
    //1.发布Content Provider的Uri地址
    private static final String AUTHORITY = "cn.androidstudy.providers.personprovider";
    public static final Uri CONTENT_URI = Uri.parse(
            "content://cn.androidstudy.providers.personprovider/persons");

    //2.注册需要匹配的Uri
    private static UriMatcher uriMatcher = new UriMatcher(UriMatcher.NO_MATCH);
    static{
        uriMatcher.addURI(AUTHORITY, "persons", 1);
        uriMatcher.addURI(AUTHORITY, "persons/#", 2);
    }
    public PersonProvider() {
    }
    @Override
    public boolean onCreate() {
        //3.实例化dbh
        helper = new DatabaseHelper(getContext());
        return false;
    }
    @Override
    public int delete(Uri uri, String selection, String[] selectionArgs) {
```

第14章 数据共享机制精讲

```java
        // 7.实现删除方法
        SQLiteDatabase db = helper.getWritableDatabase();
        int num = 0;//已经删除的记录数量
        switch (uriMatcher.match(uri)) {
            case 1:
                num = db.delete("person", selection, selectionArgs);
                break;
            case 2:
                //获取ID
                long id = ContentUris.parseId(uri);
                //在selection上增加条件_id=id
                if(selection == null){
                    selection = "_id="+id;
                }else{
                    selection = "_id="+id+" and ("+selection+")";
                }
                num = db.delete("person", selection, selectionArgs);
                break;
            default:
                break;
        }
        //通知所有的观察者，数据集已经改变
        getContext().getContentResolver().notifyChange(uri, null);
        return num;
    }
    @Override
    public String getType(Uri uri) {
        // 4.返回当前Uri所代表数据的MIME类型
        switch (uriMatcher.match(uri)) {
            case 1:
                return "vnd.android.cursor.dir/person";
            case 2:
                return "vnd.android.cursor.item/person";
        }
        return null;
    }
    @Override
    public Uri insert(Uri uri, ContentValues values) {
        // 6.实现插入方法
        SQLiteDatabase db = helper.getWritableDatabase();
        long id = db.insert("person", null, values);
        if (id>-1) {//插入数据成功
```

```java
            //构建新插入行的Uri
            Uri insertUri = ContentUris.withAppendedId(CONTENT_URI, id);
            //通知所有的观察者，数据集已经改变
            getContext().getContentResolver().notifyChange(insertUri, null);
            return insertUri;
        }
        return null;
    }
    @Override
    public Cursor query(Uri uri, String[] projection, String selection,
                String[] selectionArgs, String sortOrder) {
        // 5.实现查询
        SQLiteDatabase db = helper.getReadableDatabase();
        Cursor cursor = null;

        switch (uriMatcher.match(uri)) {
            case 1://查询所有行
                cursor = db.query(
                        "person",   //表名
                        null,       //列的数组，null代表所有列
                        selection,     //where条件
                        selectionArgs, //where条件的参数值的数组
                        null,       //分组
                        null,       //having
                        sortOrder);    //排序规则
                break;
            case 2://查询指定ID的行
                //获取ID
                long id = ContentUris.parseId(uri);
                //在selection上增加条件 _id=id
                if(selection == null){
                    selection = "_id="+id;
                }else{
                    selection = "_id="+id+" and ("+selection+")";
                }
                cursor = db.query(
                        "person",   //表名
                        null,       //列的数组，null代表所有列
                        selection,     //where条件
                        selectionArgs, //where条件的参数值的数组
                        null,       //分组
                        null,       //having
```

```java
                    sortOrder);       //排序规则
            break;
        default:
            break;
        }
        return cursor;
    }
    @Override
    public int update(Uri uri, ContentValues values, String selection,
                String[] selectionArgs) {
        // 8.实现修改方法
        SQLiteDatabase db = helper.getWritableDatabase();
        int num = 0;//已经修改的记录数量
        switch (uriMatcher.match(uri)) {
        case 1:
            num = db.update("person", values, selection, selectionArgs);
            break;
        case 2:
            //获取ID
            long id = ContentUris.parseId(uri);
            //在selection上增加条件 _id=id
            if(selection == null){
                selection = "_id="+id;
            }else{
                selection = "_id="+id+" and ("+selection+")";
            }
            num = db.update("person", values, selection, selectionArgs);
            break;
        default:
            break;
        }
        //通知所有的观察者，数据集已经改变
        getContext().getContentResolver().notifyChange(uri, null);
        return num;
    }
}
```

（3）注册Provider

在AndroidManifest.xml使用<provider>对该ContentProvider进行配置，代码如下：

```xml
<provider
    android:name=".PersonProvider"
    android:authorities="cn.androidstudy.providers.personprovider"
```

```
android:enabled="true"
android:exported="true"></provider>
```

（4）访问Provider

在MainActivity的onCreate方法中增加如下代码：

```
ContentResolver cr = getContentResolver();
//增加记录
ContentValues values = new ContentValues();
values.put("name", "Mike");
values.put("age", 20);
cr.insert(PersonProvider.CONTENT_URI, values);
values.clear();
values.put("name", "Mary");
values.put("age", 18);
cr.insert(PersonProvider.CONTENT_URI, values);
//查询所有记录
Cursor cursor = cr.query(PersonProvider.CONTENT_URI,
      null, null, null, null);
Log.i("after inserted", "-----------------------------------------");
while(cursor.moveToNext()){
   Log.i("after inserted", "id:"+cursor.getString(0)+" name:" +
         cursor.getString(1)+" age:"+cursor.getString(2));
}
cursor.close();
//修改记录
values.clear();
values.put("age", 19);
//构建的Uri为："content://cn.androidstudy.providers.personprovider/persons/2"
Uri uri = ContentUris.withAppendedId(PersonProvider.CONTENT_URI, 2);
//修改id为2的记录
cr.update(uri, values, null, null);
//查询id为2的记录
cursor = cr.query(uri, null, null, null, null);
Log.i("after updated", "-----------------------------------------");
while(cursor.moveToNext()){
   Log.i("after updated", "id:"+cursor.getString(0)+" name:"+
         cursor.getString(1)+" age:"+cursor.getString(2));
}
cursor.close();
```

```
//删除id为2的记录
cr.delete(uri, null, null);
//查询记录
cursor = cr.query(PersonProvider.CONTENT_URI,
        null, null, null, null);
Log.i("after deleted", "----------------------------------------
-----------");
while(cursor.moveToNext()){
    Log.i("after deleted", "id:"+cursor.getString(0)+" name:"+
        cursor.getString(1)+" age:"+cursor.getString(2));
}
cursor.close();
```

运行应用程序，在LogCat窗口中看到的输出如图14-2所示。

```
inserted: ------------------------------------------
inserted: id:1 name:Mike age:20
inserted: id:2 name:Mary age:18
updated: -------------------------------------------
updated: id:2 name:Mary age:19
deleted: -------------------------------------------
deleted: id:1 name:Mike age:20
```

图14-2　访问PersonProvider的结果

如果在其他应用程序中调用此Provider，因为PersonProvider.CONTENT_URI无法访问，需要定义一个新的Uri变量，如：

```
private static final Uri CONTENT_URI = Uri.parse(
        "content://cn.androidstudy.providers.personprovider/
persons");
```

然后将PersonProvider.CONTENT_URI代替即可。

14.3 监听ContentProvider中数据的变化

在使用内容提供者ContentProvider时，Android程序开发人员经常需要监听ContentProvider中数据的变化来实现对程序运行的控制。在PersonProvider的insert、update和delete方法的最后，都调用了getContext().getContentResolver().notifyChange(Uri, null)方法，该方法的作用是可以在ContentProvider发生数据变化时来通知注册在此Uri上的访问者。其中Uri表示监听的Uri，null表示发送消息给任何人。

如果ContentProvider的访问者需要得到数据变化通知，必须使用ContentObserver对数据进行监听，当监听到数据变化通知时，系统就会调用ContentObserver的onChange方法。

为了监听指定的ContentProvider的数据变化，需要通过ContentResolver向指定Uri注册ContentObserver监听器。用如下方法来注册监听器：

registerContentObserver(Uri uri,boolean notifyForDescendents,ContentObserver observer)；

notifyForDescendents：如果该参数设为true,假如Uri为"content://abc"，那么Uri为"content://abc/xyz""content://abc/xyz/foo"的数据改变时也会触发该监听器；如果参数为false，那么只有"content://abc"的数据改变时会触发该监听器。

☞【实例14-2】下面通过一个实例来演示一下如何监听ContentProvider中数据的变化。需要创建两个应用程序，第一个应用程序用来监听数据的变化，新建一个项目ContentObserver，将app的MainActivity.java文件修改如下：

```java
package cn.androidstudy.contentobserver;
import android.content.ContentResolver;
import android.database.ContentObserver;
import android.net.Uri;
import android.os.Handler;
import android.support.v7.app.AppCompatActivity;
import android.os.Bundle;
import android.util.Log;
public class MainActivity extends AppCompatActivity {
    private static final Uri CONTENT_URI = Uri.parse(
        "content://cn.androidstudy.providers.personprovider/persons");
    @Override
    protected void onCreate(Bundle savedInstanceState) {
        super.onCreate(savedInstanceState);
        setContentView(R.layout.activity_main);
        ContentResolver cr = getContentResolver();
        cr.registerContentObserver(CONTENT_URI, true, new PersonObserver(new Handler()));
    }
    private class PersonObserver extends ContentObserver {//监听
        public PersonObserver(Handler handler) {
            super(handler);
        }
        //当ContentProvier数据发生改变，则触发该函数
        @Override
        public void onChange(boolean selfChange) {
            super.onChange(selfChange);
            Log.i("监听到", "数据改变");
```

```
            }
        }
    }
```

然后创建第二个项目，在Activity上放置一个按钮，该按钮的单击事件如下：

```
public void insert(View view){
    ContentResolver cr = getContentResolver();
    Uri uri = Uri.parse("content://cn.androidstudy.providers.personprovider/persons");
    //增加记录
    ContentValues values = new ContentValues();
    values.put("name", "Mike");
    values.put("age", 20);
    cr.insert(uri, values);
}
```

先运行第一个应用程序，然后运行第二个，点击按钮时，会增加一条记录，ContentProvider就会通知所有的监听者数据已经发生改变，应用程序的onChange方法就会触发，输出结果如图14-3所示。

```
cn.androidstudy.contentobserver I/监听到：数据改变
```

图14-3 ContentProvider监听

14.4 系统ContentProvider

作为Android四大基本组件之一，Android自然对内容提供者ContentProvider非常重视。Android系统为常用数据类型提供了很多预定义的ContentProvider，例如音频、视频、图片和私人通讯录等。可在android.provider包下面找到一些Android提供的ContentProvider。可以获得这些ContentProvider，查询它们包含的数据，当然前提是已获得适当的读取权限。

Contacts Contract Content Provider提供了一个可扩展的联系人信息数据库。它允许开发人员任意地扩展为每个联系人存储数据，甚至可以为联系人管理提供可替代程序。该数据库保存的路径为"data/data/com.android.providers.contacts/databases/contacts2.db"。可以使用第13章介绍的方法来查看该数据库，常用的几张表如下。

① Data表：储存所有与Raw_Contacts相关的具体信息。表的每一条记录对应一个特定信息，如名字、电话、E-mail地址、头像和组信息等。每一个记录通过一个mimetype_id字段来表明该行所记录的数据类型。例如，如果row的数据类型是Phone.CONTENT_ITEM_TYPE，那么第一个字段应该保存电话号码，如果数据类型为Email.CONTENT_ITEM_TYPE，那么这一记录的字段应该保存邮件地址。

Data表的字段如表14-1所示。

表 14-1　Data 表的字段

字段	说明
mimetype_id	表示该行存储的信息的类型
raw_contact_id	表示该行所属的 Raw_Contact
is_primary	多个 data 数据组成一个 Raw Contact，该字段表示此 data 是否是其所属的 Raw Contact 的主 data，即其 display name 是否会作为 Raw Contact 的 display name
is_super_primary	该 data 是否是其所属的 contact 的主 data，如果 is_super_primary 为 1 则 is_primary 一定为 1
data1~data15	15 个数据字段，对于不同类型的信息，表示不同的含义，ContactsContract.CommomDataKinds 类中定义了与常用的数据类型相对应的一些类，这些类中分别定义了相应数据类型中这些字段表示的含义。一般 data1 表是主信息（如电话、E-mail 地址等），data2 表示副信息，data15 表示 Blob 数据
data_sync1~data_sync4	sync_adapter 要用的字段（sync_adapter 用于数据的同步，例如手机中的 G-mail 账户与 Google 服务器的同步）
data_version	数据的版本，用于数据的同步

② Raw_Contacts：Raw_Contacts 表中的一行存储 Data 表中一些数据行的集合及一些其他的信息，表示一个联系人某一特定账户的信息，比如 Facebook 或 Exchange 的一个联系人。

当插入一个 Raw Contact 或当一个 Raw Contact 所属的一个 Data 改变时，系统会检查这个 Raw Contact 跟其他的 Raw Contact 是否可以匹配（如果两个 Raw Contact 的 Data 包含相同的电话号码或名字），如果匹配，它们就会被综合到一起，也就是说它们会属于同一个 Contact，表现为在 Raw_Contacts 表中它们引用的 contact_id 是一样的。

③ Contacts：Contacts 表中的一行表示一个联系人，它是 Raw_Contacts 表中的一行或多行数据的组合，这些 Raw_Contacts 表中的行表示同一个人的不同的账户信息。Contacts 中的数据由系统组合 Raw_Contacts 表中的数据自动生成，不可以直接向这个表中插入数据，当一个 Raw Contact 被插入的时候，系统会首先查找 Contacts 表看是否有记录跟插入的 Raw Contact 表示同一个人，如果找到了，则把找到的这个 Contact 的 _ID 插入 Raw Contact 记录的 CONTACT_ID 字段，如果没有找到，则系统自动插入一个 Contact 记录并把它的 _ID 插入新插入的 Raw Contact 的 CONTACT_ID 列。

从 Android 2.0 SDK 开始，有关联系人 Provider 的类由 android.provider.Contacts 变成了 android.provider.ContactsContract，虽然老的 android.provider.Contacts 能用，但是在 SDK 中标记为 deprecated 的是将被放弃或不推荐的方法。

使用 ContentResolver 对通讯录中的数据进行添加、删除、修改和查询操作，需要加入读写联系人信息的权限：

```
<uses-permissionandroid:name="android.permission.READ_CONTACTS" />
<uses-permissionandroid:name="android.permission.WRITE_CONTACTS" />
```

ContactsContract.Contacts 以静态形式访问 Contacts 表中数据的常量，如列名、Uri 等。下面的代码实现了查询通讯录中联系人的 ID 和姓名，并将结果输出到 LogCat 窗口中。

```
//创建一个数组，将结果Cursor限制为所需的列
String[] projection = {
    ContactsContract.Contacts._ID,
    ContactsContract.Contacts.DISPLAY_NAME
};
```

```
Cursor cursor = getContentResolver().query(
    ContactsContract.Contacts.CONTENT_URI, projection,
    null, null, null);
while(cursor.moveToNext()){
    String id = cursor.getString(0);
    String name = cursor.getString(1);
    Log.i("Read Contact", "id:"+id+",name:"+name);
}
cursor.close();
```

ContactsContract.Data 以静态形式访问 Data 表中数据的常量，由于存储在表中的数据由 MIME 类型确定，例如，如果行的数据类型是 Phone.CONTENT_ITEM_TYPE，那么 data1 应该保存电话号码，如果数据类型为 Email.CONTENT_ITEM_TYPE，那么 data1 应该保存邮件地址。

为了简化操作，ContactsContract 定义了一些数据种类，如：ContactsContract.CommonDataKinds.Phone、ContactsContract.CommonDataKinds.Email。为了操作方便，这些类为 data1 定义了新的别名，如 Phone.NUMBER，等同于 Data.data1。

☞【实例 14-3】下面通过一个完整的实例演示一下如何来读取联系人信息，并把结果通过一个 ExpandableListView 进行显示（图 14-4、图 14-5）。

图 14-4　联系人列表　　　　　　图 14-5　联系人 zhang san 详情

在 Chapter14 中新建一个 Module——readcontactsdemo，修改 MainActivity 的布局文件如下：

```xml
<?xml version="1.0" encoding="utf-8"?>
<android.support.constraint.ConstraintLayout
    xmlns:android="http://schemas.android.com/apk/res/android"
    xmlns:app="http://schemas.android.com/apk/res-auto"
```

```xml
    xmlns:tools="http://schemas.android.com/tools"
      android:layout_width="match_parent"
    android:layout_height="match_parent" tools:context="cn.androidstudy.chapter14.MainActivity">

    <ExpandableListView
        android:id="@+id/expandableListView"
        android:layout_width="368dp"
        android:layout_height="495dp"
        android:layout_marginLeft="8dp"
        app:layout_constraintLeft_toLeftOf="parent"
        app:layout_constraintTop_toTopOf="parent"
        android:layout_marginTop="8dp" />
</android.support.constraint.ConstraintLayout>
```

添加读写联系人的权限，而该权限属于危险权限，参考12.3.3节，添加权限的动态申请代码，MainActivity的代码如下：

```java
package cn.androidstudy.chapter14;
import android.Manifest;
import android.content.pm.PackageManager;
import android.database.Cursor;
import android.os.Build;
import android.provider.ContactsContract;
import android.support.annotation.NonNull;
import android.support.v7.app.AppCompatActivity;
import android.os.Bundle;
import android.util.Log;
import android.view.Gravity;
import android.view.View;
import android.view.ViewGroup;
import android.widget.AbsListView;
import android.widget.BaseExpandableListAdapter;
import android.widget.ExpandableListView;
import android.widget.TextView;
import java.util.ArrayList;
public class MainActivity extends AppCompatActivity {
boolean isGranted = false;
private ExpandableListView elv = null;
// 定义两个List来封装系统的联系人信息、指定联系人的电话号码、E-mail等详情
ArrayList<String> names = new ArrayList<String>();
ArrayList<ArrayList<String>> details = new ArrayList<ArrayList<String>>();
    @Override
```

```java
protected void onCreate(Bundle savedInstanceState) {
    super.onCreate(savedInstanceState);
    setContentView(R.layout.activity_main);
    //0.请求权限
    grantedAndRequest();
    //1.获取控件
    elv = (ExpandableListView)findViewById(R.id.expandableListView);
    if(!isGranted) return;
    //2.获取联系人数据
    getData();
    //3.定义适配器
    MyAdapter adapter = new MyAdapter();
    //4.设置适配器
    elv.setAdapter(adapter);
}
//判断是否授权，如未授权，则申请授权
private void grantedAndRequest() {
    //针对Android 6.0及之后版本
    if (Build.VERSION.SDK_INT >= Build.VERSION_CODES.M) {
        int permission=checkSelfPermission(Manifest.permission.READ_CONTACTS);
            //和下面语句等效
        //int permission=checkSelfPermission("android.permission.WRITE_EXTERNAL_STORAGE");
            //如果已授权
        if (permission == PackageManager.PERMISSION_GRANTED) {
            isGranted = true;
        }else{
            //未授权，弹对话框，申请授权
            requestPermissions(new String[]{Manifest.permission.READ_CONTACTS}, 1);
        }
    }else{
        isGranted = true;
    }
}

/**
 * 处理权限请求结果
 * @param requestCode
 *              请求权限时传入的请求码，用于区别是哪一次请求的
 * @param permissions
```

```
 *              所请求的所有权限的数组
 * @param grantResults
 *              权限授予结果,和 permissions 数组参数中的权限一一对应,元素
值为两种情况,如下:
 *              授予:PackageManager.PERMISSION_GRANTED
 *              拒绝:PackageManager.PERMISSION_DENIED
 */
@Override
public void onRequestPermissionsResult(int requestCode, @NonNull String[] permissions, @NonNull int[] grantResults) {
    super.onRequestPermissionsResult(requestCode, permissions, grantResults);
    if(requestCode==1){
        if(grantResults[0]==PackageManager.PERMISSION_GRANTED){
            isGranted = true;
            //2.获取联系人数据
            getData();
            //3.定义适配器
            MyAdapter adapter = new MyAdapter();
            //4.设置适配器
            elv.setAdapter(adapter);
        }
    }
}
private void getData() {
    // 使用ContentResolver查找联系人数据
    Cursor cursor = getContentResolver().query(
            ContactsContract.Contacts.CONTENT_URI, null, null,
            null, null);
    // 遍历查询结果,获取系统中所有联系人
    while (cursor.moveToNext())
    {
        // 获取联系人ID
        String contactId = cursor.getString(cursor
                .getColumnIndex(ContactsContract.Contacts._ID));
        // 获取联系人的名字
        String name = cursor.getString(cursor.getColumnIndex(
                ContactsContract.Contacts.DISPLAY_NAME));
        names.add(name);
        // 使用ContentResolver查找联系人的电话号码
        Cursor phones = getContentResolver().query(
```

```java
                ContactsContract.CommonDataKinds.Phone.CONTENT_URI,null,
                ContactsContract.CommonDataKinds.Phone.CONTACT_ID
                    + " = " + contactId, null, null);
            ArrayList<String> detail = new ArrayList<String>();
            // 遍历查询结果，获取该联系人的多个电话号码
            while (phones.moveToNext())
            {
                // 获取查询结果中电话号码列中数据
                String phoneNumber = phones.getString(phones
                        .getColumnIndex(ContactsContract
                             .CommonDataKinds.Phone.NUMBER));
                detail.add("电话号码：" + phoneNumber);
            }
            phones.close();
            // 使用ContentResolver查找联系人的E-mail地址
            Cursor emails = getContentResolver().query(
                ContactsContract.CommonDataKinds.Email.CONTENT_URI,
                null,
                ContactsContract.CommonDataKinds.Email.CONTACT_ID
                    + " = " + contactId, null, null);
            // 遍历查询结果，获取该联系人的多个E-mail地址
            while (emails.moveToNext())
            {
                // 获取查询结果中E-mail地址列中数据
                String emailAddress = emails.getString(emails
                        .getColumnIndex(ContactsContract
                             .CommonDataKinds.Email.DATA));
                detail.add("邮件地址：" + emailAddress);
            }
            emails.close();
            details.add(detail);
        }
        cursor.close();
    }
    //Group：如图14-4，每个联系人就是一个组；
    //Children：如图14-5，联系人的详情
    class MyAdapter extends BaseExpandableListAdapter{
        //确定显示组的数量
        @Override
        public int getGroupCount() {
            return names.size();
        }
        //确定第 i 组孩子的数量，在本例指联系人详情
```

```java
@Override
public int getChildrenCount(int i) {
    return details.get(i).size();
}
//获取第 i 个组
@Override
public Object getGroup(int i) {
    return names.get(i);
}
//获取第 i 组, 第 i1 个孩子
@Override
public Object getChild(int i, int i1) {
    return details.get(i).get(i1);
}
//获取第 i 组的id
@Override
public long getGroupId(int i) {
    return i;
}
//获取第 i 组, 第 i1 个孩子的id
@Override
public long getChildId(int i, int i1) {
    return i1;
}

@Override
public boolean hasStableIds() {
    return true;
}
//自定义方法, 用来定义一个TextView
private TextView getTextView()
{
    AbsListView.LayoutParams lp = new AbsListView
            .LayoutParams(ViewGroup.LayoutParams.MATCH_PARENT
            , 64);
    TextView textView = new TextView(MainActivity.this);
    textView.setLayoutParams(lp);
    textView.setGravity(Gravity.CENTER_VERTICAL| Gravity.LEFT);
    textView.setPadding(64, 0, 0, 0);
    textView.setTextSize(20);
    return textView;
}
//返回第 i 组的view
```

```java
        @Override
        public View getGroupView(int i, boolean b, View view,
ViewGroup viewGroup) {
            TextView textView = getTextView();
            textView.setText(getGroup(i).toString());
            return textView;
        }
        //返回第 i 组，第 i1 个孩子的View
        @Override
        public View getChildView(int i, int i1, boolean b, View view,
ViewGroup viewGroup) {
            TextView textView = getTextView();
            textView.setText(getChild(i,i1).toString());
            return textView;
        }
        @Override
        public boolean isChildSelectable(int i, int i1) {
            return true;
        }
    }
}
```

强化训练

内容提供者ContentProvider也被称为Android四大基本组件之一，Android程序开发人员应该熟练掌握。本章主要介绍了Android系统中ContentProvider组件的功能和用法，ContentProvider的本质就像是一个"网站"，它可以把应用程序的数据按照"固定规范"暴露出来，其他应用程序就可以通过ContentProvider暴露出来的接口操作内部的数据了。

学习本章需要重点掌握三个API的用法：ContentResolver、ContentProvider和ContentObserver。其中，ContentProvider是所有ContentProvider组件的基类，ContentResolver用于操作ContentProvider提供的数据，而ContentObserver用于监听ContentProvider的数据改变。

学完本章知识后，不妨通过以下习题来巩固所学的内容吧。

一、填空题

1. Android系统采用了＿＿＿＿＿＿内核，因此继承了＿＿＿＿＿＿的严格权限管理机制。
2. Android系统中ContentResolver类提供了与ContentProvider类相同签名的四个方法：＿＿＿＿＿＿，＿＿＿＿＿＿，＿＿＿＿＿＿，＿＿＿＿＿＿。
3. Android系统提供了两个用于操作Uri的工具类，分别为＿＿＿＿＿＿和＿＿＿＿＿＿。
4. 数据存储常用的方式是文件和数据库，为了操作方便，采用＿＿＿＿＿＿数据库对数据进行存储。

5. 要操作 XML 文件中 person 节点下的 name 节点，可以构建路径：_____

6. ContentProvider 在 Android 中的作用是对外_____，也就是说用户可以通过 ContentProvider 把应用中的数据共享给其他应用访问，其他应用可以通过 ContentProvider 对用户应用中的数据进行_____。

二、选择题

1. Android 系统中（　　）用于获取 Uri 路径后面的 ID 部分。
   A. ContentUris
   B. UriMatcher
   C. ContentProvider
   D. Context

2. Data 表的字段中，（　　）表示该行存储的信息的类型。
   A. is_super_primary
   B. mimetype_id
   C. data_version
   D. is_primary

3. Android 系统中，在 PersonProvider 的 insert、update 和 delete 方法的最后，都调用了（　　）方法，该方法的作用是可以在 ContentProvider 发生数据变化时来通知注册在此 URI 上的访问者。
   A. new ContentValues()
   B. getContentResolver()
   C. registerContentObserver
   D. getContext().getContentResolver().notifyChange(Uri, null)

4. 表中的一行存储 Data 表中一些数据行的集合及一些其他的信息，表示一个联系人某一特定账户的信息的表是（　　）。
   A. Contract
   B. Contacts
   C. Raw_Contacts
   D. Data

5. Android 系统中用于返回当前 Uri 所代表数据的 MIME 类型的方法是（　　）。
   A. public Cursor query(Uri uri, String[] projection, String selection, String[] selectionArgs, String sortOrder)
   B. public int update(Uri uri, ContentValues values, String selection, String[] selectionArgs)
   C. public Uri insert(Uri uri, ContentValues values)
   D. public String getType(Uri uri)

三、操作题

1. 利用 Android Studio 开发环境创建一个新的 Android 应用程序，自定义 ContentProvider，可以通过此 ContentProvider 实现课程的增、删、改、查。

2. 利用 Android Studio 开发环境创建一个新的 Android 应用程序，访问手机中联系人信息，并实现读取联系人信息，并把结果通过一个 ExpandableListView 显示出来。

3. 利用 Android Studio 开发环境创建一个新的 Android 应用程序，学习并实现监听 ContentProvider 中数据的变化。

第15章

在线音视频的应用与管理

内容导读

多媒体功能是智能手机用户的日常基本需求之一。音频、视频不单丰富了智能手机用户的娱乐生活,还诞生了巨大的商机,从而出现了一批优秀的上市公司。本章主要介绍Android平台的多媒体应用开发方面的基础知识。多媒体主要包括音频和视频,为方便起见,本章将分开介绍音频和视频的录制与播放相关的功能。此外,本章还将探索在线音频和视频的播放等高级功能。

学习目标

- 掌握音频录制和播放的方法;
- 掌握视频录制和播放的方法。

15.1 录制音频

音频的录制是使用音频的第一步,也是Android手机用户多媒体基本需求之一。Android系统中提供了不同的音频录制方法。本节将介绍Android系统中录制音频的两种主要方法。每种方法都有它们各自的使用范围,第一种方法,使用Intent调用系统功能进行录制,该方法简单方便但灵活性不足;第二种方法,使用MediaRecorder类录制音频,该方法使用难度适中并且灵活性较强;第三种方法,使用AudioRecord类录制音频,该方法难于掌握,灵活性很强,但功能十分强大。下面主要介绍前两种方法。

15.1.1 使用Intent录制音频

录制音频最简单的方法是使用Intent调用系统已有的录制应用程序提供的录制功能。Android平台默认提供一个录音机应用程序，当然也可以使用第三方提供的其他录音应用程序。使用Intent录制音频的基础代码如下所示：

```
Intent audio_recording_intent =
        new Intent(MediaStore.Audio.Media.RECORD_SOUND_ACTION);
startActivity(audio_recording_intent);
```

其中，MediaStore.Audio.Media.RECORD_SOUND_ACTION常量表示启动录音机应用程序的动作意图。

调用startActivity(audio_recording_intent)可以启动录音机应用程序进行音频录制，但是控制权已经交给了录音机应用程序，在录制完音频后，音频数据无法返回给调用程序。

☞【实例15-1】下面给出一个较完整的示例，提供使用Intent录制音频的功能。示例运行主界面如图15-1所示，单击"录音"按钮，将打开系统自带的录音机（图15-2），单击开始"按钮"进行录音。

图15-1 录制音频　　　　图15-2 系统录音机

由于MainActivity的布局文件中只包含两个按钮，请自行设计，或者查看本书配套资源，本示例在Chapter15中的app中。

MainActivity的代码如下：

```
package cn.androidstudy.chapter15;
import android.content.Intent;
import android.media.MediaPlayer;
```

```java
import android.net.Uri;
import android.provider.MediaStore;
import android.support.v7.app.AppCompatActivity;
import android.os.Bundle;
import android.view.View;
import android.widget.Button;
public class MainActivity extends AppCompatActivity {
    Button createRecording, playRecording;
    public static int RECORD_REQUEST = 0;
    Uri audioFileUri;
    @Override
    protected void onCreate(Bundle savedInstanceState) {
        super.onCreate(savedInstanceState);
        setContentView(R.layout.activity_main);
        createRecording = (Button) this.findViewById(R.id.button);
        playRecording = (Button) this.findViewById(R.id.button2);
        //录音完成前，禁用播放功能
        playRecording.setEnabled(false);
    }
    public void record(View view){
        //启动系统自带录音机
        Intent intent = new Intent(
                MediaStore.Audio.Media.RECORD_SOUND_ACTION);
        startActivityForResult(intent, RECORD_REQUEST);
    }
    public void play(View view){
        MediaPlayer mediaPlayer = MediaPlayer.create(this, audioFileUri);
        mediaPlayer.start();
        playRecording.setEnabled(false);
    }
    @Override
    protected void onActivityResult(int requestCode, int resultCode,
Intent data) {
        //录音完成后的回调
        if (resultCode == RESULT_OK && requestCode == RECORD_REQUEST) {
            audioFileUri = data.getData();//获取录音文件
            playRecording.setEnabled(true);//允许播放
        }
    }
}
```

15.1.2 使用MediaRecorder录制音频

在介绍了使用Intent录制音频后，接下来介绍与如何使用MediaRecorder类来录制音频。MediaRecorder类是Android SDK提供的录制音频和视频的功能类，使用它可以建立功能更加完善的音频录制应用，例如可以控制录制的时长等。MediaRecorder类内部维护一个状态机来管理录制音频和视频时的各种状态。该状态机如图15-3所示，它描述了音频和视频录制过程中的各种状态和每个状态下可以调用的方法。

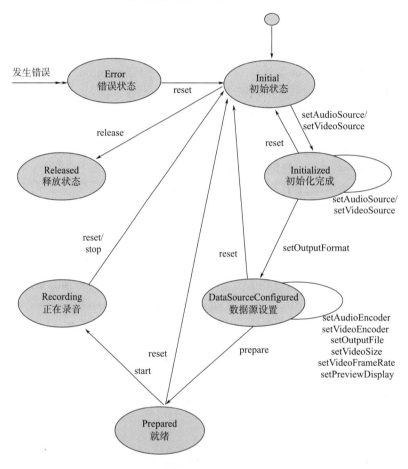

图15-3　MediaRecorder的状态机图

MediaRecorder类提供音频和视频录制的功能，这里只介绍与它的音频录制相关的API，视频录制相关的API将在15.3中进行介绍。MediaRecorder类中录制音频的主要API如表15-1所示，使用这几个主要的API就可以构建出功能较完善的音频录制应用。

表15-1　音频录制的主要API

方法名	描述
MediaRecorder.setAudioSource	设置录制的音频源
MediaRecorder.setOutputFormat	设置输出文件的格式
MediaRecorder.setOutputFile	设置音频数据的存储文件

第15章 在线音视频的应用与管理

续表

方法名	描述
MediaRecorder.setAudioEncoder	设置音频数据的编码器
MediaRecorder.prepare	开始录制前准备MediaRecorder对象
MediaRecorder.start	开始录制并存储音频数据到指定的文件
MediaRecorder.stop	停止录制
MediaRecorder.release	释放MediaRecorder对象占用的资源

录制音频的过程大致遵循以下9个步骤。

① 创建android.media.MediaRecorder对象实例，以后所有的工作都是围绕该对象展开。

```
MediaRecorder mRecorder = new MediaRecorder();
```

② 设置音频录制时采用的音频源(audio source)。这里，通过调用前面介绍的android.media.MediaRecorder对象的MediaRecorder.setAudioSource方法完成。MediaRecorder.AudioSource是MediaRecorder的内部类，它主要定义智能手机常用的音频源资源，MediaRecorder.AudioSource.MIC是最常用的音频源，代表手机的麦克风外设。除了MediaRecorder.AudioSource.MIC以外，还提供VOICE_CALL、VOICE_DOWNLINK、VOICE_UPLINK音频源，可以通过这些音频源进行语音通话的录制。

```
mRecorder.setAudioSource(MediaRecorder.AudioSource.MIC);
```

③ 设置输出音频数据的存储文件格式。这里，通过调用前面介绍的MediaRecorder.setOutputFormat方法完成。Android平台支持的音频文件格式由MediaRecorder.OutputFormat内部类定义。以下为Android支持的主要文件格式。

• MediaRecorder.OutputFormat.AMR_NB：该常量表示输出文件是AMR-NB格式的语音文件，主要对人声进行编码。

• MediaRecorder.OutputFormat.MPEG_4：该常量表示输出文件是MPEG-4格式的多媒体文件，其中，可能同时包含音频和视频信息。

• MediaRecorder.OutputFormat.THREE_GPP：该常量表示输出文件是3GPP格式的文件，其中，可能同时包含音频和视频信息。

```
mRecorder.setOutputFormat(MediaRecorder.OutputFormat.AMR_NB);
```

④ 指定输出音频数据存放的文件。这里，通过调用前面介绍的MediaRecorder. setOutputFile方法完成。该方法可以接收两种参数：文件描述符(FileDescriptor)和文件路径字符串。

```
FileDescriptor fd = ...;
mRecorder.setOutputFile(fd);
```

或者

```
String mFileName = ...;
mRecorder.setOutputFile(mFileName);
```

⑤ 设置音频编码器。这里，通过调用前面介绍的MediaRecorder. setAudioEncoder方法完成。Android平台支持的音频编码器由MediaRecorder.AudioEncoder内部类定义。最常用的音频编码器是MediaRecorder.AudioEncoder.AMR_NB。AMR_NB是自适应多速率窄带音频编

315

解码器。该编解码器主要针对语音数据进行优化，使其不适应语音之外的其他音频信息。

```
mRecorder.setAudioEncoder(MediaRecorder.AudioEncoder.AMR_NB);
```

⑥ 调用 MediaRecorder.prepare 方法，准备工作就绪，可以开始录制音频。

```
mRecorder.prepare();
```

⑦ 调用 MediaRecorder.start 方法，开始录制。

```
mRecorder.start();
```

⑧ 录制完成之后，要停止录制，需要调用 MediaRecorder.stop 方法。

```
mRecorder.stop();
```

⑨ 最后，还需要调用 MediaRecorder.release 方法，释放所占用的资源。到这里，整个录制过程就完成了。

【实例 15-2】下面给出一个完整的录制音频的程序示例，详细介绍设置、启动、停止录制音频的过程。在 Chapter15 中新建 Module——MediaRecorderDemo，布局文件同【实例 15-1】相同，修改 MainActivity.java 文件如下：

```java
package cn.androidstudy.chapter15;
import android.Manifest;
import android.content.pm.PackageManager;
import android.media.MediaPlayer;
import android.media.MediaRecorder;
import android.os.Build;
import android.os.Environment;
import android.support.annotation.NonNull;
import android.support.v7.app.AppCompatActivity;
import android.os.Bundle;
import android.util.Log;
import android.view.View;
import android.widget.Button;
import java.io.IOException;
import java.util.ArrayList;
import java.util.List;
public class MainActivity extends AppCompatActivity {
    private static final String TAG = "MainActivity";
    private MediaPlayer mPlayer = null;
    private MediaRecorder mRecorder = null;
    private static String mFileName = null;
    Button createRecording, playRecording;
    boolean isRecording,isPlaying;
    boolean isGranted =false;
    //需要动态申请的权限
```

```java
        String[] permissions = new String[]{Manifest.permission.RECORD_AUDIO
                , Manifest.permission.WRITE_EXTERNAL_STORAGE};
        //存储未获取的权限
        List<String> mPermissionList = new ArrayList<>();
        @Override
        protected void onCreate(Bundle savedInstanceState) {
            super.onCreate(savedInstanceState);
            setContentView(R.layout.activity_main);
            grantedAndRequest();
            createRecording = (Button) this.findViewById(R.id.button);
            playRecording = (Button) this.findViewById(R.id.button2);
            isRecording = false;
            isPlaying = false;
        }
        //判断是否授权，如未授权，则申请授权
        private void grantedAndRequest() {
            if (Build.VERSION.SDK_INT >= Build.VERSION_CODES.M) {//针对Android 6.0及之后版本
                mPermissionList.clear();
                for (int i = 0; i < permissions.length; i++) {
                    if (checkSelfPermission( permissions[i]) != PackageManager.PERMISSION_GRANTED) {
                        mPermissionList.add(permissions[i]);
                    }
                }
                if (mPermissionList.isEmpty()) {//未授予的权限为空，表示都授予了
                    isGranted = true;
                } else {//请求权限方法
                    String[] permissions = mPermissionList.toArray(new String[mPermissionList.size()]);//将List转为数组
                    requestPermissions(permissions, 1);
                }
            }else{
                isGranted = true;
            }
        }
        @Override
        public void onRequestPermissionsResult(int requestCode, @NonNull String[] permissions, @NonNull int[] grantResults) {
            super.onRequestPermissionsResult(requestCode, permissions, grantResults);
```

```java
        if(requestCode==1){
            isGranted = true;
            for (int i = 0; i < grantResults.length; i++) {
                if (grantResults[i] != PackageManager.PERMISSION_GRANTED) {
                    isGranted = false;
                }
            }
        }
    }
    public void record(View view){
        if(isGranted) {
            mFileName = Environment.getExternalStorageDirectory().getAbsolutePath();
            mFileName += "/audiorecord.3gp";
        }else {
            return;
        }
        if(isRecording){//正在录音
            stopRecording();
            createRecording.setText("录音");
            isRecording = false;
        }else {
            startRecording();
            createRecording.setText("停止");
            isRecording = true;
        }
    }
    private void startRecording() {
        mRecorder = new MediaRecorder();
        mRecorder.setAudioSource(MediaRecorder.AudioSource.MIC);
        mRecorder.setOutputFormat(MediaRecorder.OutputFormat.THREE_GPP);
        mRecorder.setOutputFile(mFileName);
        mRecorder.setAudioEncoder(MediaRecorder.AudioEncoder.AMR_NB);
        try {
            mRecorder.prepare();
        } catch (IOException e) {
            Log.e(TAG, "prepare() failed");
        }
        mRecorder.start();
    }
    private void stopRecording() {
        mRecorder.stop();
```

```java
            mRecorder.release();
            mRecorder = null;
        }
    public void play(View view){
        if(!isGranted) return;
        if(isPlaying){
            stopPlaying();
            playRecording.setText("播放");
            isPlaying = false;
        }else {
            startPlaying();
            playRecording.setText("停止");
            isPlaying = true;
        }
    }
    private void startPlaying() {
        mPlayer = new MediaPlayer();
        try {
            mPlayer.setDataSource(mFileName);
            mPlayer.prepare();
            mPlayer.start();
        } catch (IOException e) {
            Log.e(TAG, "prepare() failed");
        }
    }
    private void stopPlaying() {
        mPlayer.release();
        mPlayer = null;
    }
    @Override
    public void onPause() {
       super.onPause();
        if (mRecorder != null) {
           mRecorder.release();
           mRecorder = null;
        }
      if (mPlayer != null) {
         mPlayer.release();
         mPlayer = null;
      }
    }
}
```

由于进行录音需要授予"android.permission.RECORD_AUDIO"权限，录音文件存放到SD卡，故需要授予"android.permission.WRITE_EXTERNAL_STORAGE"权限，故在AndroidManifest.xml文件添加如下两个权限：

```
<uses-permission android:name="android.permission.WRITE_EXTERNAL_STORAGE"/>
<uses-permission android:name="android.permission.RECORD_AUDIO"/>
```

由于Android6.0之后危险权限需要动态申请，请读者结合前面的介绍，认真理解并掌握多个权限的申请。

15.2 应用音频

Android平台提供了强大的媒体播放功能，它支持相当广泛的音频格式，例如MP3（.mp3）、3GPP（.3gp）、OGG（.ogg）和WAVE（.wav）等。本节将首先介绍Android平台支持的常见音频格式，然后再介绍播放音频的两种主要方法。同音频录制类似，第一种方法，使用Intent调用系统功能进行播放，该方法简单方便但灵活性不足；第二种方法，使用MediaPlayer播放音频，该方法使用难度适中并且灵活性较强。

15.2.1 常见的音频格式

Android支持多种音频格式和编解码器，下面介绍几种常见的音频格式。

① AMR：自适应多速率编解码器，包括AMR窄带AMR-NB和AMR宽带AMR-WB，文件扩展名是".3gp"（audio/3gpp）或".amr"（audio/amr）。AMR是3GPP使用的基本音频编解码标准。AMR主要用于手机上的语音通话应用，并得到手机厂商的广泛支持。该编码标准适用于简单的语音编码，不适用于处理更复杂的音频数据，例如音乐等。

② AAC：全称是Advanced Audio Coding，是一种专为声音数据设计的文件压缩格式，与MP3不同，它采用了全新的算法进行编码，更加高效，具有更高的"性价比"。利用AAC格式，可使人感觉声音质量没有明显下降的前提下，资源占用更少。Android除了支持AAC外，还支持新添加到AAC规范中的高效AAC(High Efficiency AAC)格式。

③ MP3：是一种音频压缩技术，其全称是动态影像专家压缩标准音频层面3（Moving Picture Experts Group Audio Layer Ⅲ），简称为MP3。它被设计用来大幅度地降低音频数据量。利用MP3技术，将音乐以1∶10甚至1∶12的压缩率压缩成容量较小的文件，而对于大多数用户来说，重放的音质与最初的不压缩音频相比没有明显的下降。它是在1991年由位于德国埃尔朗根的研究组织Fraunhofer-Gesellschaft的一组工程师发明和标准化的。用MP3形式存储的音乐就叫作MP3音乐，能播放MP3音乐的机器就叫作MP3播放器。MP3是目前互联网上使用最广泛的音频编解码器之一。

④ OGG：全称是OGG Vorbis，是一种新的音频压缩格式，类似于MP3的音乐格式。但有一点不同的是，它是完全免费、开放和没有专利限制的。OGG Vorbis有一个特点是支持多声道。Vorbis是这种音频压缩机制的名字，而OGG则是一个计划的名字，该计划意图设计一

个完全开放性的多媒体系统。OGG Vorbis文件的扩展名是".ogg"。这种文件的设计格式是非常先进的,创建的OGG文件可以在未来的任何播放器上播放,因此,这种文件格式可以不断地进行大小和音质的改良,而不影响旧有的编码器或播放器。

15.2.2 使用Intent播放音频

播放音频最简单的方法是使用Intent调用系统已有的音乐播放应用程序提供的播放功能。Android平台默认提供一个音乐播放应用程序,当然也可以使用第三方提供的其他音乐播放应用程序。使用Intent播放音频的基础代码如下所示:

```
Intent audio_playing_intent = 
        new Intent(android.content.Intent.ACTION_VIEW);
audio_playing_intent.setDataAndType(audioFileURI, "audio/mpeg");
startActivity(audio_playing_intent);
```

其中,android.content.Intent.ACTION_VIEW常量表示启动音乐播放应用程序的动作意图。然后设置Intent对象的参数:播放音乐的URI和MIME类型。

调用startActivity(audio_playing_intent)可以启动音乐播放应用程序进行音频播放,但是控制权已经交给了音乐播放应用程序。

MIME代表多用途Internet邮件扩展(Multipurpose Internet Mail Extension),最初被用于帮助电子邮件客户端发送和接收附件。现在它的使用范围已经超出电子邮件,扩展到了许多其他通信协议,包括HTTP或Web服务。当解析一个Intent时,Android使用MIME类型来帮助确定应该处理该Intent的应用程序。每种文件类型都有一个特定的MIME类型。使用至少两部分(由斜杠分隔开)来指定类型。第一部分是一般的类型,例如"audio";第二部分是具体的类型,例如"mpeg"。一般类型"audio"和具体类型"mpeg"构成一个MIME类型"audio/mpeg",这通常用于MP3文件的MIME类型。

☞【实例15-3】下面给出一个较完整的示例,提供音频播放的功能。示例运行主界面见图15-4,单击按钮则开始播放MP3,如图15-5所示。

图15-4　播放音频　　　　　　　　　　图15-5　播放MP3

在Chapter15中，新建一个Module——AudioPlayerDemo，在布局文件activity_main.xml文件中添加一个Button，并给按钮添加单击事件，修改后的MainActivity.java文件如下：

```java
package cn.androidstudy.chapter15;
import android.Manifest;
import android.content.Intent;
import android.content.pm.PackageManager;
import android.net.Uri;
import android.os.Build;
import android.os.Environment;
import android.support.annotation.NonNull;
import android.support.v7.app.AppCompatActivity;
import android.os.Bundle;
import android.view.View;
import java.io.File;
import java.util.ArrayList;
import java.util.List;
public class MainActivity extends AppCompatActivity {
    boolean isGranted =false;
    //需要动态申请的权限
    String[] permissions = new String[]{Manifest.permission.RECORD_AUDIO
            , Manifest.permission.WRITE_EXTERNAL_STORAGE};
    //存储未获取的权限
    List<String> mPermissionList = new ArrayList<>();
    @Override
    protected void onCreate(Bundle savedInstanceState) {
        super.onCreate(savedInstanceState);
        setContentView(R.layout.activity_main);
        grantedAndRequest();
    }
    //判断是否授权，如未授权，则申请授权
    private void grantedAndRequest() {
        if (Build.VERSION.SDK_INT >= Build.VERSION_CODES.M) {//针对Android6.0及之后版本
            mPermissionList.clear();
            for (int i = 0; i < permissions.length; i++) {
                if (checkSelfPermission( permissions[i]) != PackageManager.PERMISSION_GRANTED) {
                    mPermissionList.add(permissions[i]);
                }
            }
            if (mPermissionList.isEmpty()) {//未授予的权限为空，
```

表示都授予了

```
            isGranted = true;
        } else {//请求权限方法
            String[] permissions = mPermissionList.toArray(new String[mPermissionList.size()]);//将List转为数组
            requestPermissions(permissions, 1);
        }
    }else{
        isGranted = true;
    }
}
@Override
public void onRequestPermissionsResult(int requestCode, @NonNull String[] permissions, @NonNull int[] grantResults) {
    super.onRequestPermissionsResult(requestCode, permissions, grantResults);
    if(requestCode==1){
        isGranted = true;
        for (int i = 0; i < grantResults.length; i++) {
            if (grantResults[i] != PackageManager.PERMISSION_GRANTED) {
                isGranted = false;
            }
        }
    }
}
public void playByIntent(View view){
    Intent intent = new Intent(android.content.Intent.ACTION_VIEW);

    File sdcard = Environment.getExternalStorageDirectory();
    File audioFile = new File(sdcard.getPath()
            + "/Download/hjbj.mp3");
    intent.setDataAndType(Uri.fromFile(audioFile), "audio/mp3");
    startActivity(intent);
}
```

在Android6.0及之前的版本中运行没有问题,但是从Android 7.0开始,不再允许在app中把file:// Uri暴露给其他app,否则应用会抛出"File Uri Exposed Exception"。原因在于Google认为使用file:// Uri存在一定的风险。例如,文件是私有的,其他app无法访问该文件,或者其他app没有申请READ_EXTERNAL_STORAGE运行的权限。解决方案是使用FileProvider生成content:// Uri来替代file:// Uri。

① 声明FileProvider。

首先在清单文件中申明FileProvider：

```xml
<manifest>
  ...
  <application>
    ...
    <provider
      android:name="android.support.v4.content.FileProvider"
      android:authorities="cn.androidstudy.fileprovider"
      android:exported="false"
      android:grantUriPermissions="true">
      <meta-data
        android:name="android.support.FILE_PROVIDER_PATHS"
        android:resource="@xml/file_paths" />
    </provider>
    ...
  </application>
</manifest>
```

其中，android:name是固定写法；android:authorities可自定义，是用来标识该<provider>的唯一标识，建议结合包名来保证authority的唯一性；android:exported必须设置成false，否则运行时会报错"java.lang.SecurityException: Provider must not be exported"；android:grantUriPermissions用来控制共享文件的访问权限；<meta-data>节点中的android:resource指定了共享文件的路径，此处的file_paths即是该<provider>对外提供文件的目录的配置文件，存放在res/xml下。

② 添加file_paths.xml文件。

文件格式如下：

```xml
<paths>
  <files-path name="name" path="path"/>
  ...
</paths>
```

其中，根元素<paths>是固定的，内部元素可以是以下节点：
- <files-path name="name" path="path" /> 对应getFilesDir。
- <cache-path name="name" path="path" /> 对应getCacheDir。
- <external-path name="name" path="path" /> 对应Environment.getExternalStorageDirectory。
- <external-files-path name="name" path="path" /> 对应getExternalFilesDir。
- <external-cache-path name="name" path="path" /> 对应getExternalCacheDir。

将MP3文件拖放到模拟器中，默认存储到SD卡的Download文件夹，定义file_paths.xml文件如下：

```xml
<?xml version="1.0" encoding="utf-8"?>
<paths>
  <external-path name="download" path="Download"/>
</paths>
```

③ 在 Java 代码中使用 FileProvider。
修改 playByIntent 方法如下：

```java
public void playByIntent(View view){
    Intent intent = new Intent(android.content.Intent.ACTION_VIEW);
    File sdcard = Environment.getExternalStorageDirectory();
    File audioFile = new File(sdcard.getPath()
            + "/Download/hjbj.mp3");
    Uri data;
    // 判断版本大于等于7.0
    if (Build.VERSION.SDK_INT >= Build.VERSION_CODES.N) {
        // "cn.androidstudy.fileprovider"即是在清单文件中配置的authorities
        data = FileProvider.getUriForFile(this, "cn.androidstudy.fileprovider", audioFile);
        // 给目标应用一个临时授权
        intent.addFlags(Intent.FLAG_GRANT_READ_URI_PERMISSION);
    } else {
        data = Uri.fromFile(audioFile);
    }
    intent.setDataAndType(data, "audio/mp3");
    startActivity(intent);
}
```

15.2.3 使用 MediaPlayer 播放音频

在介绍了如何使用 Intent 播放音频后，接下来介绍如何使用 MediaPlayer 类来播放音频。MediaPlayer 类是 Android SDK 提供的播放音频和视频的功能类，这里仅使用其音频播放功能，使用它可以建立功能更加完善的音频播放应用。

MediaPlayer 播放音频的最简单情况是播放与应用程序本身一起打包的音频文件。音频文件放置在应用程序的原始资源中。具体操作是在项目的 res 文件夹中创建一个新文件夹，命名为 raw，把音频文件放入该文件夹中，ADT 将自动更新 R.java 文件（位于 gen 文件夹中），为音频文件生成资源 ID，使用 R.raw.file_name_without_extension 语法访问音频文件。

播放与应用程序一起打包的音频文件非常简单。使用 MediaPlayer 类的静态方法 create 实例化一个 MediaPlayer 对象，传入上下文 this 以及音频文件的资源 ID。

```java
MediaPlayer mediaPlayer =
    MediaPlayer.create(this, R.raw.audio_file_name_without_extension);
```

由于调用 MediaPlayer 的静态方法 create 创建 MediaPlayer 对象成功后，系统自动调用 prepare 方法，不再需要手动调用，MediaPlayer 对象已经处于 Prepared 状态，因此，只需调用 MediaPlayer 对象的 start 方法即可播放音频文件。

```
mediaPlayer.start();
```

　　MediaPlayer类内部维护一个状态机来管理播放音频和视频中的各种状态，该状态机如图15-6所示(摘自Android API参考手册)，它描述了音频和视频播放过程中的各种状态和每个状态下可以调用的方法。

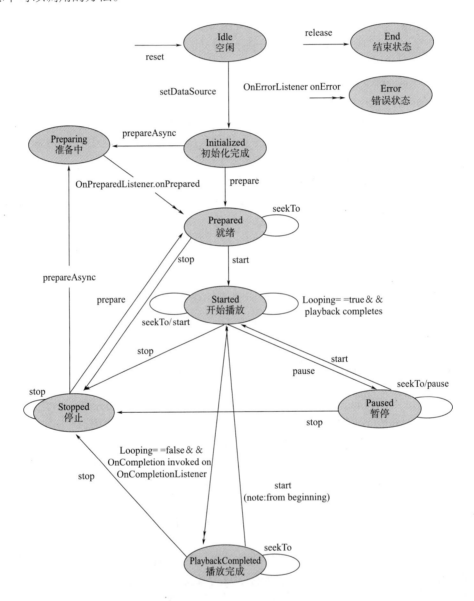

图15-6　MediaPlayer的状态机图

　　这里需要着重强调的是MediaPlayer是基于状态的。当写代码时，必须始终注意MediaPlayer所处的状态，因为MediaPlayer的方法都有其可以正常执行的有效状态。如果在一个错误的状态执行一个方法时，系统会抛出异常或者产生不可预料的行为。

　　前面所述的MediaPlayer的状态机图明确指出哪些方法可以把MediaPlayer从一个状态迁移到另一个状态。例如，当新建一个MediaPlayer对象时(使用new操作)，它处于Idle状态。

第15章 在线音视频的应用与管理

在 Idle 状态时，通过调用 setDataSource 方法初始化 MediaPlayer 对象，迁移到 Initialized 状态。之后，使用 prepare 或 prepareAsync 方法准备 MediaPlayer 对象。当 MediaPlayer 对象准备就绪时，它进入 Prepared 状态，这时可以调用 start 方法来播放媒体文件。此时，MediaPlayer 对象处于 Started 状态，正如图 15-6 所示，可以通过调用 start、pause 和 seekTo 方法，在 Started、Paused 和 PlaybackCompleted 三个状态之间进行迁移。当调用 stop 方法停止播放后，就不能再调用 start 方法播放媒体文件了，除非再次调用 prepare 方法准备 MediaPlayer 对象。

再强调一次，编写媒体播放代码时，一定注意 MediaPlayer 对象的状态，因为在错误的状态调用方法会造成许多错误。

使用 MediaPlayer 播放时，需要注意不要在应用程序的 UI 主线程中准备 MediaPlayer。调用 prepare 方法通常会花费一定时间，因为它需要获取并解码媒体数据。因此，只要是任何需要花费一定时间执行的方法，都不要在应用程序的主 UI 线程中调用。如果那样做的话，将会挂起 UI 主线程直到该方法执行完毕，这是一个非常差的用户体验，会造成应用程序不响应（Application Not Responding）错误。即使是很短的媒体数据，加载可能很快，但是记住任何花费超过 100ms 的操作都会造成 UI 界面发生明显的停顿并给用户特别明显的印象——应用程序响应很慢。

为了避免挂起 UI 主线程，创建一个新的线程准备 MediaPlayer，当准备就绪时，通知主 UI 线程。其实，我们不用自己编写新的线程逻辑，MediaPlayer 本身提供了一个非常方便的方法可以完成该任务，即 prepareAsync 方法。该方法在后台准备媒体数据，准备就绪后立刻返回。当媒体数据准备好后，MediaPlayer.OnPreparedListener 接口的 onPrepared 回调方法会自动被调用，以便继续后续处理。我们可以通过 MediaPlayer 的 setOnPreparedListener 方法注册监听器。

MediaPlayer 可能会耗尽系统资源，因此，我们应该采取措施避免 MediaPlayer 一直占用系统资源。当使用 MediaPlayer 播放完毕时，我们应该总是调用 release 方法，确保任何分配给它的系统资源被合适地释放。例如，如果应用的 Activity 收到 onStop 回调方法的调用，我们必须在其中释放 MediaPlayer，因为当该 Activity 不和用户进行交互时，MediaPlayer 还占用系统资源已经没有意义，除非正在后台播放。当 Activity 被继续或重新启动时，我们需要创建一个新的 MediaPlayer，并再次准备它，之后才能继续播放。

```
mediaPlayer.release();
mediaPlayer = null;
```

下面代码实现自定义 MediaPlayer 进行音频资源的播放，为了简单起见，在【实例 15-3】的按钮下面添加一个 Button，并在 MainActivity.java 中添加如下代码：

```java
public void playByMediaplayer(View view){
    mediaPlayer = new MediaPlayer();
    File sdcard = Environment.getExternalStorageDirectory();
    File audioFile = new File(sdcard.getPath()
            + "/Download/hjbj.mp3");
    try {
        mediaPlayer.setDataSource(audioFile.getPath());
        mediaPlayer.prepare();
        mediaPlayer.start();
    } catch (Exception e) {
```

```
        Toast.makeText(this, "", Toast.LENGTH_SHORT).show();
    }
}
public void onStop() {
    super.onStop();
    mediaPlayer.stop();
    mediaPlayer.release();
}
```

15.3 录制视频

Android 程序开发人员也要对录制视频有所掌握。本章前面向读者介绍了音频录制功能，接下来本节将向读者主要讲解录制视频的两种主要方法。每种方法都有它们各自的使用范围，第一种方法，使用 Intent 调用系统摄像头进行录制，该方法简单方便但灵活性不足；第二种方法，使用 MediaRecorder 录制视频，该方法使用难度适中并且灵活性较强。

15.3.1 使用 Intent 录制视频

正如音频录制一样，Android 平台上录制视频的最简单方式是通过 Intent 来调用系统已有的摄像应用程序。Android 平台默认提供一个 Camera 应用程序，当然也可以使用第三方提供的其他视频录制应用程序。使用 Intent 录制视频的基础代码如下所示：

```
Intent video_recording_intent =
        new Intent(MediaStore.ACTION_VIDEO_CAPTURE);
startActivity(video_recording_intent);
```

其中，MediaStore.ACTION_VIDEO_CAPTURE 常量表示启动视频录制应用程序的动作意图。调用 startActivity(video_recording_intent) 可以启动视频录制应用程序进行视频录制，但是控制权已经交给了视频录制应用程序，在录制完视频后，视频数据无法返回给调用程序。

调用 startActivityForResult(video_recording_intent) 可以把录制的视频数据返回给调用程序。用户 Android 手机上可能有多个视频应用程序将该字符串常量注册为一个 Intentf 过滤器，这时系统会弹出提示，以供用户选择使用哪个应用程序来执行该项操作。

```
Intent video_recording_intent =
        new Intent(MediaStore.ACTION_VIDEO_CAPTURE);
startActivityForResult (video_recording_intent, VIDEO_CAPTURED);
```

其中，VIDEO_CAPTURED 是我们定义的常量，当 Camera 应用程序调用 onActivityResult 方法将结果返回给 Activity 时，可以使用该常量进行确认。

onActivityResult方法返回给Activity的Intent中包含录制的视频文件的URI，该文件由Camera应用程序创建。

```
protected void onActivityResult(int requestCode, int resultCode, Intent video_data){
        if (requestCode == VIDEO_CAPTURED && resultCode == RESULT_OK) {
            Uri videoFileUri = video_data.getData();
        }
    }
```

【实例15-4】下面给出一个较完整的示例，提供使用Intent录制视频的功能。

新建一个Module——VideoCaptureDemo，在布局文件中添加两个按钮和一个VideoView，示例运行时界面如图15-7所示。

图15-7　录制视频

修改MainActivity.java文件如下：

```
package cn.androidstudy.chapter15;
import android.content.Intent;
import android.net.Uri;
import android.support.v7.app.AppCompatActivity;
import android.os.Bundle;
import android.view.View;
import android.widget.Button;
import android.widget.VideoView;
public class MainActivity extends AppCompatActivity implements View.OnClickListener{
```

```java
//创建Activity的VIDEO_CAPTURED常量，将在调用onActivityResult时返回
public static int VIDEO_CAPTURED = 1;
/*本Activity的两个按钮，一个用于触发Intent，即recordVideoBtn，另一个用于
 * 播放录制的视频，即playVideoBtn
 */
Button recordVideoBtn;
Button playVideoBtn;
//使用VideoView播放视频
VideoView videoView;
Uri videoFileUri;
@Override
protected void onCreate(Bundle savedInstanceState) {
    super.onCreate(savedInstanceState);
    setContentView(R.layout.activity_main);
    recordVideoBtn = (Button) this
            .findViewById(R.id.button);
    playVideoBtn = (Button) this.findViewById(R.id.button2);
    recordVideoBtn.setOnClickListener(this);
    playVideoBtn.setOnClickListener(this);
    //初始化时设置playVideoBtn为不可用状态，因为这时还没有录制好视频
    playVideoBtn.setEnabled(false);
    videoView = (VideoView) this.findViewById(R.id.videoView);
}

@Override
public void onClick(View view) {
    if (view == recordVideoBtn) {
        //按下录制按钮，开始录制视频
        Intent captureVideoIntent = new Intent(
                android.provider.MediaStore.ACTION_VIDEO_CAPTURE);
        startActivityForResult(captureVideoIntent, VIDEO_CAPTURED);
    } else if (view == playVideoBtn) {
        //按下播放按钮，播放录制的视频
        videoView.setVideoURI(videoFileUri);
        videoView.start();
    }
}
/*当Camera应用程序返回时，将调用onActivityResult回调方法。
    首先检查requestCode是否为传入startActivityForResult的值VIDEO_
CAPTURED和resultCode是否为常量RESULT_OK，然后获取录制的视频文件的URI，接
着启用playVideoBtn按钮，从而用户可以单击它来播放视频。
 */
```

第15章 在线音视频的应用与管理

```
    protected void onActivityResult(int requestCode, int
resultCode, Intent data) {
        if (resultCode == RESULT_OK && requestCode == VIDEO_CAPTURED) {
            videoFileUri = data.getData();
            playVideoBtn.setEnabled(true);
        }
    }
}
```

15.3.2 使用MediaRecorder录制视频

在介绍了使用Intent录制视频后,接下来介绍如何使用MediaRecorder来录制视频。MediaRecorder是Android SDK提供的录制音频和视频的功能类,使用它可以建立功能更加完善的视频录制应用。MediaRecorder内部维护一个状态机来管理录制音频和视频中的各种状态,该状态机如图15-3所示,它描述了音频和视频录制过程中的各种状态和每个状态下可以调用的方法。

为了将MediaRecorder用于视频录制,必须采用与音频录制相似的步骤,同时加上与视频相关的特殊步骤。

首先需要创建MediaRecorder对象,然后依次进行后续的操作。

```
MediaRecorder video_recorder = new MediaRecorder();
```

(1) 设置音频和视频源

创建MediaRecorder对象后,需要设置音频和视频源。可以使用setAudioSource方法设置音频源,传入一个想要使用的音频源常量,设置方法已在15.1节音频录制中介绍过,此处不再重复叙述。为了设置视频源,可以使用setVideoSource方法。可能的视频源的值定义在MediaRecorder.VideoSource类的常量上,其中只包含两个常量:CAMERA和DEFAULT。其实这两个常量表示含义一样,都是指设备上的主摄像头。

```
video_recorder.setVideoSource(MediaRecorder.VideoSource.DEFAULT);
```

(2) 输出格式

设置音频和视频源之后,可以使用MediaRecorder的setOutputFormat方法设置输出格式。传入要使用的格式。

```
video_recorder.setOutputFormat(MediaRecorder.OutputFormat.
DEFAULT);
```

可能的格式定义在MediaRecorder.OutputFormat中的常量上。
- DEFAULT:使用默认的输出格式。默认的输出格式根据设备的不同而不同。
- MPEG_4:指定音频和视频被录制在一个MPEG-4格式的文件中,扩展名是".mp4"。MPEG-4文件通常包含H.264、H.263或MPEG-4 Part 2编码的视频,以及AAC或MP3编码的音频。MPEG-4文件被广泛用于许多其他在线视频技术或消费电子设备上。
- THREE_GPP:指定音频和视频将被录制到一个3GPP格式的文件中,扩展名是

331

".3gp"。3GPP文件通常包含使用H.264、H.263或MPEG-4 Part 2编码的视频和使用AMR或AAC编码的音频。

（3）设置音频和视频编解码器

设置输出格式之后，需要指定想要使用的音频和视频编解码器。可以使用MediaRecorder的setVideoEncoder方法设置视频编解码器。

```
video.setVideoEncoder(MediaRecorder.VideoEncoder.DEFAULT);
```

可以使用的编解码器定义在MediaRecorder.VideoEncoder中的常量上：

- DEFAULT：使用默认的视频编解码器。多数情况下是H.263，Android设备上必须唯一支持的编解码器。
- H263：指定H.263为视频编解码器。H.263是在1995年发布的编解码器，专门为低比特率视频传输而开发。它是许多早期Internet视频技术的基础，如Flash和RealPlayer早期使用的技术。在Android中它是必须支持的编码，因此可以可靠地使用。
- H264：指定H.264为视频编解码器。H.264是当前最先进的编解码器，广泛应用于各种技术，从BlueRay到Flash。
- MPEG_4_SP：指定视频编解码器为MPEG-4 SP。MPEG-4 SP是MPEG-4 Part 2 Simple Profile，它发布于1999年，为需要低比特率视频且不需要大的处理器能力的技术而开发。

对于音频部分，可以使用setAudioEncoder方法设置音频编解码器，设置方法已在15.1节音频录制中介绍过，此处不再重复叙述。

（4）设置音频和视频比特率

使用MediaRecorder的setVideoEncodingBitRate方法设置视频编码比特率。视频的低比特率设置在256000位/秒（256kbps）范围之内，而高比特率在3000000位/秒（3Mbps）范围之内。

```
video_recorder.setVideoEncodingBitRate(150000);
```

使用MediaRecorder的setAudioEncodingBitRate方法设置音频编码比特率。8000位/s（8kbps）是一个非常低的比特率，适合于在慢速网络上实时传输的音频，而196000位/秒（196kbps）在MP3文件中很常见。

```
video_recorder.setAudioEncodingBitRate (8000);
```

（5）设置音频采样率

和编码比特率一样，音频采样率对于音频的质量也非常重要。可以使用MediaRecorder的setAudioSamplingRate方法设置音频采样率。采样率以"Hz"为单位，表示每秒采样的数量。采样率越高，则在录制音频文件中可以表示的音频频率的范围越大。一个低端的采样率8000Hz适合于录制低质量的音频，而高端的采样率48000Hz可用于DVD和许多其他高质量的视频格式。

```
video_recorder. setAudioSamplingRate (8000);
```

（6）设置音频通道

可以使用setAudioChannels方法指定将要录制的音频的通道数量。目前，音频大都限制为大多数Android设备上的单一通道麦克风，因此使用一个以上的通道不会有益处。对于通道数量，一般是单声道为一个通道，立体声为两个通道。

第15章 在线音视频的应用与管理

```
video_recorder. setAudioChannels (1);
```

（7）设置视频帧速率

可以使用setVideoFrameRate方法来控制每秒录制的视频帧数。每秒12~15帧通常足以表示运动。具体使用的实际帧率取决于设备的能力。

```
video_recorder. setVideoFrameRate(15);
```

（8）设置视频大小

可以通过setVideoSize方法并传入宽高值来控制录制的视频的宽度和高度。标准大小的范围是176×144至640×480，许多设备甚至支持更高的分辨率。

```
video_recorder. setVideoSize (640, 480);
```

（9）设置最大文件大小

使用setMaxFileSize方法设置最大文件大小，单位为字节。

```
video_recorder. setMaxFileSize (10000000); //10MB
```

为了确定是否已达到最大文件大小，需要在Activity中实现MediaRecorder.OnInfoListener，同时在MediaRecorder中注册它，然后系统会调用onInfo方法检查what参数是否等于MediaRecorder.MEDIA_RECORDER_INFO_FILESIZE_REACHED。

（10）设置持续时间

使用setMaxDuration方法设置最长持续时间，单位为毫秒。

```
video_recorder. setMaxDuration(10000); //10s
```

为了确定是否已达到最长持续时间，需要在Activity中实现MediaRecorder.OnInfoListener，同时在MediaRecorder中注册它，当已达到最长持续时间时就会触发onInfo方法，检查what参数是否等于MediaRecorder.MEDIA_RECORDER_INFO_MAX_DURATION_REACHED。

（11）概要

MediaRecorder有一个setProfile方法，接受CamcorderProfile实例作为参数。使用该方法允许根据预设值设置整个配置变量集。其中，CamcorderProfile.QUALITY_LOW指低质量视频捕获设置，CamcorderProfile.QUALITY_HIGH指高质量视频捕获设置。

（12）输出文件

使用setOutputFile方法设置输出文件的位置。

```
video_recorder. setOutputFile ("/sdcard/video_recorded.mp4"); //10s
```

（13）预览表面

由于是视频录制，录制过程中需要看到画面。因此，需要为MediaRecorder指定一个取景器以预览要录制的图像。需要使用SurfaceView和SurfaceHolder.Callback，我们将在实例中进行介绍。

(14)准备录制

设置好 MediaRecorder 实例后,就可以使用 prepare 方法准备 MediaRecorder 了。

```
video_recorder.prepare();
```

(15)开始录制

MediaRecorder 实例准备好后,就可以开始录制了。

```
video_recorder.start();
```

(16)停止录制

录制过程中,可以通过 stop 方法停止录制。

```
video_recorder.stop();
```

(17)释放资源

最后,不要忘记需要释放占用的资源。

```
video_recorder.release;
```

【实例 15-5】下面代码实现 MediaRecorder 进行视频的录制。

在 Chapter15 中新建 Module——MediaRecorderVideo,修改 activity_main.xml 文件如下:

```xml
<?xml version="1.0" encoding="utf-8"?>
<android.support.constraint.ConstraintLayout
    xmlns:android="http://schemas.android.com/apk/res/android"
    xmlns:app="http://schemas.android.com/apk/res-auto"
    xmlns:tools="http://schemas.android.com/tools" android:layout_width="match_parent"
    android:layout_height="match_parent" tools:context="cn.androidstudy.chapter15.MainActivity">
    <SurfaceView
        android:id="@+id/surfaceView"
        android:layout_width="0dp"
        android:layout_height="0dp"
        android:layout_marginLeft="8dp"
        app:layout_constraintLeft_toLeftOf="parent"
        android:layout_marginRight="8dp"
        app:layout_constraintRight_toRightOf="parent"
        app:layout_constraintHorizontal_bias="0.0"
        app:layout_constraintBottom_toBottomOf="parent"
        android:layout_marginBottom="8dp"
        app:layout_constraintTop_toTopOf="parent"
        android:layout_marginTop="8dp" />
</android.support.constraint.ConstraintLayout>
```

在配置文件中添加权限：

```
<uses-permission android:name="android.permission.WRITE_EXTERNAL_STORAGE"/>
<uses-permission android:name="android.permission.RECORD_AUDIO"/>
<uses-permission android:name="android.permission.CAMERA"/>
```

修改MainActivity文件如下：

```java
package cn.androidstudy.chapter15;
import android.Manifest;
import android.content.pm.ActivityInfo;
import android.content.pm.PackageManager;
import android.media.CamcorderProfile;
import android.media.MediaRecorder;
import android.os.Build;
import android.support.annotation.NonNull;
import android.support.v7.app.AppCompatActivity;
import android.os.Bundle;
import android.util.Log;
import android.view.SurfaceHolder;
import android.view.SurfaceView;
import android.view.View;
import android.view.Window;
import android.view.WindowManager;
import java.io.IOException;
import java.util.ArrayList;
import java.util.List;
// 实现OnClickListener以便响应单击启动和停机录制按钮，实现
// SurfaceHolder.Callback以处理和图像预览的操作
public class MainActivity extends AppCompatActivity implements View.OnClickListener,
        SurfaceHolder.Callback {
    private static final String TAG = "MainActivity";
    MediaRecorder recorder;
    SurfaceHolder holder;
    // 该布尔量recording用于表示当前是否正在录制
    boolean recording = false;
    boolean isGranted =false;
    //需要动态申请的权限
    String[] permissions = new String[]{Manifest.permission.RECORD_AUDIO
            ,Manifest.permission.CAMERA
            , Manifest.permission.WRITE_EXTERNAL_STORAGE};
    //存储未获取的权限
```

```java
        List<String> mPermissionList = new ArrayList<>();
        @Override
        protected void onCreate(Bundle savedInstanceState) {
            super.onCreate(savedInstanceState);
            // 将以全屏和横屏模式运行
            requestWindowFeature(Window.FEATURE_NO_TITLE);
            getWindow().setFlags(WindowManager.LayoutParams.FLAG_FULLSCREEN,
                    WindowManager.LayoutParams.FLAG_FULLSCREEN);
            setRequestedOrientation(ActivityInfo.SCREEN_ORIENTATION_LANDSCAPE);
            setContentView(R.layout.activity_main);
            grantedAndRequest();
            // 实例化MediaRecorder对象
            recorder = new MediaRecorder();
            // 初始化MediaRecorder对象
            initRecorder();
            // 获取SurfaceView和SurfaceHolder的引用,同时注册该Activity为
            // SurfaceHolder.Callback
            SurfaceView cameraView = (SurfaceView) findViewById(R.id.surfaceView);
            holder = cameraView.getHolder();
            holder.addCallback(this);
            holder.setType(SurfaceHolder.SURFACE_TYPE_PUSH_BUFFERS);
            // 设置SurfaceView为可单击,注册监听器
            cameraView.setClickable(true);
            cameraView.setOnClickListener(this);
        }
        //判断是否授权,如未授权,则申请授权
        private void grantedAndRequest() {
            if (Build.VERSION.SDK_INT >= Build.VERSION_CODES.M) {//针对Android6.0及之后版本
                mPermissionList.clear();
                for (int i = 0; i < permissions.length; i++) {
                    if (checkSelfPermission( permissions[i]) != PackageManager.PERMISSION_GRANTED) {
                        mPermissionList.add(permissions[i]);
                    }
                }
                if (mPermissionList.isEmpty()) {//未授予的权限为空,表示都授予了
                    isGranted = true;
                } else {//请求权限方法
                    String[] permissions = mPermissionList.toArray(new String[mPermissionList.size()]);//将List转为数组
```

第15章 在线音视频的应用与管理

```java
            requestPermissions(permissions, 1);
        }
    }else{
        isGranted = true;
    }
}
@Override
    public void onRequestPermissionsResult(int requestCode, @NonNull String[] permissions, @NonNull int[] grantResults) {
        super.onRequestPermissionsResult(requestCode, permissions, grantResults);
        if(requestCode==1){
            isGranted = true;
            for (int i = 0; i < grantResults.length; i++) {
                if (grantResults[i] != PackageManager.PERMISSION_GRANTED) {
                    isGranted = false;
                }
            }
        }
    }
// 处理所有和MediaRecorder设置有关的操作
private void initRecorder() {
    recorder.setAudioSource(MediaRecorder.AudioSource.DEFAULT);
    recorder.setVideoSource(MediaRecorder.VideoSource.DEFAULT);
    CamcorderProfile cpHigh = CamcorderProfile
            .get(CamcorderProfile.QUALITY_HIGH);
    recorder.setProfile(cpHigh);
    recorder.setOutputFile("/sdcard/videocapture_example.mp4");
    recorder.setMaxDuration(50000); // 50s
    recorder.setMaxFileSize(5000000); // Approximately 5MB
}
@Override
public void surfaceCreated(SurfaceHolder surfaceHolder) {
    Log.v(TAG, "surfaceCreated");
    prepareRecorder();
}
private void prepareRecorder() {
    // 设置MediaRecorder的预览表面
    recorder.setPreviewDisplay(holder.getSurface());
    try {
        recorder.prepare();
    } catch (IllegalStateException e) {
```

337

```java
            e.printStackTrace();
            finish();
        } catch (IOException e) {
            e.printStackTrace();
            finish();
        }
    }

    @Override
    public void surfaceChanged(SurfaceHolder surfaceHolder, int i, int i1, int i2) {
    }
    // 当销毁表面时,如果正在录制,那么停止录制。这可能会在Activity不可见时发生。
    @Override
    public void surfaceDestroyed(SurfaceHolder surfaceHolder) {
        Log.v(TAG, "surfaceDestroyed");
        if (recording) {
            recorder.stop();
            recording = false;
        }
        recorder.release();
        finish();
    }
    @Override
    public void onClick(View view) {
        if(!isGranted){
            return;
        }
        if (recording) {
            recorder.stop();
            recording = false;
            Log.v(TAG, "Recording Stopped");
            initRecorder();
            prepareRecorder();
        } else {
            recording = true;
            recorder.start();
            Log.v(TAG, "Recording Started");
        }
    }
}
```

15.4 应用视频

目前随着iPhone和Android智能手机的流行，视频播放功能已经成为智能手机的标配，而且智能手机用户对视频的播放要求也越来越高。市场决定资源配置，Android程序开发人员对此也应该掌握并进一步研究。接下来本节将探讨Android的视频播放功能，将重点对读者介绍在Android上播放视频的各种方法，以及所支持的视频格式。

15.4.1 常见的视频格式

在深入介绍如何播放视频的具体机制之前，先来了解一下可以播放的视频类型。Android支持多种视频格式和编解码器，并且支持的类型还在不断增加，下面介绍几种常见的视频格式。

① H.263：自适应多速率编解码器，包括AMR窄带AMR-NB和AMR宽带AMR-WB，文件扩展名是".3gp"(audio/3gpp)或".amr"(audio/amr)。AMR是3GPP使用的基本音频编解码标准。AMR主要应用于手机上的语音通话，并得到手机厂商的广泛支持。该编码标准适用于简单的语音编码，不适用于处理更复杂的音频数据，例如音乐等。

② H.264 AVC：全称是Advanced Audio Coding，一种专为声音数据设计的文件压缩格式，与MP3不同，它采用了全新的算法进行编码，更加高效，具有更高的"性价比"。利用AAC格式，可使人感觉声音质量没有明显下降的前提下，资源占用更小。Android除了支持AAC外，还支持新添加到AAC规范中的高效AAC(High Efficiency AAC)格式。

③ MPEG-4 SP：是一种音频压缩技术，其全称是动态影像专家压缩标准音频层面3 (Moving Picture Experts Group Audio Layer 3)，简称为MP3。它被设计用来大幅度地降低音频数据量。利用 MPEG Audio Layer 3 的技术，将音乐以1∶10甚至1∶12的压缩率压缩成容量较小的文件，而对于大多数用户来说，重放的音质与最初的不压缩音频相比没有明显的下降。它是在1991年由位于德国埃尔朗根的研究组织Fraunhofer-Gesellschaft的一组工程师发明和标准化的。用MP3形式存储的音乐就叫作MP3音乐，能播放MP3音乐的机器就叫作MP3播放器。MP3是目前互联网上使用最广泛的音频编解码器之一。

④ VP8：全称是OGG Vorbis，是一种新的音频压缩格式，类似于MP3的音乐格式。但有一点不同的是，它是完全免费、开放和没有专利限制的。OGG Vorbis有一个特点是支持多声道。Vorbis是这种音频压缩机制的名字，而Ogg则是一个计划的名字，该计划意图设计一个完全开放的多媒体系统。Ogg Vorbis文件的扩展名是".ogg"。这种文件的设计格式是非常先进的。创建的". ogg"文件可以在未来的任何播放器上播放，因此，这种文件格式可以不断地进行大小和音质的改良，而不影响旧有的编码器或播放器。

15.4.2 使用Intent播放视频

正如本章已经探讨的播放音频的功能，Android可以很容易地通过使用Intent调用内置的媒体播放器来实现简单的视频播放功能。

为了通过创建Intent来调用内置的媒体播放器应用程序的播放功能，可以使用Intent.

ACTION_VIEW来构建Intent，并通过setDataAndType方法传入视频文件的URI和MIME类型。

使用Intent播放视频的核心代码如下：

```java
public void playByIntent(View view){
    Intent intent = new Intent(android.content.Intent.ACTION_VIEW);
    // Download "Test_Movie_iPhone.m4v from
    // http://www.mobvcasting.com/android/video/Test_Movie_iPhone.m4v
    // and save to the root of your device's SDCard.
    Uri data = Uri.parse(Environment.getExternalStorageDirectory().getPath()
            + "/Download/Test_Movie_iPhone.m4v");
    intent.setDataAndType(data, "video/mp4");
    startActivity(intent);
}
```

完整的代码请查看本书配套资源。

15.4.3 使用VideoView播放视频

VideoView是一个带有视频播放功能的视图，可以直接在布局中使用，使用起来非常简单。在15.3.1中已经使用VideoView进行了录制视频的播放，如果要播放SD卡的文件，步骤如下：

① 通过Uri类的parse方法获取视频文件的URI，该视频文件放置于设备SD卡的Download目录下。

```java
Uri videoUri = Uri.parse(Environment.getExternalStorageDirectory().getPath()
        + "/Download/Test_Movie_iPhone.m4v");
```

② 通过VideoView的setVideoView方法设置视频URI，并调用start方法播放该视频文件。

```java
videoView.setVideoURI(videoUri);
videoView.start();
```

然而，VideoView控制视频播放的功能相对较少，他只有start和pause方法。为了提供更多的控制，可以实例化一个MediaController，并通过setMediaController方法把它设置为VideoView的控制器。

默认的MediaController有后退(rewind)、暂停(pause)、播放(play)和快进(fast forward)按钮，还有一个清除和控制条组合空间，可以用来定位到视频中的任何一个位置。

☞【实例15-6】下面是对前面VideoView示例的更新，在通过setContentView方法设置内容视图后，通过调用setMediaController方法设置MediaController为VideoView的控制器。

```java
@Override
protected void onCreate(Bundle savedInstanceState) {
    super.onCreate(savedInstanceState);
```

第15章 在线音视频的应用与管理

```java
        setContentView(R.layout.activity_main);
        grantedAndRequest();
        videoView = (VideoView)findViewById(R.id.videoView);
        videoView.setMediaController(new MediaController(this));
    }
    public void playByVideoView(View view){
        Uri videoUri = Uri.parse(Environment.getExternalStorageDirectory().getPath()
                + "/Download/Test_Movie_iPhone.m4v");
        videoView.setVideoURI(videoUri);
        videoView.start();
    }
```

完整的代码请查看Chapter15/VideoPlayerDemo。

15.4.4 使用MediaPlayer播放视频

15.2.3节音频播放介绍了MediaPlayer类，同样，MediaPlayer类也可以通过类似的方式用于视频播放。与使用Intent和VideoView播放视频相比，将MediaPlayer用于视频播放能够为播放视频文件提供更大的灵活性。事实上，在通过Intent和VideoView播放视频时，处理视频播放的内部机制也有由MediaPlayer完成的。

使用SurfaceView+MediaPlayer，这种方式效果比较好，SurfaceView从Android 1.0就有了，十分好用。一般来说，UI对刷新都需要在UI线程中完成，但是，SurfaceView可以在非UI线程中完成刷新，这样一来就很方便了，比如在线播放，就不需要自己去写Handler来实现两个线程之间的通信了，直接可以在非UI线程中播放视频。

步骤如下：

① 调用mediaPlayer.setDataSource方法设置要播放的资源，可以是文件、文件路径或者URL。

② 调用MediaPlayer.setDisplay(holder)设置surfaceHolder，surfaceHolder可以通过surfaceview的getHolder方法获得。

③ 调用MediaPlayer.prepare来准备。

④ 调用MediaPlayer.start来播放视频。

为了简单起见，在VideoPlayerDemo中添加一个新的Activity——CustomVideoPlayer，在其布局文件中添加一个SurfaceView，然后修改CustomVideoPlayer文件如下：

```java
package cn.androidstudy.chapter15;
import android.media.MediaPlayer;
import android.os.Environment;
import android.support.constraint.ConstraintLayout;
import android.support.v7.app.AppCompatActivity;
import android.os.Bundle;
import android.util.Log;
import android.view.Display;
import android.view.SurfaceHolder;
```

```java
    import android.view.SurfaceView;
    import java.io.IOException;
    public class CustomVideoPlayer extends AppCompatActivity
implements
            MediaPlayer.OnCompletionListener, MediaPlayer.OnErrorListener,
MediaPlayer.OnInfoListener,
            MediaPlayer.OnPreparedListener, MediaPlayer.OnSeekCompleteListener,
MediaPlayer.OnVideoSizeChangedListener,
            SurfaceHolder.Callback{
        private static final String TAG = "CustomVideoPlayer";
        Display currentDisplay;
        SurfaceView surfaceView;
        SurfaceHolder surfaceHolder;
        MediaPlayer mediaPlayer;
        int videoWidth = 0;
        int videoHeight = 0;
        boolean readyToPlay = false;
        @Override
        protected void onCreate(Bundle savedInstanceState) {
          super.onCreate(savedInstanceState);
          setContentView(R.layout.activity_custom_video_player);
          surfaceView = (SurfaceView) this.findViewById(R.id.surfaceView);
          surfaceHolder = surfaceView.getHolder();
          surfaceHolder.addCallback(this);
          mediaPlayer = new MediaPlayer();
          mediaPlayer.setOnCompletionListener(this);
          mediaPlayer.setOnErrorListener(this);
          mediaPlayer.setOnInfoListener(this);
          mediaPlayer.setOnPreparedListener(this);
          mediaPlayer.setOnSeekCompleteListener(this);
          mediaPlayer.setOnVideoSizeChangedListener(this);
          String filePath = Environment.getExternalStorageDirectory().getPath()
                  + "/Download/Test_Movie_iPhone.m4v";
          try {
              mediaPlayer.setDataSource(filePath);
          } catch (IllegalArgumentException e) {
              Log.v(TAG, e.getMessage());
              finish();
          } catch (IllegalStateException e) {
              Log.v(TAG, e.getMessage());
              finish();
          } catch (IOException e) {
```

```java
            Log.v(TAG, e.getMessage());
            finish();
        }
        currentDisplay = getWindowManager().getDefaultDisplay();
    }
    @Override
    public void onCompletion(MediaPlayer mediaPlayer) {
        Log.d(TAG, "onCompletion: ");
    }
    @Override
    public boolean onError(MediaPlayer mediaPlayer, int i, int i1) {
        Log.d(TAG, "onError: ");
        if (i == MediaPlayer.MEDIA_ERROR_SERVER_DIED) {
            Log.v(TAG, "Media Error, Server Died " + i1);
        } else if (i == MediaPlayer.MEDIA_ERROR_UNKNOWN) {
            Log.v(TAG, "Media Error, Error Unknown" + i1);
        }
        return false;
    }
    @Override
    public boolean onInfo(MediaPlayer mediaPlayer, int whatInfo, int extra) {
        if (whatInfo == MediaPlayer.MEDIA_INFO_BAD_INTERLEAVING) {
            Log.v(TAG, "Media Info, Media Info Bad Interleaving " + extra);
        } else if (whatInfo == MediaPlayer.MEDIA_INFO_NOT_SEEKABLE) {
            Log.v(TAG, "Media Info, Media Info Not Seekable " + extra);
        } else if (whatInfo == MediaPlayer.MEDIA_INFO_UNKNOWN) {
            Log.v(TAG, "Media Info, Media Info Unknown " + extra);
                } else if (whatInfo == MediaPlayer.MEDIA_INFO_VIDEO_TRACK_LAGGING) {
            Log.v(TAG, "MediaInfo, Media Info Video Track Lagging " + extra);
        }
        return false;
    }
    @Override
    public void onPrepared(MediaPlayer mediaPlayer) {
        videoWidth = mediaPlayer.getVideoWidth();
        videoHeight = mediaPlayer.getVideoHeight();
        if (videoWidth > currentDisplay.getWidth()
                || videoHeight > currentDisplay.getHeight()) {
            float heightRatio = (float) videoHeight
                    / (float) currentDisplay.getHeight();
```

```java
            float widthRatio = (float) videoWidth
                    / (float) currentDisplay.getWidth();
            if (heightRatio > 1 || widthRatio > 1) {
                if (heightRatio > widthRatio) {
                    videoHeight = (int) Math.ceil((float) videoHeight
                            / (float) heightRatio);
                    videoWidth = (int) Math.ceil((float) videoWidth
                            / (float) heightRatio);
                } else {
                    videoHeight = (int) Math.ceil((float) videoHeight
                            / (float) widthRatio);
                    videoWidth = (int) Math.ceil((float) videoWidth
                            / (float) widthRatio);
                }
            }
        }
        surfaceView.setLayoutParams(new ConstraintLayout.LayoutParams(videoWidth,
                videoHeight));
        mediaPlayer.start();
    }
    @Override
    public void onSeekComplete(MediaPlayer mediaPlayer) {
        Log.d(TAG, "onSeekComplete: ");
    }

    @Override
    public void onVideoSizeChanged(MediaPlayer mediaPlayer, int i, int i1) {
        Log.d(TAG, "onVideoSizeChanged: ");
    }

    @Override
    public void surfaceCreated(SurfaceHolder surfaceHolder) {
        Log.v(TAG, "surfaceCreated Called");

        mediaPlayer.setDisplay(surfaceHolder);

        try {
            mediaPlayer.prepare();
        } catch (IllegalStateException e) {
            Log.v(TAG, e.getMessage());
```

```java
            finish();
        } catch (IOException e) {
            Log.v(TAG, e.getMessage());
            finish();
        }
    }

    @Override
    public void surfaceChanged(SurfaceHolder surfaceHolder, int i,
int i1, int i2) {
        Log.d(TAG, "surfaceChanged: ");
    }

    @Override
    public void surfaceDestroyed(SurfaceHolder surfaceHolder) {
        Log.d(TAG, "surfaceDestroyed: ");
    }
}
```

修改AndroidManifest.xml文件，设置CustomVideoPlayer为启动界面，代码如下：

```xml
<?xml version="1.0" encoding="utf-8"?>
<manifest xmlns:android="http://schemas.android.com/apk/res/android"
    package="cn.androidstudy.chapter15">
    <uses-permission android:name="android.permission.READ_EXTERNAL_STORAGE" />
    <application
        android:allowBackup="true"
        android:icon="@mipmap/ic_launcher"
        android:label="@string/app_name"
        android:roundIcon="@mipmap/ic_launcher_round"
        android:supportsRtl="true"
        android:theme="@style/AppTheme">
        <activity android:name=".MainActivity">

        </activity>
        <activity android:name=".CustomVideoPlayer">
            <intent-filter>
                <action android:name="android.intent.action.MAIN" />
                <category android:name="android.intent.category.LAUNCHER" />
            </intent-filter>
        </activity>
    </application>
</manifest>
```

强化训练

过去、现在甚至将来很长时间,多媒体功能都是智能手机用户的日常基本需求之一,Android程序开发人员都应该对智能手机用户的相关需求进行研究。本章主要对基于Android平台的音视频录制和播放功能进行介绍,重点讲解了两种方法,使用Intent进行录制和播放与使用系统API进行录制和播放。此外,本章并没有探讨在线音视频的播放,但是通过本章介绍的知识,读者完全有能力自己开发包含在线播放功能的应用程序。

学完本章知识后,不妨通过以下习题来巩固所学的内容吧。

一、填空题

1. 录制音频的三种主要方法,每种方法都有它们各自的使用范围,第一种方法,_____,该方法简单方便但灵活性不足;第二种方法,_____,该方法使用难度适中并且灵活性较强;第三种方法,_____,该方法难于掌握,灵活性很强,但功能十分强大。

2. Android中录制音频最简单的方法是_____。

3. Android中MediaRecorder类是_____提供的录制音频和视频的功能类,使用它可以建立功能更加完善的_____,例如可以控制录制的时长等。

4. AMR是自适应多速率编解码器,包括_____和_____,文件扩展名是".3gp" (audio/3gpp)或".amr" (audio/amr)。

5. MediaPlayer播放音频的最简单情况是_____。音频文件放置在应用程序的_____中。

6. Android平台上录制视频的最简单方式是通过_____来调用系统已有的摄像应用程序。Android平台默认提供一个_____应用程序。

7. _____是一个带有视频播放功能的视图,可以直接在布局中使用,使用起来非常简单。

8. MediaPlayer类也可以通过类似的方式用于视频播放。与使用_____和_____播放视频相比,将MediaPlayer用于视频播放能够为播放视频文件提供更大的灵活性。

9. 多媒体主要包括_____和_____。

二、选择题

1. (   ) 是开始录制前准备MediaRecorder的对象。
   A. MediaRecorder.setAudioSource
   B. MediaRecorder.setOutputFormat
   C. MediaRecorder.prepare
   D. MediaRecorder.release

2. 录制音频的过程,(   ) 的步骤是通过调用MediaRecorder. setOutputFile方法完成。
   A. 设置音频录制时采用的音频源
   B. 指定输出音频数据存放的文件
   C. 设置输出音频数据的存储文件格式
   D. 停止录制

3. （　）音频格式是一种新的音频压缩格式，类似于 MP3 的音乐格式。

   A. MP3

   B. AMR

   C. AAC

   D. Ogg

4. MediaRecorder.VideoEncoder 中的常量（　）是为需要低比特率视频且不需要大的处理器能力的技术而开发。

   A. MPEG_4_SP

   B. H264

   C. H263

   D. DEFAULT

5. 音频采样率对于音频的质量也非常重要。可以使用 MediaRecorder 的（　）方法设置音频采样率。

   A. setAudioEncoder

   B. setVideoEncodingBitRate

   C. setAudioChannels

   D. setAudioSamplingRate

6. （　）是自适应多速率编解码器，包括 AMR 窄带 AMR-NB 和 AMR 宽带 AMR-WB。

   A. H.263

   B. H.264 AVC

   C. MPEG-4 SP

   D. VP8

三、操作题

1. 利用 Android Studio 开发环境创建一个新的 Android 应用程序，实现音频的播放。要求从 MediaStore 中获取音频的信息，并使用 ListView 进行显示；在点击歌曲时，实现该音频的播放功能，要求实现音频的顺序播放、随机播放、单曲循环播放功能。

2. 利用 Android Studio 开发环境创建一个新的 Android 应用程序，提供使用 Intent 录制音频的功能。单击"录音"按钮，可打开系统自带的录音机，单击开始按钮进行录音。

3. 利用 Android Studio 开发环境创建一个新的 Android 应用程序——对 VideoView 示例的更新，在通过 setContentView 方法设置内容视图后，通过调用 setMediaController 方法设置 MediaController 为 VideoView 的控制器。

第16章

网络编程精讲

内容导读

　　智能手机是便携式的移动通信设备,访问互联网是其基本功能之一。作为优秀的智能手机操作系统的代表,Android强大优秀的网络功能是不容置疑的,Android程序开发人员自然应该用心学习。本章主要介绍Android平台下进行网络编程的相关知识,内容包括Android网络编程的常用接口,基于HTTP协议的GET请求和POST请求的两种网络编程方式,套接字Socket的基本概念以及服务器和客户端网络编程,WebView的常用组件及方法和基于WebView的浏览器开发方法。

学习目标

- 熟悉Android网络编程的方法;
- 掌握使用HttpURLConnection访问网络的方法;
- 掌握使用Socket进行网络通信的方法;
- 掌握WebView的使用。

16.1　Android网络编程基础

　　作为优秀的智能手机操作系统,Android的网络编程是Android程序开发人员值得花时间学习和研究的。ISO的OSI七层网络模型和事实标准TCP/IP五层网络模型这里就不再一一介绍了，读者需要的话可以查阅相关书籍，这里主要讲解基于Android的网络编程。Android是基于Linux内核的操作系统，继承了Linux优秀的联网功能。在网络通信应用的开发中，Android平台有以下三种网络编程接口可以使用。

（1）标准的Java网络接口

Java的标准网络接口java.net.*中提供了网络编程的相关类，主要包括流操作、数据包、套接字Socket以及HTTP协议处理，通过设置参数，能够完成连接服务器、和服务器进行数据交换等网络操作，有Java开发基础的人员可以使用这些包快速地创建Android平台下的网络应用程序。

（2）Apache网络接口

HTTP协议是目前互联网中应用最为广泛的网络协议，虽然标准的Java网络接口提供了对HTTP协议的支持，但是灵活性不足。HttpClient是Apache下的子项目，是一个功能强大并全面支持HTTP协议的客户端编程工具包，它实际是对java.net.*进行了封装和扩展。Android系统中已经集成了HttpClient的开源包org.apache.http.*，可以在Android中直接使用HttpClient访问网络。

（3）Android网络接口

Android的网络接口android.net.*是通过对Apache网络接口进行封装来实现的，它同样提供了对HTTP协议的支持。该包下的类常常用来开发Wi-Fi连接、手机邮件处理等Android系统特有的网络应用。

在Android平台下进行网络应用程序的开发，必须遵守Android系统的规范，否则应用程序将无法访问服务器或出错。下面的这些规范适用于本章的实例，以后不再逐一描述。

第一，Android的网络程序需要为其添加网络访问权限，否则将无法连接服务器，必须在项目的AndroidManifest.xml文件的Manifest标签中加入如下指令：

```
<uses-permission android:name="android.permission.INTERNET"/>
```

第二，在Android系统中，主线程的任务是UI操作和交互，为了提升用户感受和响应度，不允许主线程进行网络和磁盘读写等响应周期长的操作，如果将访问网络的代码放在主线程中，将会出现异常信息：

```
"android.os.NetworkOnMainThreadException"。
```

解决该问题有两种方法。第一种方法是在主线程中引入系统提供的开发工具StrictMode类，该类用来捕获磁盘访问或者网络访问中与主线程交互产生的潜在问题，提示开发者对其进行修复。在主线程的onCreate方法中加入下面的代码。

```
StrictMode.setThreadPolicy(new StrictMode.ThreadPolicy.
Builder()
    .detectDiskReads()
    .detectDiskWrites()
    .detectNetwork()
    .penaltyLog()
    .build());
StrictMode.setVmPolicy(new StrictMode.VmPolicy.Builder()
    .detectLeakedSqlLiteObjects()
```

```
        .penaltyLog()
        .penaltyDeath()
        .build());
```

加入上述代码后，运行程序依然会有异常抛出，但程序能够正常执行。由于该方法并不符合Android平台的规范和要求，不建议使用。

第二种方法是启动一个新的线程去执行访问网络的代码。

```
new Thread(new Runnable() {
    public void run() {
            ……;
            网络任务;
            ……;
    }
}).start();
```

本章的所有实例将采用第二种方法。

一般情况下，在网络任务完成以后，主线程需要进行UI交互和更新，在网络通信异常的情况下，主线程需要给出提示或者警告，因此执行网络任务的线程需要使用Handler消息通信机制向主线程发送消息，通知主线程网络任务的状态。

16.2 基于HTTP协议的网络编程

HTTP协议是超文本传输协议(Hyper Text Transfer Protocol)的缩写，是用于WWW服务器与本地浏览器之间传输超文本的传输协议，其最大优点是简单、实用。HTTP协议是一个应用层协议，由请求和响应构成，是一个标准的客户端和服务器模型，其客户端的实现程序主要是Web浏览器。Android平台提供了大量使用HTTP协议进行网络编程的接口，比较重要的是标准的Java接口和HttpClient接口，本节主要针对这些接口的网络编程方式进行介绍。

16.2.1 HTTP介绍

HTTP协议是Internet上使用最为广泛的协议，在TCP/IP体系中处于应用层，是一个标准的客户端服务器模型，由一个客户端程序和一个Web服务器端的服务程序构成，客户端程序通过统一资源定位器URL向服务器发送HTTP请求报文，服务程序在80号端口监听客户端的请求并向客户端发送HTTP响应报文。HTTP协议永远都是客户端发起请求，服务器回送响应，是一个无状态的协议。

HTTP协议中定义了八种方法和服务器进行交互，其中GET和HEAD方法是服务器必须实现的，也是最常用的。GET方法请求的数据会附在URL之后，以"?"分割URL和传输数据，参数之间以"&"相连，如"login.action？name＝Administrator＆password＝123456"。GET方法提交的数据量跟URL的长度有直接关系，不同的浏览器对URL的长度要求不一样。POST把提交的数据放置在HTTP协议的包体中，理论上对数据长度没有要求。POST的安全性要比GET的安全性高，通过GET提交数据，如果对数据不做修改和加密，用户名和密码将以明文的形式出现在URL上。HTTP协议默认采用的是GET方法。

目前广泛使用的Web服务器是Microsoft公司的IIS和Apache软件基金会的Tomcat。由于本节的实例需要，首先架设一个Tomcat服务器。搭建Tomcat服务器的步骤如下：

① 进入Apache的官方网站http://tomcat.apache.org，在该网站可以免费获取各个版本的Tomcat软件，根据操作系统的版本选择对应的下载。

② 将下载的压缩包解压缩到本地磁盘任意目录，以D:\Tomcat为例。

③ 配置环境变量。打开环境变量窗口，找到CLASSPATH变量，在变量值的最后添加如下内容：

```
;D:\tomcat\lib\jsp-api.jar;D:\tomcat\lib\servlet-api.jar
```

④ 进入D:\Tomcat\bin目录，双击执行startup.bat启动Tomcat。

⑤ 打开浏览器，输入网址"http://localhost:8080"，若出现图16-1则代表安装成功。

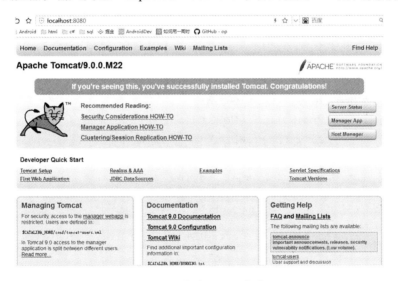

图16-1 启动Tomcat成功

16.2.2 使用HttpURLConnection访问网络

HttpURLConnection类位于Java.NET包中，它是Java中提供的访问网络资源的接口，使用这些接口可以方便地编写网络应用程序。由于该类是一个抽象类，需要使用URL对象的openConnection方法实例化。创建一个HttpURLConnection对象的代码如下：

```
URL myUrl = new URL(stringURL);
HttpURLConnection myConn =(HttpURLConnection)myUrl.openConnection();
```

创建了HttpURLConnection对象之后，就可以使用该对象向服务器发送GET请求和POST请求。

（1）GET请求

HttpURLConnection对象默认使用GET请求，下面通过实例来说明。

【实例16-1】在Android Studio中新建一个名称为Chapter16的Android项目，该项目

向服务器发送 GET 请求，从服务器获取一个文本文件的内容，并将该文件的内容显示在屏幕的 EditText 控件中。

① 在 Tomcat\webapps 目录下新建一个文件夹 AndroidTest，将文本文件 Message.txt 复制到该文件夹下面作为服务器资源。

② 修改项目的布局文件 activity_main.xml，使用相对布局管理器，添加一个向服务器发送 GET 请求的按钮控件和一个显示文本内容的 EditText 控件。

③ 在 MainActivity 类中，创建要使用的变量，同时本例中使用了消息通信机制 Handler，执行网络任务的线程向主线程发送消息，通知主线程更新 EditText 控件，需要重写 Handler 对象的 handleMessage 方法，代码如下：

```java
    private static final int SUCCESS = 0;
    private static final int FAILURE = 1;
    private EditText et;
    private Button btn;
    private String txt;
    //统一资源定位器URL字符串，192.168.1.120为服务器地址
    String stringURL = "http://192.168.1.120:8080/AndroidTest/Message.txt";
    private  Handler mHandler = new Handler(){
        //重写handleMessage方法
        public void handleMessage (Message msg){
            switch(msg.what) {
                case SUCCESS:
                    et.setText(txt);
                    break;
                case FAILURE:
                    et.setText("Download the file Failure!");
                    break;
            }
        }
    };
```

④ 重写 onCreate 方法，获取布局管理器中的对象，并为按钮对象添加事件监听器，调用自定义的 GetURLResources，代码如下：

```java
    protected void onCreate(Bundle savedInstanceState) {
        super.onCreate(savedInstanceState);
        setContentView(R.layout.activity_main);
        et = (EditText)findViewById(R.id.editText);
        btn = (Button)findViewById(R.id.button);
        //为按钮添加事件监听器
        btn.setOnClickListener(new View.OnClickListener() {
            @Override
            public void onClick(View v) {
                GetURLResources();
```

            }
        });
}
```

⑤ 编写无返回值的方法GetURLResources，该方法启动一个新线程创建一个HTTP连接，并用GET请求从服务器获取文本文件内容，如果获取成功，则向主线程发送what值为0的消息，否则发送what值为1的消息（what为sendEmptyMessage的参数）。代码如下：

```java
public void GetURLResources(){
    //创建一个新线程，读取服务器资源
    new Thread(new Runnable(){
        public void run(){
            try{
                URL myUrl = new URL(stringURL);
                //创建一个HttpURLConnection对象，打开链接
                HttpURLConnection myConn =(HttpURLConnection)myUrl.openConnection();
                //设置连接超时
                myConn.setConnectTimeout(3000);
                //获取输入流，得到读取的内容
                InputStreamReader in = new InputStreamReader(myConn.getInputStream());
                BufferedReader buffer = new BufferedReader(in);
                String inputLine = null;
                StringBuffer pageBuffer = new StringBuffer();
                while((inputLine = buffer.readLine())!= null){
                    pageBuffer.append(inputLine +"\n");
                }
                //设置字符的编码格式
                txt =new String(pageBuffer.toString().getBytes("UTF-8"));
                //去掉最后一行的换行符
                txt = txt.substring(0,txt.length()-1);
                mHandler.sendEmptyMessage(0);
                in.close();
                buffer.close();
                //关闭连接
                myConn.disconnect();
            }
            catch(Exception e){
                mHandler.sendEmptyMessage(1);
                e.printStackTrace();
            }
        }
    }).start();
}
```
```

启动Tomcat服务器，运行本实例，结果如图16-2所示。

图16-2 利用GET请求获取文件内容实例图

（2）POST请求

由于HttpURLConnection对象默认使用GET请求，在发送POST请求时，需要调用setRequestMethod方法来进行指定，同时POST请求还必须进行一些必要的设置，常用的设置如表16-1所示。

表 16-1 POST常用设置

方法	描述
SetRequestMethod(String)	设置连接方式
SetDoInput(Boolean)	设置输入是否允许
setDoOutput(boolean)	设置输出是否允许
setUseCaches(boolean)	是否使用Cache
setInstanceFollowRedirects(boolean)	是否使用HTTP重定向
setRequestProperty(String, String)	配置请求属性

☞【实例16-2】在图16-2中，在"获取文件内容"按钮右侧添加一个新的按钮"POST"，向服务器发送POST请求，POST请求提交的数据是参数username以及对应的值Android User，服务器获取参数username的值，并向客户端返回欢迎信息在屏幕的EditText控件中显示。

① 编写一个UrlPost.jsp文件，代码如下所示，将该文件保存在文件夹Tomcat\webapps\AndroidTest\下面。

```
<%@ page contentType="text/html;charset=GBK"%>
```

```
<%
        request.setCharacterEncoding("GBK");
        String name = (String)request.getParameter("username");
        if (name != null) {
            out.print("Welcome: "+name);
        }
%>
```

② 修改项目的布局文件 activity_main.xml，在"获取文件内容"按钮右侧添加一个新的按钮"POST"。

③ 在 MainActivity 类中创建要使用的变量，代码如下：

```
private Button btnPost;
//统一资源定位器URL字符串
String postURL = "http://192.168.1.120:8080/AndroidTest/UrlPost.jsp";
```

④ 修改 onCreate 方法，获取布局管理器中的对象，并为按钮对象添加事件监听器，调用自定义的 GetURLResourcesByPost，代码如下：

```
btnPost = (Button)findViewById(R.id.button2);
//为按钮添加事件监听器
btnPost.setOnClickListener(new View.OnClickListener() {
    @Override
    public void onClick(View v) {
        GetURLResourcesByPost();
    }
});
```

⑤ 编写无返回值的方法 GetURLResourcesByPost，该方法启动一个新线程创建一个 HTTP 连接，并用 POST 请求向服务器发送数据同时从服务器接收数据。如果接收数据成功，则向主线程发送 what 值为 0 的消息；如果服务器响应异常，发送 what 值为 1 的消息。代码如下：

```
public void GetURLResourcesByPost(){
   new Thread(new Runnable(){
        public void run(){
            try{
                String Param = "username=" + "Android User";     //发送的数据
                byte[] PostData = Param.getBytes();
                URL myUrl = new URL(stringURL);
                //创建一个HttpURLConnection对象，打开链接
HttpURLConnection myConn = (HttpURLConnection)myUrl.openConnection();
                myConn.setConnectTimeout(3000);              //设置连接超时
                myConn.setDoInput(true);                     //设置输入允许
                myConn.setDoOutput(true);                    //设置输出允许
```

```java
            myConn.setRequestMethod("POST");        //设置POST请求
            myConn.setUseCaches(false);             //POST请求不能使用Cache
            myConn.setInstanceFollowRedirects(true);   //允许HTTP重定向
            myConn.setRequestProperty("Content-Type", "application/
x-www-form-urlencoded");                            //配置请求设置
            myConn.connect();
            //发送数据
    DataOutputStream out = new DataOutputStream(myConn.getOutputStream());
            out.write(PostData);
            out.flush();
            out.close();
            if(myConn.getResponseCode()==200){       //判断连接状态
             InputStreamReader in = new InputStreamReader(myConn.
getInputStream());
                BufferedReader buffer = new BufferedReader(in);
                String inputLine = null;
                StringBuffer pageBuffer = new StringBuffer();
                while((inputLine = buffer.readLine())!= null){
                   pageBuffer.append(inputLine +"\n");
                }
                //设置字符编码格式
    txt =   new String(pageBuffer.toString().getBytes("UTF-8"));
                mHandler.sendEmptyMessage(0);
                in.close();
                buffer.close();
                myConn.disconnect();
            }
            else{
                mHandler.sendEmptyMessage(1);
            }
        }
        catch(Exception e){
            mHandler.sendEmptyMessage(1);
            e.printStackTrace();
        }
      }
    }).start();
  }
```

启动Tomcat服务器，运行本实例，结果如图16-3所示。

第16章 网络编程精讲

图16-3 POST请求接收信息

16.2.3 使用HttpClient访问网络

一般情况下，对于比较复杂的网络访问操作，使用Java编写的标准接口程序烦琐，工作量比较大。在Android系统中，最常用的是Apache提供的HttpClient，它对Java.NET进行了抽象和封装，GET请求封装成了HttpGet类，Post请求封装成了HttpPost类，服务器的响应封装成了HttpResponse类，发送和接收HTTP报文的实体封装成了HttpEntity类。在Android6.0版本直接删除了HttpClient类库，如果仍想使用则解决方法如下：

如果使用的是Eclipse，则在libs中加入org.apache.http.legacy.jar，这个jar包在sdk\platforms\android-25\optional目录中。

如果使用的是Android Studio，则在相应的module下的build.gradle中加入：

```
android {
    useLibrary 'org.apache.http.legacy'
}
```

下面分别介绍利用HttpClient完成的GET请求和POST请求。

（1）GET请求

使用HttpClient发送GET请求的步骤大致如下：
① 利用DefaultHttpClient类创建HttpClient对象。

```
HttpClient httpclient = new DefaultHttpClient();
```

② 利用HttpGet类生成HttpGet对象。

```
HttpGet httpGet = new HttpGet(stringURL);
```

③ 如果需要向服务器发送请求参数，生成HTTP请求报文的HttpEntity对象，使用HttpGet的setEntity方法设置发送请求参数。

```
httpGet.setEntity(httpentity);
```

④ 调用HttpClient的execute方法发送GET请求，该方法返回一个服务器的响应HttpResponse对象。

```
HttpResponse httpresponse = httpclient.execute(httpGet);
```

⑤ 调用HttpResponse对象的getEntity方法得到服务器HTTP响应报文的HttpEntity对象，该对象包含了服务器的应答信息。

☞【实例16-3】在Chapter16中新建一个Module——HttpClientGet，该项目向服务器发送GET请求，从服务器获取图片，并在客户端屏幕的ImageView控件中显示。

① 将图片文件Flower.png保存在文件夹Tomcat\webapps\AndroidTest中。

② 修改项目的布局文件activity_httpclient_get.xml，使用相对布局管理器，添加一个向服务器发送GET请求的按钮控件、一个用于显示图片信息的ImageView控件和一个显示状态信息的EditText控件。

③ 在MainActivity类中，创建要使用的变量，重写Handler对象的handleMessage方法，如果接收到的消息的what值为0，将ImageView控件的对象设置为消息中的图片对象，代码如下：

```
private static final int SUCCESS = 0;
private static final int FAILURE = 1;
private Button btn;
private TextView tv;
private ImageView  iv;
//统一资源定位器URL字符串
String stringURL = "http://192.168.1.120:8080/AndroidTest/Flower.png";
private  Handler mHandler = new Handler(){
      //重写handleMessage方法
      public void handleMessage (Message msg) {
           switch(msg.what) {
                case SUCCESS:
                     tv.setText("Download the Picture Success!");
                     //设置图片控件中显示的图片对象
                     iv.setImageBitmap((Bitmap) msg.obj);
                     break;
                case FAILURE:
                     tv.setText("Download the Picture Failure!");
                     break;
           }
      }
};
```

④ 重写onCreate方法，获取布局管理器中的对象，并为按钮对象添加事件监听器，调用自定义的GetURLResources，代码如下：

```java
protected void onCreate(Bundle savedInstanceState) {
    super.onCreate(savedInstanceState);
    setContentView(R.layout.activity_main);
    btn = (Button)findViewById(R.id.button);
    tv = (TextView)findViewById(R.id.textView);
    iv = (ImageView)findViewById(R.id.imageView);
    //为按钮添加事件监听器
    btn.setOnClickListener(new View.OnClickListener() {
        @Override
        public void onClick(View v) {
            GetURLResources();
        }
    });
}
```

⑤ 编写无返回值的方法GetURLResources，该方法启动一个新线程使用HttpClient向服务器发送GET请求，从服务器获取图片，如果获取成功，则向主线程发送what值为0，并且包含一个图片对象的消息，否则发送what值为1的消息。代码如下：

```java
public void GetURLResources(){
    new Thread(new Runnable(){
        public void run(){
            try{
                HttpClient httpclient = new DefaultHttpClient();    //创建httpclient对象
                HttpGet httpGet = new HttpGet(stringURL);           //创建连接
                httpclient.getParams().setParameter(CoreConnectionPNames.CONNECTION_TIMEOUT, 3000);
                //设置超时
                HttpResponse httpresponse = httpclient.execute(httpGet); //执行GET请求
                //判断连接状态
                if(httpresponse.getStatusLine().getStatusCode() == HttpStatus.SC_OK){
                    HttpEntity httpentity = httpresponse.getEntity();    //获取响应实体
                    InputStream in = httpentity.getContent();            //获取输入流
                    Bitmap bmp = BitmapFactory.decodeStream(in);
                    mHandler.obtainMessage(0, bmp).sendToTarget();       //发送包含对象的消息
                    in.close();
                }
```

```
        }
        catch(Exception e){
            mHandler.obtainMessage(1).sendToTarget();           //连接服
务器失败消息
            e.printStackTrace();
        }
    }
}).start();
}
```

启动Tomcat服务器，运行本实例，结果如图16-4所示。

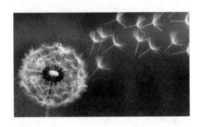

图16-4　GET获取图片信息

（2）POST请求

POST请求和GET请求的步骤基本一致，大致流程如下。

① 利用DefaultHttpClient类创建HttpClient对象。

② 利用HttpPost类生成HttpPost对象。

③ 如果需要向服务器发送请求参数，生成HTTP请求报文的HttpEntity对象，使用HttpPost的setEntity方法设置发送请求参数。

④ 调用HttpClient的execute方法发送POST请求，该方法返回一个服务器的响应HttpResponse对象。

⑤ 调用HttpResponse对象的getEntity方法得到服务器HTTP响应报文的HttpEntity对象，该对象包含了服务器的应答信息。

下面通过实例来说明POST请求的过程。

☞【实例16-4】在Chapter16中新建一个Module——HttpClientPost，该项目模拟登录的

过程，客户端输入用户名和密码，向服务器发送POST请求，如果验证通过，则打开一个新的Activity。

① 编写一个HttpClientPost.jsp文件，代码如下所示，将该文件保存在文件夹Tomcat\webapps\AndroidTest中。

```jsp
<%@ page contentType="text/html;charset=GBK"%>
<%
request.setCharacterEncoding("GBK");
String name = (String)request.getParameter("username");
String pass = (String)request.getParameter("password");
if (name.equals("Android User") && pass.equals("123456")) {
    out.print("ok");
}
else{
    out.print("error");
}
%>
```

② 修改项目的布局文件activity_main.xml，添加一个向服务器发送POST请求的按钮控件、两个用于输入用户名和密码的EditText控件。新建一个布局文件activity_response.xml，添加一个TextView控件，用于显示欢迎信息。

③ 在MainActivity类中，创建要使用的变量，重写Handler对象的handleMessage方法，如果接收到的消息的what值为0，打开一个新的Activity，代码如下：

```java
private static final int SUCCESS = 0;
private static final int FAILURE = 1;
private static final int ServerError = 2;
private static final int LinkFailure = 3;
private Button btn;
private EditText editname;
private EditText editpass;
private String repeat;
//统一资源定位器URL字符串
String stringURL="http://192.168.1.120:8080/AndroidTest/HttpClientPost.jsp";
private  Handler mHandler = new Handler(){
        //重写handleMessage方法
        public void handleMessage (Message msg) {
            switch(msg.what) {
              case SUCCESS:
                    //打开一个新的Activity
                    Intent _intent = new Intent(MainActivity.this,ResponseActivity.class);
                    startActivity(_intent);
```

```
                break;
            case FAILURE:
                Toast.makeText(getApplicationContext(),"Login failed!",
Toast.LENGTH_SHORT).show();
                break;
            case ServerError:
    Toast.makeText(getApplicationContext(),"The response of Server
is error! ",
            Toast.LENGTH_SHORT).show();
            break;
            case LinkFailure:
                Toast.makeText(getApplicationContext(),"Link Server failed! ",
                Toast.LENGTH_SHORT).show();
                break;
            }
        }
    };
```

④ 重写 onCreate 方法，获取布局管理器中的对象，并为按钮对象添加事件监听器，调用自定义的 PostMessage，代码如下：

```
    protected void onCreate(Bundle savedInstanceState) {
        super.onCreate(savedInstanceState);
        setContentView(R.layout.activity_httpclient_post);
        btn = (Button)findViewById(R.id.button);
        editname = (EditText)findViewById(R.id.editText2);
        editpass = (EditText)findViewById(R.id.editText3);
        //为按钮添加事件监听器
        btn.setOnClickListener(new View.OnClickListener() {
            @Override
            public void onClick(View v) {
                PostMessage();
            }
        });
    }
```

⑤ 编写无返回值的方法 PostMessage，该方法启动一个新线程使用 HttpClient 向服务器发送 POST 请求，POST 请求提交的数据是参数 username 和 password 以及对应的值，服务器获取参数的值，判断用户名和密码是否合法，向客户端返回响应报文。客户端接收到响应报文后，根据报文内容向主线程发送消息。如果登录成功，则向主线程发送 what 值为 0 的消息；登录失败，则向主线程发送 what 值为 1 的消息；如果服务器响应异常，发送 what 值为 2 的消息；如果服务器连接异常，发送 what 值为 3 的消息。代码如下：

```
    public void PostMessage(){
```

```
new Thread(new Runnable(){
    public void run(){
        try{
            HttpClient httpclient = new DefaultHttpClient();//创建HttpClient对象
            HttpPost httppost = new HttpPost(stringURL); //创建连接
            String name = editname.getText().toString(); //获取用户名
            String pass = editpass.getText().toString(); //获取密码
            List<NameValuePair> params = new ArrayList<NameValuePair>();
            params.add(new BasicNameValuePair("username", name));
            //添加参数
            params.add(new BasicNameValuePair("password", pass));
            //添加参数
            //创建HttpEntity对象，添加参数并设置编码格式
            HttpEntity httpentity = new UrlEncodedFormEntity(params, HTTP.UTF_8);
            //设置POST请求的参数实体
            httppost.setEntity(httpentity);
            httpclient.getParams().setParameter(CoreConnectionPNames.CONNECTION_TIMEOUT, 3000);
            //设置网络延时
            HttpResponse httpresponse = httpclient.execute(httppost);
            //发送POST请求
            //判断连接状态
            if(httpresponse.getStatusLine().getStatusCode() == HttpStatus.SC_OK){
                httpentity = httpresponse.getEntity();      //获取响应实体
                repeat = EntityUtils.toString(httpentity);
                repeat = repeat.replace("\r\n", "");
                //去掉响应字符串的回车和换行
                if(repeat.equals("ok")){
                    mHandler.sendEmptyMessage(0); //验证成功消息
                }
                else{
                    mHandler.sendEmptyMessage(1); //验证失败消息
                }
            }
            else{
                mHandler.sendEmptyMessage(2);       //服务器响应异常消息
            }
        }
        catch(Exception e){
            mHandler.sendEmptyMessage(3);           //连接服务器失败消息
            e.printStackTrace();
```

```
            }
        }
    }).start();
}
```

启动 Tomcat 服务器，运行本实例如图 16-5 所示，登录成功后打开一个新的 Activity，登录失败如图 16-6 所示。

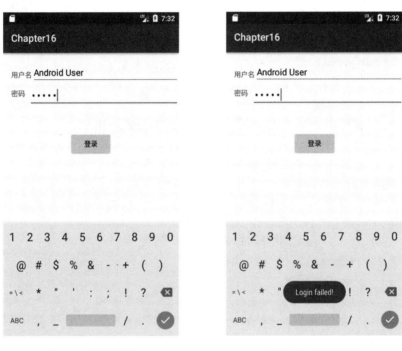

图16-5　登录界面　　　　　　图16-6　登录失败

16.3　基于 Socket 的网络编程

套接字（Socket）是一个抽象层，应用程序可以通过它发送或接收数据，可对其进行像对文件一样的打开、读写和关闭等操作。套接字允许应用程序将 I/O 插入到网络中，并与网络中的其他应用程序进行通信。实际上套接字（Socket）是 IP 地址与端口的组合。在 HTTP 通信中，客户端需要首先发起连接请求，和服务器建立连接后接收服务器的响应信息，在请求结束之后连接主动释放，这种连接称为短连接，是 HTTP 协议中最常使用的方式，这种方式能够使服务器的资源充分的利用，但是也带来了一定的问题，如果客户端需要服务器更新的数据，必须再次发送连接请求，服务器更新的数据不会主动发送给客户端，在一些比较复杂的网络通信应用中，需要客户端和服务器建立连接后能够相互发送数据传递消息，这就需要使用基于 Socket 的网络编程来完成。

16.3.1 套接字Socket

　　TCP/IP协议是互联网的基础,其核心部分是传输层的TCP协议和UDP协议以及网际层的IP协议,通常在操作系统实现,网络应用程序一般不直接使用TCP/IP协议,而是使用一组调用TCP/IP协议的应用程序接口函数,这就是Socket,也称为套接字,通过Socket可以很容易地编写网络应用程序完成不同主机之间的相互通信。

　　说明:TCP/IP本身并没有对应用程序接口函数进行标准化,不同的操作系统提供的应用程序接口函数形式不一样。

　　Socket有两种方式:基于TCP协议的面向连接的流Socket和基于UDP协议的面向无连接的数据包Socket。流Socket在发送数据之前,客户端的Socket和服务器Socket建立连接,效率不高,但可以保证数据安全有序地到达接收方,通信完成后需要关闭Socket来断开连接;数据包Socket在发送数据之前,不需要建立连接,效率高但不保证数据安全有序地到达接收方。推荐使用流Socket。

　　Java.NET中提供了两个类,ServerSocket和Socket。ServerSocket类用于创建在服务器的指定端口进行监听的对象,监听客户端的连接请求;Socket类用于创建在服务器端和客户端相互连接的对象,该对象已经封装了输入流和输出流,一旦客户端的Socket对象和服务器端的Socket对象建立了连接,只要把数据写出到相关联的Socket对象的输出流中就可以完成发送,读取相关联的Socket对象的输入流就可以完成接收。

　　ServerSocket类常用的构造函数如下。

- ServerSocket(int port):创建绑定到指定端口进行监听的对象,参数port是端口号。
- ServerSocket(int port, int backlog):创建绑定到指定端口进行监听的对象,参数port是端口号,参数backlog是请求等待队列的最大长度。

　　服务器端和客户端必须约定好使用的通信端口,端口不一致将不能通信。在选择端口号的时候需要注意,计算机中的端口号在0~65535之间,其中0~1023为系统保留,用户程序尽量不要使用,否则可能影响系统服务。

　　Socket类常用的构造函数如下。

- Socket(Int address, int port):创建一个连接指定IP地址和端口号的流Socket,参数address是IP地址,参数port是端口号。
- Socket(String host, int port):创建一个连接指定主机名和端口号的流Socket,参数host是主机名或者域名,参数port是端口号。

16.3.2 Socket编程

　　客户端和服务器端要完成一次通信,首先客户端必须知道远程主机的IP地址(或者主机名)和端口号,其次客户端要发送一个连接请求以建立连接。下面分别介绍服务器端和客户端的网络通信的步骤。

（1）服务器端

① 指定端口实例化一个ServerSocket。

```
ServerSocket serversocket = new ServerSocket(5678);
```

② 调用ServerSocket的accept方法,该方法是一个阻塞方法,调用该方法会使服务器在

指定的端口监听客户端的连接请求,如果没有连接请求,该方法被阻塞,如果有连接请求,该方法返回一个Socket对象,该对象用于和客户端的Socket建立连接。

```
Socket socket = serversocket.accept();
```

③ 调用Socket对象的getInputStream方法获取输入流,接收客户端的Socket发送的信息,调用getOutputStream方法获取输出流,向客户端Socket发送信息。

```
DataInputStream  in = new DataInputStream(socket.getInputStream());
DataOutputStream out = new DataOutputStream(socket.getOutputStream());
```

④ 在通信完成后关闭流和Socket。

```
socket.close();
```

(2)客户端

① 通过IP地址和端口号实例化一个Socket对象,该对象会向指定IP和端口的远程主机发送连接请求。

```
Socket socket = new Socket("192.168.1.120", 5678);
```

② 调用Socket对象的getOutputStream方法获取输出流,向服务器的Socket对象发送信息,调用getInputStream方法获取输入流,接收服务器的Socket发送的信息。

```
DataInputStream  in = new DataInputStream(socket.getInputStream());
DataOutputStream out = new DataOutputStream(socket.getOutputStream());
```

③ 在通信完成后关闭流和Socket。

```
socket.close();
```

☞【实例16-5】本例的服务器端是PC,运行在Java SE平台下,客户端是Android系统,通过Socket完成信息传送。

① 在Eclipse中,新建一个Java项目SocketServer,main方法中的代码如下:

```
try{
    ServerSocket serversocket = new ServerSocket(5678);//实例化ServerSocket
    System.out.println("Listening...");
    while(true){
        Socket socket = serversocket.accept();//监听端口,得到服务器Socket
        System.out.println("Client Connected...");
        //得到输入流
        DataInputStream in = new DataInputStream(socket.getInputStream());
        String InStr = in.readUTF();            //获取客户端发送的信息
        System.out.println(InStr);
        DataOutputStream out = new DataOutputStream(socket.getOutputStream()); //得到输出流
        out.writeUTF("Link Server Success!");//向客户端发送信息
```

```
            out.flush();
            in.close();
            out.close();
            socket.close();                           //关闭Socket
        }
    }
    catch(Exception e){
        e.printStackTrace();
    }
```

② 在Chapter16中新建一个Module——SocketClient，修改项目的布局文件，添加一个EditText控件用来显示服务器发送的消息。

③ 在MainActivity类中，创建要使用的变量，重写onCreate方法，获取布局管理器中的对象，同时创建Socket对象，和服务器交换信息，代码如下：

```
private EditText et;
@Override
protected void onCreate(Bundle savedInstanceState) {
    super.onCreate(savedInstanceState);
    setContentView(R.layout.activity_main);
    et = (EditText)findViewById(R.id.editText);      //获得EditText对象
    new Thread(new Runnable() {
    public void run() {
      try{
          Socket socket = new Socket("192.168.1.120", 5678); //创建Socket对象
          //获得输出流
          DataOutputStream out = new DataOutputStream(socket.getOutputStream());
          out.writeUTF("This is a Message Send by Client !");  //发送信息out.flush;
          DataInputStream in = new DataInputStream(socket.getInputStream());
          String str = in.readUTF();                 //读取服务端发来的消息
          et.setText(str);                           //设置EditText对象
          out.close();
          in.close();
          socket.close();                            //关闭Socket
      }
      catch(Exception e){
          e.printStackTrace();
      }
    }
}).start();
}
```

首先运行服务器端程序,然后启动客户端程序,连接成功后,服务器端的运行结果如图16-7所示,客户端的运行结果如图16-8所示。

图16-7 连接成功服务器端界面

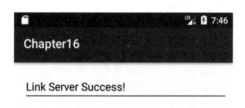

图16-8 连接成功后客户端界面

16.4 基于WebView的简单浏览器

浏览器是Internet时代最优秀的产品之一。随着PC操作系统的普及、Internet的全球连接及人们对信息需求的爆炸式增长,为浏览器的诞生和发展提供了强大的动力。Android平台内置了开源的WebKit浏览器引擎。WebKit有着很强大的网络功能,不仅能够浏览网页、收发电子邮件,还支持音视频节目的在线播放。通过WebView组件能够方便地使用WebKit来开发支持JavaSript、AJAX等功能的浏览器。WebView组件常用的方法如表16-2所示。

表16-2 WebView常用的方法

方法	描述
loadUrl(String url)	加载URL指定的页面
reload()	刷新当前加载的页面
stopLoading()	停止正在加载的页面
goBack()	返回至当前页面上一次加载的页面
goForwoard()	前进至当前页面下一次加载的页面

在实际应用开发过程中,一般还需要使用WebView组件中的WebSettings对象、WebViewClient对象与WebChromeClient对象设置浏览器的特性,以及用重写事件的方法来丰富浏览器的功能。下面对它们进行简单的介绍。

WebSettings对象使用WebView对象的getSettings方法可以得到,该对象主要是对

WebView的属性和配置进行设置，常用的方法如表16-3所示。

表16-3　WebSettings常用方法

方法	描述
setJavaSciptEnabled(boolean flag)	是否支持JavaScipt脚本
setBlockNetworkImage(boolean flag)	是否显示网络图片
setBuiltInZoomControls(boolean flag)	是否支持页面缩放
setDefaultTextEncodingName(String Encode)	设置解码时默认编码
setDefaultFontSize(int size)	设置默认的字体大小
setSupportMultipleWindows(boolean flag)	是否支持多屏幕显示
setPluginsEnabled(boolean flag)	是否支持插件

WebView对象主要是负责页面的解析和渲染工作，如果需要丰富浏览器的功能，就必须用到WebViewClient对象与WebChromeClient对象。WebViewClient的方法会在影响页面渲染的动作发生时被调用，例如页面开始加载、页面加载完毕、资源加载、页面中的URL请求等；WebChromeClient的方法在一些影响页面的交互动作发生时被调用，例如WebView的打开和关闭、页面加载进展、弹出JavaScipt确认框和警告框等。

（1）WebViewClient常用的方法

• public void onPageStarted(WebView view,String url,Bitmap favicon)：该方法在页面加载时会调用。

• public void onPageFinished(WebView view,String url)：该方法会在页面加载完毕时会调用。

• public void onLoadResource(WebView view,String url)：该方法在加载页面的资源时会调用。

• public boolean shouldOverrideUrlLoading(WebView view,String url)：该方法在点击页面内的链接时调用，重写此方法并返回true表明点击页面内的链接还是在当前的WebView里打开，不跳转到系统默认的浏览器。

（2）WebChromeClient常用的方法

• public void onProgressChanged(WebView view, int progress)：该方法在加载进度改变时会调用。

• public Boolean onCreateWindow(WebView view, boolean dialog, boolean userGesture, Message resultMsg)：该方法在创建WebView的时候会调用。

• public void onCloseWindow(WebView window)：该方法在关闭WebView的时候会调用。

• public void onReceivedIcon(WebView view, Bitmap icon)：该方法在网页图标更改时会调用。

• public void onReceivedTitle(WebView view, String title)：该方法在网页标题更改时会调用。

下面通过一个实例利用WebView组件开发一个简单的浏览器。

【实例16-6】在Chapter16中新建一个Module——WebViewExplorer，该项目实现一个具有打开、前进、后退功能，支持JavaScipt的简单浏览器。

① 修改项目的布局文件activity_main.xml，添加三个按钮控件，分别完成前进、后退和打开功能，添加一个EditText控件，用于输入URL地址，添加一个WebView控件显示网页的

内容，代码如下：

```xml
<?xml version="1.0" encoding="utf-8"?>
<android.support.constraint.ConstraintLayout
    xmlns:android="http://schemas.android.com/apk/res/android"
    xmlns:app="http://schemas.android.com/apk/res-auto"
    xmlns:tools="http://schemas.android.com/tools" android:layout_width="match_parent"
    android:layout_height="match_parent" tools:context="cn.androidstudy.chapter16.MainActivity">
    <Button
        android:id="@+id/button"
        android:layout_width="wrap_content"
        android:layout_height="wrap_content"
        android:layout_marginLeft="0dp"
        android:layout_marginTop="0dp"
        android:text="前进"
        app:layout_constraintLeft_toLeftOf="parent"
        app:layout_constraintTop_toTopOf="parent" />
    <Button
        android:id="@+id/button2"
        android:layout_width="wrap_content"
        android:layout_height="wrap_content"
        android:text="后退"
        app:layout_constraintTop_toTopOf="parent"
        app:layout_constraintLeft_toRightOf="@+id/button" />
    <Button
        android:id="@+id/button3"
        android:layout_width="wrap_content"
        android:layout_height="wrap_content"
        android:text="GO"
        app:layout_constraintRight_toRightOf="parent"
        app:layout_constraintTop_toTopOf="parent" />
    <EditText
        android:id="@+id/editText"
        android:layout_width="0dp"
        android:layout_height="wrap_content"
        android:layout_marginTop="0dp"
        android:ems="10"
        android:inputType="textPersonName"
        app:layout_constraintLeft_toRightOf="@+id/button2"
        app:layout_constraintRight_toLeftOf="@+id/button3"
        app:layout_constraintTop_toTopOf="parent"
```

```xml
        app:layout_constraintHorizontal_bias="0.0" />
    <WebView
        android:id="@+id/webView"
        android:layout_width="0dp"
        android:layout_height="0dp"
        android:layout_marginLeft="0dp"
        android:layout_marginTop="0dp"
        app:layout_constraintLeft_toLeftOf="parent"
        app:layout_constraintTop_toBottomOf="@+id/button2"
        android:layout_marginRight="0dp"
        app:layout_constraintRight_toRightOf="parent"
        app:layout_constraintBottom_toBottomOf="parent"
        android:layout_marginBottom="0dp"
        app:layout_constraintVertical_bias="0.0"
        app:layout_constraintHorizontal_bias="0.0" />
</android.support.constraint.ConstraintLayout>
```

② 在MainActivity类中创建要使用的变量：

```java
private WebView webview;
private EditText editurl;
private Button btnForward;
private Button btnBack;
private Button btnGo;
```

③ 重写onCreate方法，首先获取布局管理器中的对象，然后创建WebViewClient对象并进行浏览器基本属性的配置，重写WebChromeClient的onProgressChanged方法，实现打开网页时的进度条功能，重写WebViewClient的shouldOverrideUrlLoading方法，实现在页面内点击链接时仍然使用当前的WebView打开，最后为打开、前进、后退添加监听器实现对应功能。代码如下：

```java
@Override
protected void onCreate(Bundle savedInstanceState) {
    super.onCreate(savedInstanceState);
    getWindow().requestFeature(Window.FEATURE_PROGRESS);
    setContentView(R.layout.activity_main);
    editurl = (EditText)findViewById(R.id.editText);
    btnForward = (Button)findViewById(R.id.button);
    btnBack = (Button)findViewById(R.id.button2);
    btnGo = (Button)findViewById(R.id.button3);
    webview = (WebView)findViewById(R.id.webView);
    WebSettings browserSetting = webview.getSettings();
    //创建WebSettings对象
    browserSetting.setSupportMultipleWindows(false);    //不支持对窗口
```

```java
browserSetting.setJavaScriptEnabled(true);        //支持JavaScript脚本
webview.setWebChromeClient(new WebChromeClient(){
    @Override
    //重写onProgressChanged方法，实现打开页面时的进度条
    public void onProgressChanged(WebView view, int progress){
        //设置打开页面时滚动条的文字
        MainActivity.this.setTitle("Loading...");
        //设置滚动条的进度
        MainActivity.this.setProgress(progress * 100);
        if(progress == 100)
            MainActivity.this.setTitle(R.string.app_name);
    }
});
webview.setWebViewClient(new WebViewClient() {
    @Override
    //重写shouldOverrideUrlLoading方法，保证在当前WebView中打开页面的链接
    public boolean shouldOverrideUrlLoading(WebView view, String url) {
        //加载页面内点击链接的请求页面
        view.loadUrl(url);
        return true;
    }
});
//前进按钮添加监听器
btnForward.setOnClickListener(new View.OnClickListener() {
    @Override
    public void onClick(View v) {
        webview.goForward();        //返回当前页面下一次打开的页面
    }
});
//后退按钮添加监听器
btnBack.setOnClickListener(new View.OnClickListener() {
    @Override
    public void onClick(View v) {
        webview.goBack();           //返回当前页面上一次打开的页面
    }
});
//打开按钮添加监听器
btnGo.setOnClickListener(new View.OnClickListener() {
    @Override
    public void onClick(View v) {
        //获取EditText中的URL地址
        String url = editurl.getText().toString().trim();
```

```java
            if(URLUtil.isNetworkUrl(url)){        //判断URL的是否正确
                webview.loadUrl(url);              //打开当前的链接
            }
            else{
                Toast.makeText(MainActivity.this, "The NetAddress is Error!", Toast.LENGTH_SHORT).show();       //URL不正确,给出提示
                editurl.requestFocus();
            }
        }
    });
    //EditText添加监听器
    editurl.setOnKeyListener(new View.OnKeyListener() {
        @Override
        public boolean onKey(View v, int keyCode, KeyEvent event) {
            if(keyCode == KeyEvent.KEYCODE_ENTER) {        //是否是Enter键
                String url = editurl.getText().toString().trim();
                if(URLUtil.isNetworkUrl(url)){
                    webview.loadUrl(url);                  //打开当前的链接
                    return true;
                }
                else{
                    Toast.makeText(MainActivity.this, "The NetAddress is Error!", Toast.LENGTH_SHORT).show();      //URL不正确,给出提示
                    editurl.requestFocus();
                }
            }
            return false;
        }
    });
}
```

运行本实例,加载页面的运行结果如图16-9所示。

强化训练

网络功能是智能手机的最基本功能之一,Android网络编程也是Android程序开发人员学习的重点之一。本章主要介绍了Android平台下进行网络编程的方法,首先介绍了基于HTTP协议访问网络的两种方法,一种是基于标准的Java接口,另一种是基于Apache的HttpClient,后者在前者的基础上进行了进一步的封装和完善,在比较复杂的网络通信应用中更加方便高效;然后介绍基于Socket的网络通信,它是网络通信的基础;最后介绍了

图16-9 浏览器浏览网页

WebView组件以及该组件中常用的对象及方法，利用该组件可以方便地实现一个具有基本功能的浏览器。

学完本章知识后，不妨通过以下习题来巩固所学的内容吧。

一、填空题

1. Android平台有三种网络编程接口可以使用，分别是_____，_____，_____。

2. HTTP协议是_____的简称，是Internet上使用最为广泛的协议，在TCP/IP体系中处于_____，是一个标准的客户端服务器模型。

3. HTTP协议中定义了八种方法和服务器进行交互，其中_____和_____方法是服务器必须实现的，也是最常用的。

4. _____类位于Java.NET包中，它是Java中提供的访问网络资源的接口，使用这些接口可以方便地编写网络应用程序。

5. 在Android系统中，最常用的是Apache提供的_____，它对Java.NET进行了抽象和封装，GET请求封装成了_____类，POST请求封装成了_____类，服务器的响应封装成了_____类，发送和接收HTTP报文的实体封装成了_____类。

6. TCP/IP协议是互联网的基础，其核心部分是传输层的_____和_____以及网际层的_____，通常在操作系统实现。

7. 网络应用程序一般不直接使用TCP/IP协议，而是使用一组调用TCP/IP协议的应用程序接口函数，这就是Socket，也称为_____，通过Socket可以很容易地编写网络应用程序完成不同主机之间的_____。

8. Java.NET中提供了两个类，_____和_____。

9. ServerSocket类常用的构造函数是_____，_____。

10. 客户端和服务器端要完成一次通信，首先客户端必须知道_____（或者主机名）和_____，其次客户端要发送一个连接请求以建立连接。

二、选择题

1. 在实际应用开发过程中，一般还需要使用WebView组件中的（　　）对象、WebViewClient对象与WebChromeClient对象设置浏览器的特性，以及用重写事件的方法来丰富浏览器的功能。

　　A. getSettings
　　B. WebSettings
　　C. JavaScipt
　　D. WebViewExplorer

2. （　　）是Apache下的子项目，是一个功能强大并全面支持HTTP协议的客户端编程工具包，它实际是对java.net.*进行了封装和扩展。

　　A. StrictMode
　　B. android.net.*
　　C. org.apache.http.*
　　D. HttpClient

3. POST常用设置（　　）是配置请求属性。

　　A. SetDoInput(Boolean)
　　B. setUseCaches(boolean)
　　C. setRequestProperty(String, String)
　　D. setInstanceFollowRedirects(boolean)

4. POST请求和GET请求的步骤基本一致，下列属于POST请求的步骤的是（　　）。

　　A. 利用HttpGet类生成HttpGet对象
　　B. 利用DefaultHttpClient类创建HttpClient对象
　　C. 调用HttpClient的execute方法发送POST请求，该方法返回一个服务器的响应HttpResponse对象
　　D. 调用HttpClient的execute方法发送GET请求，该方法返回一个服务器的响应HttpResponse对象

5. 调用HttpResponse对象的（　　）方法得到服务器HTTP响应报文的HttpEntity对象，该对象包含了服务器的应答信息。

　　A. onCreate
　　B. getEntity
　　C. GetURLResources
　　D. module

6. Android中调用Socket对象的（　　）方法获取输出流，向服务器的Socket对象发送信息。

　　A. getOutputStream
　　B. DataInputStream
　　C. DataOutputStream
　　D. getInputStream

7. Android中WebView常用的方法（　　）是刷新当前加载的页面。

　　A. goBack
　　B. loadUrl（String url）
　　C. stopLoading

D. reload

8. Android中WebSettings常用方法（　　）是设置解码时的默认编码。
A. setDefaultFontSize(int size)
B. setDefaultTextEncodingName(String Encode)
C. setSupportMultipleWindows(boolean flag)
D. setPluginsEnabled(boolean flag)

三、操作题

1. 利用Android Studio开发环境创建一个新的Android应用程序，实现加载网络图片。要求使用HttpURLConnection加载指定URL的图片，并显示到ImageView中。

2. 利用Android Studio开发环境创建一个新的Android应用程序，实现下载文件。要求使用HttpURLConnection下载指定URL的文件，并保存到手机SD卡中。

3. 利用Android Studio开发环境创建一个新的Android应用程序，实现登录功能。要求实现登录窗口，使用HttpURLConnection的POST请求方式，将用户名和密码传递回服务器端进行验证。

第17章

定位服务精讲

内容导读

定位和地图已经成为智能手机不可或缺的功能，也是很多智能手机app常见功能，例如外卖、物流、导航等，里面的商机也是显而易见的。因为地图功能与用户所在国家密切相关，Android自身并不提供地图功能，智能手机app一般介入第三方地图开发包实现相关功能。本章主要介绍在Android平台下开发基于GPS的定位服务应用程序和基于Baidu Map服务的应用程序，内容包括通过GPS获取用户位置、处理位置变化事件以及利用Baidu Map API开发融合了地图服务的应用程序。

学习目标

- 了解定位服务的相关类；
- 熟悉百度地图的开发环境搭建；
- 掌握使用百度地图进行定位的方法。

17.1 定位服务相关类

通过智能手机可以实时地获取定位信息，例如用户所在的经度、纬度、高度、方向等。如果用户的位置发生变化，还可以接到变化后的相关信息。在Android系统中，可以通过GPS和Network两种方式主动获取用户的位置。通过GPS定位只能在户外，费电而且响应慢，但是精度高；通过Network定位对用户位置没有要求，省电而且响应快，但是精度低。它们被称为位置服务提供者(LocationProvider)，开发人员可以指定两种位置服务提供者中的一种来开发具有定位服务的应用程序，也可以设定定位的条件，由系统选择两种位置服务提供者中的一种来完成定位服务。

定位服务的实现需要使用下面的类和接口。

（1）Criteria类

该类用来对定位的条件经行设置，系统根据设定的定位条件选择位置服务提供者。在定位条件中，用户最关心的是耗电量和精度，该类中设置了常量来进行耗电量和精度的描述，具体如表17-1所示。

表17-1　耗电量和精度常量

常量	描述	常量	描述
ACCURACY_HIGH	精度高	POWER_HIGH	耗电量高
ACCURACY_MEDIUM	精度中等	POWER_MEDIUM	耗电量中等
ACCURACY_LOW	精度低	POWER_LOW	耗电量低

除了耗电量和精度之外，还能对是否产生费用、海拔高度和速度等条件进行设置，该类常用的方法如表17-2所示。

表17-2　Criteria类常用的方法

方法	描述
setAccuracy(int accuracy)	设置精确度
setPowerRequirement (boolean powerRequirement)	设置用电量
setAltitudeRequired (boolean altitudeRequired)	设置是否需要高度信息
setBearingRequired (boolean bearingRequired)	设置是否需要方位信息
setSpeedAccuracy (int accuracy)	设置是否需要速度信息
setCostAllowed (boolean costAllowed)	设置是否产生费用

（2）Location类

该类封装了位置服务提供者描述当前设备的一些物理数据，包括经纬度、高度、速度、方向等数据。通过该类定义的一系列Get方法，可以返回这些数据供应用程序使用，该类定义的常用方法如表17-3所示。

表17-3　Location类常用方法

方法	描述
public float getAccuracy ()	返回定位的精确度
public double getAltitude ()	返回设备的高度数据
public float getBearing ()	返回设备方向
public double getLatitude ()	返回设备的经度
public double getLongitude ()	返回设备的纬度
public float getSpeed ()	返回设备的速度

（3）LocationProvider类

该类用来描述位置服务提供者，设置位置提供者的一些属性，这些属性一般采用系统默认的参数。

（4）LocationManager类

该类是实现设备的定位、追踪，是定位服务重要的类。需要注意的是该类不能被实例化，它是通过getSystemService方法来获得的，下面的代码能够获得一个LocationManager的实例：

```
LocationManager manager = (LocationManager)getSystemService(LOCATION_SERVICE);
```

该类中定义了两个用于描述位置服务提供者的字符串常量：GPS_PROVIDER表示使用GPS方式，NETWORK_PROVIDER表示使用网络方式。通过这些字符串常量可以直接指定位置服务提供者，也可以通过调用该类的getBestProvider方法，通过Criteria类设定的定位条件由系统确定最合适的位置服务提供者，该方法的返回值是字符串，表示位置服务提供者。该方法如下所示：

```
public string getBestProvider(Criteria criteria, boolean enabledOnly);
```

确定了位置服务提供者之后，就可以使用表示位置服务提供者的字符串作为参数调用该类的getLastKnownLocation方法来获取当前设备的定位信息，该方法返回一个Location对象，通过该对象我们就能够得到设备的诸如经纬度、速度、高度等信息。该方法如下所示：

```
public Location getLastKnownLocation(String provider);
```

通过前面的步骤我们只能主动获得设备的定位信息，如果需要在定位信息或者状态发生变化的时候主动通知系统，就需要为LocationManager添加一个LocationListener监听器，调用该类的requestLocationUpdates方法就可以添加一个监听器，该方法如下所示：

```
public void requestLocationUpdates(String provider, long minTime, float minDistance, LocationListener listener);
```

该方法的第二个参数是位置更新的最短时间间隔，第三个参数是位移变化的最短距离。

（5）LocationListener监听器

在LocationListener监听器中定义了4个方法，实现监听器需要重写这几个方法。这4个方法如表17-4所示。

表17-4　LocationListener监听器方法

方法	描述
onLocationChanged(Location location)	设备位置变化时调用该方法
onProviderDisabled(String Provider)	设备禁用时调用该方法
onProviderEnabled(String Provider)	设备启用时调用该方法
onStatusChanged(String Provider,int status,Bundle extras)	当设备状态变化时调用该方法

17.2 定位实例

本节将通过实例来讲解 Android 平台下的定位服务开发。Android 平台下的定位服务开发的一般步骤如下。

① 通过 getSystemService 方法实例化一个 LocationManager 类的对象：

```
LocationManager manager = (LocationManager)getSystemService(LOCATION_SERVICE);
```

② 如果不需要设置定位条件，直接转入第④步，否则实例化一个 Criteria 类的对象，设置查询条件：

```
Criteria criteria = new Criteria();
criteria.setAccuracy(Criteria. ACCERACY_HIGH);    //设置精度
criteria.setPowerRequirement(Criteria.POWER_LOW);  //设置耗电量
criteria.setAltitudeRequired(false);         //设置是否需要海拔高度
criteria.setBearingRequired(false);          //设置是否需要方向
criteria.setSpeedRequired(false);           //设置是否需要速度
criteria.setCostAllowed(false);            //设置是否允许产生费用
```

③ 通过 LocationManager 类的 getBestProvider 方法得到位置服务提供者：

```
String provider = manager.getBestProvider(criteria, true);
```

④ 通过 LocationManager 类的 getLastKnownLocation 方法来获取当前设备的定位信息：

```
Location location = manager.getLastKnownLocation(provider);
```

如果没有设定定位条件，使用 LocationManager 的常量 GPS_PROVIDER 或者 NETWORK_PROVIDER 作为参数：

```
Location location = manager.getLastKnownLocation(LocationManager.GPS_PROVIDER);
```

需要注意的是模拟器不提供网络定位服务，无法使用 NETWORK_PROVIDER。

⑤ 由于定位设备的定位需要一定的时间，在使用得到的 Location 对象时，可能出现 Location 对象为 null 的情况，导致信息不正确。应该对 Location 对象进行判断，然后再获取定位信息进行应用。

⑥ 为 LocationManager 绑定 LocationListener 监听器并重写对应的方法：

```
manager.requestLocationUpdates(provider,10000,10, new LocationListener(){
  //重写对应的方法
});
```

扫一扫 看视频

☞【实例17-1】在 Android Studio 中新建一个名称为 Chapter17 的 Android 的项目，获取更新后的经度和纬度信息，在 EditText 中进行显示。

① 在项目的 AndroidManifest.xml 文件的 Manifast 标签中加入如下指令：

第17章 定位服务精讲

```
<uses-permission android:name="android.permission.ACCESS_FINE_LOCATION" />
<uses-permission android:name="android.permission.ACCESS_COARSE_LOCATION" />
```

② 修改项目的布局文件activity_main.xml，使用线性布局管理器，添加一个显示经度和纬度的EditText控件。

③ 在MainActivity类中，创建要使用的变量，重写onCreate方法，代码如下：

```java
public class MainActivity extends AppCompatActivity {
    private LocationManager locationmanager;
    private Location location;
    private EditText edittext;
    boolean isGranted = false;
    //需要动态申请的权限
    String[] permissions = new String[]{Manifest.permission.ACCESS_FINE_LOCATION
            , Manifest.permission.ACCESS_COARSE_LOCATION};
    //存储未获取的权限
    List<String> mPermissionList = new ArrayList<>();
    @Override
    protected void onCreate(Bundle savedInstanceState) {
        super.onCreate(savedInstanceState);
        setContentView(R.layout.activity_main);
        grantedAndRequest();
        edittext = (EditText) findViewById(R.id.editText);
        //获取LocationManager的实例
        locationmanager = (LocationManager) getSystemService(LOCATION_SERVICE);
        if (ActivityCompat.checkSelfPermission(this,
Manifest.permission.ACCESS_FINE_LOCATION) != PackageManager.PERMISSION_GRANTED && ActivityCompat.checkSelfPermission(this,
Manifest.permission.ACCESS_COARSE_LOCATION) != PackageManager.PERMISSION_GRANTED) {
            return;
        }
        //使用GPS作为位置服务提供者，获得包含位置信息的Location对象
        location = locationmanager.getLastKnownLocation(LocationManager.GPS_PROVIDER);
        //更新EditText的内容
        update(location);
        //添加LocationListener监听器
        locationmanager.requestLocationUpdates(LocationManager.GPS_PROVIDER, 10000, 5, new LocationListener() {
            @Override
            //当位置信息发生变化时，更新EditText的内容
            public void onLocationChanged(Location newLocation) {
```

```java
                update(newLocation);
            }
            @Override
            //当定位设备停止服务时，更新EditText的内容
            public void onProviderDisabled(String provider) {
                edittext.setText("定位设备" + provider + "停止服务");
            }
            @Override
            //当定位设备启动服务时，获取Location对象，更新EditText的内容
            public void onProviderEnabled(String provider) {
                location = locationmanager.getLastKnownLocation(provider);
                update(location);
            }
            @Override
             public void onStatusChanged(String provider, int status, Bundle extras) {
            }
        });
    }
    //判断是否授权，如未授权，则申请授权
    private void grantedAndRequest() {
        if (Build.VERSION.SDK_INT >= Build.VERSION_CODES.M) {//针对Android6.0及之后版本
            mPermissionList.clear();
            for (int i = 0; i < permissions.length; i++) {
                if (checkSelfPermission( permissions[i]) != PackageManager.PERMISSION_GRANTED) {
                    mPermissionList.add(permissions[i]);
                }
            }
            if (mPermissionList.isEmpty()) {//未授予的权限为空，表示都授予了
                isGranted = true;
            } else {//请求权限方法
                String[] permissions = mPermissionList.toArray(new String[mPermissionList.size()]);//将List转为数组
                requestPermissions(permissions, 1);
            }
        }else{
            isGranted = true;
        }
    }
    @Override
```

第17章 定位服务精讲

```java
public void onRequestPermissionsResult(int requestCode, @NonNull String[] permissions, @NonNull int[] grantResults) {
    super.onRequestPermissionsResult(requestCode, permissions, grantResults);
    if(requestCode==1){
        isGranted = true;
        for (int i = 0; i < grantResults.length; i++) {
            if (grantResults[i] != PackageManager.PERMISSION_GRANTED) {
                isGranted = false;
            }
        }
    }
}
```

④ 编写无返回值的方法update，该方法判断Location是否为空，如果不为空，获取Location对象中的经度和纬度并在EditText中显示，代码如下：

```java
public void update(Location location){
    if (location != null){
        edittext.setText("当前位置:" + "\n");
        edittext.append("纬度:" + location.getLatitude() + "\n");
        edittext.append("经度:" + location.getLongitude());
    }
    else {
        edittext.setText("定位失败");
    }
}
```

模拟器中没有GPS设备，可以通过DDMS中的模拟器控制台向模拟器发送模拟的GPS数据，模拟器控制台如图17-1所示。启动应用程序后，由于无法得到定位信息，会出现如图17-2所示的状态，单击SEND按钮，当模拟器接收到模拟数据后，文本框将显示出模拟的经度和纬度，如图17-3所示。

图17-1　模拟器控制台

图17-2　定位失败　　　　　　　　　图17-3　定位成功

17.3　Baidu Map使用

百度地图是为用户提供包括智能路线规划、智能精准导航（驾车、步行、骑行）、实时路况等相关出行服务的地图服务，是目前最常用的地图服务之一。通过该地图提供的服务，用户可以方便地查找周边商家信息、行车及公交路线。开发融合了地图服务的Android应用程序是程序员必须掌握的一项技术。百度公司提供的百度地图 Android SDK是一套基于 Android4.0及以上版本的应用程序接口。程序开发人员可以使用该套 SDK 开发适用于 Android系统移动设备的地图应用。不过要使用百度的地图服务，必须先申请经过验证的 Baidu Map的API KEY。

17.3.1　申请Map API KEY

API KEY 的申请地址为：http://lbsyun.baidu.com/apiconsole/key。申请与配置步骤如下。

（1）登录百度账号

访问 API 控制台页面，若未登录百度账号，将会进入百度账号登录页面，如图17-4所示。

图17-4　登录百度账号

（2）登录 API 控制台

登录账号后会跳转到 API 控制台服务，具体如图17-5所示。

第17章 定位服务精讲

图17-5 API控制台服务

（3）创建应用

点击"创建应用"，进入创建应用页面（图17-6），输入应用名称，将应用类型改为"Android SDK"。

图17-6 创建应用

（4）配置应用

在应用类型选为"Android SDK"后，需要配置应用的安全码，如图17-7所示。

图17-7 配置应用的安全码

（5）获取安全码

安全码的组成规则为："Android签名证书的SHA1值+包名"。例如：

```
SHA1:BB:0D:AC:74:D3:21:E1:43:67:71:9B:62:91:AF:A1:66:6E:44:5D:75
包名：com.baidumap.demo
```

Android应用获取包名(packagename)，使用 Android Studio 开发，需要在文件build.gradle中查询 applictionId，如图17-8。

```
android {
    compileSdkVersion 28
    buildToolsVersion "28.0.2"
    defaultConfig {
        applicationId "cn.androidstudy.chapter17"
        minSdkVersion 21
        targetSdkVersion 28
        versionCode 1
        versionName "1.0"
```

图17-8　获取包名

Android签名证书的SHA1值分为发布版和开发版，下面以开发版SHA1为例来说明如何获取Android签名证书的SHA1值。

第一步，找到 debug.keystore 文件，该文件一般放在C:\Users\用户名\.android下。

第二步，进入上述目录，按下Shift键和鼠标右键，选择"在此处打开命令窗口"命令，并输入命令 "keytool -v -list -keystore debug.keystore"。

第三步，输入秘钥口令"android"，看到图17-9就代表成功了。

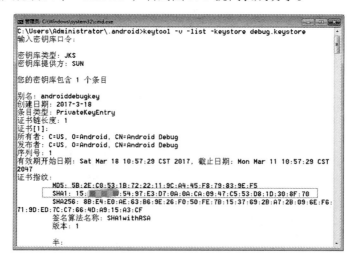

图17-9　获取Android签名证书的SHA1值

发布版SHA1可以使用相同的方法从自定义的签名文件中获取。

（6）成功创建AK

在输入安全码后，点击"提交"完成应用的配置工作，将会得到一个创建的AK（图17-10），请妥善保管所申请的AK。到这里，就可以使用新AK来完成开发工作了。

第17章 定位服务精讲

图17-10 创建密钥

到此，已经成功创建了百度地图的Map API key，使用复制功能将该密钥保存好，在后面的开发中需要用到。

17.3.2 开发和测试环境搭建

打开http://developer.baidu.com/map/index.php?title=androidsdk/sdkandev-download网址下载SDK（图17-11），可以全部下载，也可以自定义下载。从V2.3.0之后的版本，SDK的开发包以可定制的形式提供下载，用户可以根据自己的项目需要勾选相应的功能下载对应的SDK开发包。

图17-11 百度地图SDK下载

点击"自定义下载"，打开开发资源下载平台（图17-12），选择需要的资源后，点击"开发包"按钮进行下载。

图17-12 开发资源下载平台

387

将下载的压缩包解压缩，主要文件如图17-13所示。

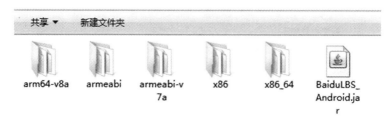

图17-13　查看下载的文件

下面通过一个示例演示如何在Android项目中使用百度地图。

☞【实例17-2】在Chapter17中新建一个Module——MapTest，该项目将一张全球地图显示在手机屏幕上。

① 添加文件。将图17-13中的"BaiduLBS_Android.jar"文件复制到项目的libs文件夹；在src/main/文件夹下新建一个文件夹jniLibs，将下载的其他文件复制到jniLibs文件夹中，如图17-14所示。

图17-14　添加JAR包

② 修改项目的AndroidManifest.xml文件，在AndroidManifest.xml中添加开发密钥和所需权限，如图17-15所示。

③ 修改项目的布局文件activity_main.xml，在布局文件中添加地图控件，代码如下所示：

```xml
<?xml version="1.0" encoding="utf-8"?>
<android.support.constraint.ConstraintLayout
    xmlns:android="http://schemas.android.com/apk/res/android"
    xmlns:app="http://schemas.android.com/apk/res-auto"
    xmlns:tools="http://schemas.android.com/tools" android:layout_
```

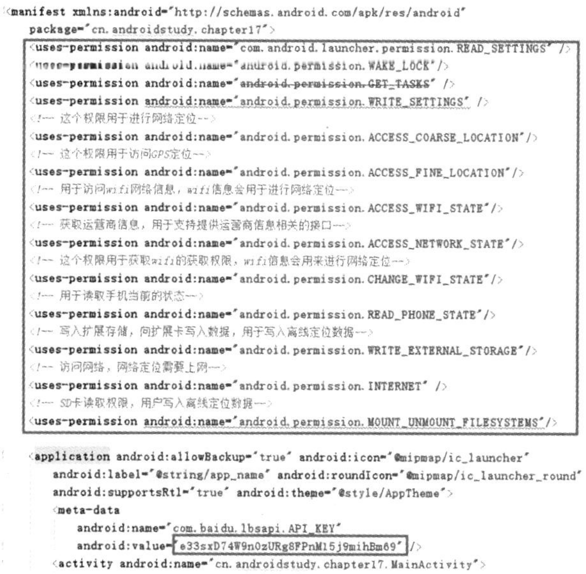

```
<manifest xmlns:android="http://schemas.android.com/apk/res/android"
    package="cn.androidstudy.chapter17">
    <uses-permission android:name="com.android.launcher.permission.READ_SETTINGS" />
    <uses-permission android:name="android.permission.WAKE_LOCK"/>
    <uses-permission android:name="android.permission.GET_TASKS" />
    <uses-permission android:name="android.permission.WRITE_SETTINGS" />
    <!-- 这个权限用于进行网络定位-->
    <uses-permission android:name="android.permission.ACCESS_COARSE_LOCATION"/>
    <!-- 这个权限用于访问GPS定位-->
    <uses-permission android:name="android.permission.ACCESS_FINE_LOCATION"/>
    <!-- 用于访问wifi网络信息，wifi信息会用于进行网络定位-->
    <uses-permission android:name="android.permission.ACCESS_WIFI_STATE"/>
    <!-- 获取运营商信息，用于支持提供运营商信息相关的接口-->
    <uses-permission android:name="android.permission.ACCESS_NETWORK_STATE"/>
    <!-- 这个权限用于获取wifi的获取权限，wifi信息会用来进行网络定位-->
    <uses-permission android:name="android.permission.CHANGE_WIFI_STATE"/>
    <!-- 用于读取手机当前的状态-->
    <uses-permission android:name="android.permission.READ_PHONE_STATE"/>
    <!-- 写入扩展存储，向扩展卡写入数据，用于写入离线定位数据-->
    <uses-permission android:name="android.permission.WRITE_EXTERNAL_STORAGE"/>
    <!-- 访问网络，网络定位需要上网-->
    <uses-permission android:name="android.permission.INTERNET" />
    <!-- SD卡读取权限，用户写入离线定位数据-->
    <uses-permission android:name="android.permission.MOUNT_UNMOUNT_FILESYSTEMS"/>

    <application android:allowBackup="true" android:icon="@mipmap/ic_launcher"
        android:label="@string/app_name" android:roundIcon="@mipmap/ic_launcher_round"
        android:supportsRtl="true" android:theme="@style/AppTheme">
        <meta-data
            android:name="com.baidu.lbsapi.API_KEY"
            android:value="e33sxD74W9n0zURg8FPnM15j9mihBm69" />
        <activity android:name="cn.androidstudy.chapter17.MainActivity">
```

图17-15　添加开发密钥和所需权限

```
width="match_parent"
        android:layout_height="match_parent"  tools:context="cn.
androidstudy.chapter17.MainActivity">
    <com.baidu.mapapi.map.MapView
        android:id="@+id/bmapView"
        android:layout_width="fill_parent"
        android:layout_height="fill_parent"
        android:clickable="true" />
</android.support.constraint.ConstraintLayout>
```

④ 修改MainActivity类，代码如下：

```
package cn.androidstudy.chapter17;
import android.location.Location;
import  android.location.LocationListener;
import android.location.LocationManager;
import  android.support.v7.app.AppCompatActivity;
import  android.os.Bundle;
import com.baidu.mapapi.SDKInitializer;
import com.baidu.mapapi.map.BaiduMap;
import  com.baidu.mapapi.map.BitmapDescriptorFactory;
import  com.baidu.mapapi.map.MapView;
import com.baidu.mapapi.map.MyLocationConfiguration;
import com.baidu.mapapi.map.MyLocationData;

public class MainActivity extends AppCompatActivity {
```

```java
        MapView mMapView = null;
        BaiduMap mBaiduMap;

        @Override
        protected void onCreate(Bundle savedInstanceState) {
            super.onCreate(savedInstanceState);
            //在使用SDK各组件之前初始化Context信息,传入ApplicationContext
            //注意该方法要再setContentView方法之前实现
            SDKInitializer.initialize(getApplicationContext());
            setContentView(R.layout.activity_main);
            //获取地图控件引用
            mMapView = (MapView) findViewById(R.id.bmapView);
            //mBaiduMap = mMapView.getMap();


        }
        @Override
        protected void onDestroy() {
            super.onDestroy();
            //在Activity执行onDestroy时执行mMapView.onDestroy(),实现地图生命周期管理
            mMapView.onDestroy();
        }
        @Override
        protected void onResume() {
            super.onResume();
            //在Activity执行onResume时执行mMapView. onResume (),实现地图生命周期管理
            mMapView.onResume();
        }
        @Override
        protected void onPause() {
            super.onPause();
            //在Activity执行onPause时执行mMapView. onPause (),实现地图生命周期管理
            mMapView.onPause();
        }
    }
```

运行该项目,如果开发环境、测试环境、密钥、权限、代码都正确无误的话,将会在屏幕上看到如图17-16所示的地图,使用鼠标左右拖拽会看到不同国家和地区的地图,使用"+"和"−"可以对地图进行放大和缩小。

百度地图将地图的类型分为两种:普通矢量地图和卫星图。可以使用如下语句进行切换:

```
//获取地图控件引用
mMapView = (MapView) findViewById(R.id.bmapView);
mBaiduMap = mMapView.getMap();
//普通地图
mBaiduMap.setMapType(BaiduMap.MAP_TYPE_NORMAL);
//卫星地图
mBaiduMap.setMapType(BaiduMap.MAP_TYPE_SATELLITE);
```

请大家自行测试。

图17-16　地图显示

17.4 地图定位

上一节的实例仅仅是一个全球地图的显示。本节利用GPS定位服务确定设备所在的位置，并将该位置显示在地图上，完成地图定位的功能。下面对MapTest进行修改，使其打开后，自动进行定位，然后将地图的中心设为当前位置。

① 添加权限。定位可以通过GPS或者Wi-Fi，添加如下权限：

```
<uses-permission android:name="android.permission.ACCESS_COARSE_LOCATION"/>
<uses-permission android:name="android.permission.ACCESS_FINE_LOCATION"/>
```

```xml
<uses-permission android:name="android.permission.ACCESS_NETWORK_STATE"/>
<uses-permission android:name="android.permission.READ_PHONE_STATE"/>
<uses-permission android:name="android.permission.MOUNT_UNMOUNT_FILESYSTEMS"/>
```

② 初始化LocationClient类。在MainActivity类中定义LocationClient对象：

```java
public LocationClient mLocationClient = null;
```

然后在OnCreate方法的最后添加：

```java
if(isGranted) {
   //声明LocationClient类
   mLocationClient = new LocationClient(getApplicationContext());
mLocationClient.registerLocationListener(myListener);//注册监听函数
   mLocationClient.start();//开始定位
}
```

由于使用危险权限，请参照前文介绍添加权限的动态申请代码。

③ 由于定位是一个耗时的操作，需要异步运行，定位完成后通知主线程，在前面的代码的第二行注册监听函数。BDLocation类封装了SDK的定位结果，在BDLocationListener的onReceiveLocation方法中获取。定义myListener的代码如下所示：

```java
private BDLocationListener myListener = new BDLocationListener() {
   @Override
   public void onReceiveLocation(BDLocation location) {
       //定位sdk获取位置后回调
       if (null != location && location.getLocType() != BDLocation.TypeServerError) {
           MapStatus.Builder builder = new MapStatus.Builder();
           //根据经纬度设置地图操作的中心点
           builder.target(new LatLng(location.getLatitude(),location.getLongitude()));
           mBaiduMap.setMapStatus(MapStatusUpdateFactory.newMapStatus(builder.build()));
       }
   }
}
```

BDLocation类常用方法如下：

```
location.getTime();  //服务端输出本次结果的时间
location.getLocType();  //定位类型 BDLocation.TypeGpsLocation——GPS定位
                //BDLocation.TypeNetWorkLocation——网络定位
location.getLocTypeDescription(); // 定位类型说明
```

```
location.getLatitude();    //纬度
location.getLongitude();   //经度
location.getRadius();  //误差半径
location.getCountryCode();//国家码，null代表没有信息
location.getCountry();//国家名称
location.getCityCode();//城市编码
location.getCity();//城市
location.getDistrict();//区
location.getStreet();//街道
location.getAddrStr();//地址信息
location.getDirection();//获取方向
```

有兴趣的话，可以自行测试。

④ 程序退出时，注销监听并停止定位，代码如下：

```
@Override
protected void onStop() {
    super.onStop();
    if(mLocationClient!=null) {
        mLocationClient.unRegisterLocationListener(myListener);
//注销掉监听
        mLocationClient.stop();  //停止定位
    }
}
```

由于模拟器不支持GPS，建议大家使用真机进行测试，定位地图的结果如图17-17所示。

图17-17　地图定位

强化训练

定位和地图是目前很多智能手机app的常见功能。本章介绍了Android平台进行定位的主要方式，以及完成定位服务常用的类和接口，还介绍了如何实现百度提供的地图服务，包括Map Key的获得、开发平台和测试平台的配置，最后结合GPS和Baidu Map实现了地图定位，将这两种功能融合到应用程序中，会增加Android系统的很多特色。

学完本章知识后，不妨通过以下习题来巩固所学的内容吧。

一、填空题

1. 在Android系统中，可以通过_____和_____两种方式主动获取用户的位置，它们被称为位置服务提供者（LocationProvider）。
2. Android中定位服务的实现需要使用的类和接口：_____，_____，_____，_____，_____。
3. 模拟器中没有GPS设备，可以通过_____中的模拟器控制台向模拟器发送模拟的GPS数据。
4. 如果用Baidu的地图服务，必须先申请经过验证Baidu Map的_____。
5. 百度地图将地图的类型分为两种：_____和_____。
6. GPS定位只能在户外用，费电而且_____，但是_____。
7. 通过Network定位对用户位置没有要求，省电而且_____，但是_____。

二、选择题

1. Criteria类常用的方法中，（ ）设置是否需要速度信息。
   A. setPowerRequirement (boolean powerRequirement)
   B. setBearingRequired (boolean bearingRequired)
   C. setSpeedAccuracy (int accuracy)
   D. setAccuracy(int accuracy)
2. Location类常用方法（ ）返回设备的纬度。
   A. public double getLongitude ()
   B. public float getBearing ()
   C. public double getAltitude ()
   D. public float getSpeed ()
3. (   ) 类实现设备的定位、追踪，是定位服务的重要的类，需要注意的是该类不能被实例化。
   A. LocationProvide
   B. Location
   C. Criteria
   D. LocationManager
4. LocationListener监听器方法中（ ）在设备启用时被调用。
   A. onProviderEnabled(String Provider)
   B. onLocationChanged(Location location)

C. onStatusChanged(String Provider,int status,Bundle extras)
D. onProviderDisabled(String Provider)
5. 实例化一个Criteria类的对象，设置的查询条件中，（ ）设置是否需要方向。
A. criteria.setAltitudeRequired(false)
B. criteria.setAltitudeRequired(false)
C. criteria.setCostAllowed(false)
D. criteria.setBearingRequired(false)

三、操作题

1. 利用Android Studio开发环境创建一个新的Android应用程序，实现定位功能。要求使用百度地图进行定位，不仅输出用户的位置的经纬度，还要输出用户所在的省、市、区、街道，验证是否正确。注意：由于GPS在室内无法使用，建议在户外进行测试。
2. 利用Android Studio开发环境创建一个新的Android应用程序，实现地图定位标注功能。要求在百度地图上标注用户的当前位置。
3. 用百度地图的服务，必须先申请经过验证的API KEY。练习申请Map API KEY。
4. 搭建Android地图开发和测试环境。
5. 利用Android Studio开发环境创建一个新的Android应用程序，利用GPS定位服务确定设备所在的位置，并将该位置显示在地图上，完成地图定位的功能。

第18章

Android应用项目的设计与开发

内容导读

本章将介绍如何使用Android技术开发一个卓卓音乐盒，重点介绍了Android如何获取本地及网络音乐数据，并进行数据的展示，实现实时网络图片加载、Android UI布局、UI界面的动态更新、数据全局共享处理、界面数据交互等。通过对该实例的学习，读者可以熟悉移动应用的开发和设计的全过程，能够在Android开发技术上提升一个新的台阶。

学习目标

- 了解移动项目的开发流程；
- 熟悉移动网应用的开发和设计的全过程；
- 掌握本地及网络数据的获取、网络图片的加载等技术的应用。

18.1 系统概述

随着网络信息技术的进步以及智能终端的大面积普及，移动信息化的趋势越来越明显，移动应用开始逐渐蚕食传统互联网的市场份额。现代网络服务提供商已经开始大规模进入移动应用战圈，网络服务早已不是守候在电脑前查找相关信息的模式。现代网络服务即时、当

第18章 Android应用项目的设计与开发

面、便捷，随时随地利用移动终端查询信息成为用户使用的焦点。用户可以随时随地地浏览信息、进行网上交易、通过移动终端在线查询信息的需求也越来越迫切。音乐播放器作为手机终端最基础的功能，其质量直接影响着用户对手机的体验。

在基于Android系统的众多应用中，音乐播放器是每个手机都必备的应用软件。现今生活节奏快，生活压力大，在日常休息之余，欣赏音乐是舒缓压力的方式之一。本章以开发卓卓音乐盒项目为实例来说明项目分析、设计和实现的全过程。

18.1.1 项目总体需求

近年来随着移动终端技术的发展，智能手机已成为人们生活中不可或缺的一部分，而它的作用也不再局限于通信，它在更多场景下满足了用户不同的娱乐需求并方便着人们的生活。一部小小的手机里就集成了地图、新闻、天气、游戏等诸多功能，而在线音乐播放功能更是在诸多场景中得到了广泛的青睐和应用。Android是目前主流的移动操作系统，也是全球移动操作系统中占有市场份额最大的系统，基于Android开发的应用软件也越来越多。

18.1.2 项目功能分析

随着手机的功能的不断增加，音乐播放功能已经是时下各大手机厂商手机方案中的标配。总结设计思路和对市场的调研，音乐播放器需要提供以下核心功能。

① 播放音乐文件。
② 播放音乐文件时实现暂停、播放。
③ 播放音乐文件时支持前一首或后一首音乐文件与当前音乐文件的切换。
④ 播放音乐文件时支持进度条显示并显示已播放时间。
⑤ 可以设置播放模式。
⑥ 显示手机或网络上音乐文件列表。
⑦ 支持歌词文件同步。

18.1.3 运行环境

操作系统：Android 操作系统。
支持环境：Android6.0以上版本。
开发环境：Android Studio 3.5.1，Java JDK 8.0。

18.2 系统框架设计

根据上一节的系统功能分析，系统主要功能为音乐播放列表展示、音乐播放控制。播放列表分为本机（图18-1）和网络（图18-2），播放控制是相同的。可以使用Fragment显示不同的播放列表，Fragment之间的切换通过侧滑菜单（图18-3）进行。

实现侧滑菜单，可以通过添加DrawerLayout布局实现，在此推荐使用Android Studio的项目模板Navigation Drawer Activity（图18-4），创建项目即可包含侧滑菜单。

图18-1 本机音乐

图18-2 网络音乐

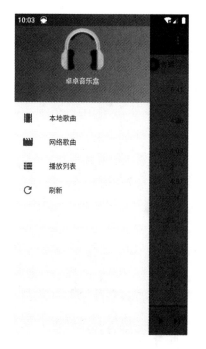

图18-3 侧滑菜单

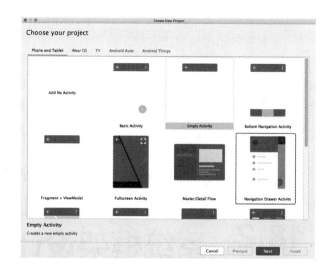

图18-4 选择项目模板

打开activity_main.xml，可以看到侧滑菜单的布局：

```
<android.support.design.widget.NavigationView
    android:id="@+id/nav_view"
    android:layout_width="wrap_content"
```

```
android:layout_height="match_parent"
android:layout_gravity="start"
android:fitsSystemWindows="true"
app:headerLayout="@layout/nav_header_main"
app:menu="@menu/activity_main_drawer" />
```

通过修改 app:headerLayout 和 app:menu 对应的文件，即可得到图 18-3 所示的效果。

通过 MainActivity 展示内容，直接修改 content_main.xml 布局即可。该布局（图 18-5）主要包括两部分：FrameLayout 和播放控制部分。FrameLayout 对应红色矩形选中部分，用来显示不同的 Fragment；播放控制部分为矩形框以外，可以采用不同的布局实现。

图18-5　侧滑菜单　　　　　　　　　图18-6　项目结构

在设计实现的时候采用分层的思想，分层的好处在于代码清晰、结构分明，有利于修改、维护和复用，每个功能模块都有自己对应的页面、业务处理、数据获取、数据一系列独立的模块结构。

根据分层的思想，不同层的文件放入不同的包中。如图 18-6 所示，根据类的不同把它们分到不同的包中，方便进行管理与维护。

18.3　本地歌曲列表

卓卓音乐盒启动后，默认情况下会打开本地歌曲列表，主要任务：
① 获取本机音乐列表；

② 使用 ListView 组件进行数据显示。

18.3.1 创建 Fragment

首先，在项目的 ui 包中创建 Fragment——LocalMusicFragment，其布局如图 18-7 所示，下方为 ListView 组件，上方为三个 TextView 组件，用来从不同的视角查看数据。

图18-7　本地音乐布局

其次，在 MainActivity 的 onCreate 方法中添加代码，指定加载的 Fragment 为本机音乐 Fragment。代码如下：

```
// 加载本地音乐Fragment
getSupportFragmentManager().beginTransaction()
    .replace(R.id.container, LocalMusicFragment.newInstance())
    .commitNow();
```

18.3.2 获取本机音乐列表

（1）创建实体类：Music

为了便于存储音乐数据，在 bean 包中创建一个实体类 Music，代码如下：

```
public class Music {
    private String song_id;//歌曲编号
    private String title;//歌名
    private String album_id;//专辑id
    private String album_title;//专辑名
    private int file_duration;//歌曲时长
```

```
private String artist_name;//歌手名
private String lrclink;//歌词url
private String pic_big;//专辑图片150×150
private String pic_small;//专辑图片90×90
private String pic_huge;//专辑图片500×500
private boolean isLocal;//区分本地和网络歌曲
private String url;//存储歌曲的实际地址，用于播放歌曲
（省略get、set）
}
```

（2）创建工具类：LocalMusicApi

在util包中创建工具类LocalMusicApi，用来封装对本地音乐的常用操作。通过前面的学习，掌握了读取本机音乐的两种方法：对手机文件进行遍历，找出所有的音乐文件；使用内容提供者MediaStore。由于第一种方法效率较低，在此采用第二种方法。

MediaStore这个类是Android系统提供的多媒体数据库，Android中多媒体信息都可以从这里提取，包括了多媒体数据库的所有信息，例如音频、视频和图像。Android把所有的多媒体数据接口进行了封装，所有的数据库不用自己创建，直接用ContentResolver去调用那些已经封装好的接口就可以进行数据库操作了。操作视频和图像的方法和操作音频类似，只需要将URL部分改动就可以了。

```
/**
 * 获取手机上的所有音乐，添加过滤条件：时长>10s，去除手机中的铃声
 * @param resolver ContentResolver对象
 * @return 音乐列表
 */
public static List<Music> getMusic(ContentResolver resolver) {
    List<Music> list = new ArrayList<>();

    Cursor cursor = resolver.query(MediaStore.Audio.Media.EXTERNAL_CONTENT_URI,
            new String[]{MediaStore.Audio.Media._ID,
                    MediaStore.Audio.Media.TITLE,
                    MediaStore.Audio.Media.ALBUM_ID,
                    MediaStore.Audio.Media.ALBUM,
                    MediaStore.Audio.Media.DURATION,
                    MediaStore.Audio.Media.ARTIST,
                    MediaStore.Audio.Media.DATA
            },
            MediaStore.Audio.Media.DURATION + ">=10000", null, null);
    while (cursor.moveToNext()) {
        Music music = new Music();
        music.setSong_id(cursor.getString(0));
        music.setTitle(cursor.getString(1));
        music.setAlbum_id(cursor.getString(2));
```

```java
            music.setAlbum_title(cursor.getString(3));
            music.setFile_duration(cursor.getInt(4) / 1000);
            music.setArtist_name(cursor.getString(5));
            music.setUrl(cursor.getString(6));
            music.setLocal(true);
            list.add(music);
        }
        cursor.close();
        return list;
    }
```

（3）动态申请权限

运行getMusic方法，需要授予读写SD卡的权限，打开AndroidManifest.xml文件，添加如下代码：

```xml
<uses-permission android:name="android.permission.READ_EXTERNAL_STORAGE"/>
<uses-permission android:name="android.permission.WRITE_EXTERNAL_STORAGE"/>
```

如果运行在Android6.0之下的版本，可以获取手机上的音乐列表，但是在Android 6.0及之后的版本，需要进行动态授权（参考12.3.3节）。为了简化动态申请权限，可以对此进行封装，定义BaseActivity类，代码如下：

```java
public class BaseActivity extends AppCompatActivity {
    protected final String TAG = getClass().getSimpleName().replace("Activity", "Act");
    private SparseArray<OnPermissionResultListener> listenerMap = new SparseArray<>();

    /**
     * 权限请求结果监听者
     */
    public interface OnPermissionResultListener {
        /**
         * 权限被允许
         */
        void onAllow();

        /**
         * 权限被拒绝
         */
        void onReject();
    }

    /**
```

第18章 Android应用项目的设计与开发

```java
 * 镜像权限申请
 * @param onPermissionResultListener 申请权限结果回调
 */
public void checkPermissions(final String[] permissions,
OnPermissionResultListener onPermissionResultListener) {
    if (Build.VERSION.SDK_INT < 23 || permissions.length == 0)
{// android6.0已下不需要申请,直接为"同意"
        if (onPermissionResultListener != null)
            onPermissionResultListener.onAllow();
    } else {
        int size = listenerMap.size();
        if (onPermissionResultListener != null) {
            listenerMap.put(size, onPermissionResultListener);
        }
        ActivityCompat.requestPermissions(this, permissions, size);
    }
}

/**
 * 显示提示"跳转到应用权限设置界面"的dialog
 * @param permission 具体的某个权限,用于展示dialog的内容文字
 */
 private void showTipDialog(String permission, final
OnPermissionResultListener onPermissionResultListener) {
    new AlertDialog.Builder(this)
            .setMessage("打开App设置")
            .setTitle("提示")
            .setPositiveButton("确定", new DialogInterface.OnClickListener() {
                @Override
                public void onClick(DialogInterface dialog, int which) {
                    toAppDetailSetting();
                }
            })
            .setNegativeButton("取消", new DialogInterface.OnClickListener() {
                @Override
                public void onClick(DialogInterface dialog, int which) {
                    onPermissionResultListener.onReject();
                }
            })
            .create().show();
}
```

```java
/**
 * 跳转系统的app应用详情页
 */
protected void toAppDetailSetting() {
    Intent localIntent = new Intent();
    localIntent.addFlags(Intent.FLAG_ACTIVITY_NEW_TASK);
    localIntent.setAction("android.settings.APPLICATION_DETAILS_SETTINGS");
    localIntent.setData(Uri.fromParts("package", getPackageName(), null));
    startActivity(localIntent);
}

@Override
protected void onDestroy() {

    listenerMap.clear();
    listenerMap = null;
    super.onDestroy();
}

@Override
  public void onRequestPermissionsResult(int requestCode, @NonNull String[] permissions, @NonNull int[] grantResults) {
    super.onRequestPermissionsResult(requestCode, permissions, grantResults);
    OnPermissionResultListener onPermissionResultListener = listenerMap.get(requestCode);
    if(onPermissionResultListener != null) {
        listenerMap.remove(requestCode);
        // 循环判断权限，只要有一个拒绝了，则回调onReject()。 全部允许时才回调onAllow()
        for (int i = 0; i < grantResults.length; i++) {
            if (grantResults[i] == PackageManager.PERMISSION_DENIED) {// 拒绝权限
                // 对于 ActivityCompat.shouldShowRequestPermissionRationale
                // 1：用户拒绝了该权限，没有勾选"不再提醒"，此方法将返回true
                // 2：用户拒绝了该权限，有勾选"不再提醒"，此方法将返回 false
                // 3：如果用户同意了权限，此方法返回false
                if (!ActivityCompat.shouldShowRequestPermissionRationale(this, permissions[i])) {
```

第18章 Android应用项目的设计与开发

```
                            // 拒绝选了"不再提醒",一般提示跳转到权限设置页面
                            showTipDialog(permissions[i], onPermission
ResultListener);
                        } else {
                            onPermissionResultListener.onReject();
                        }
                        return;
                    }
                }
                onPermissionResultListener.onAllow();
            }
        }

        public void showMsg(String msg){
            Toast.makeText(this, msg, Toast.LENGTH_SHORT).show();
        }
    }
```

然后,让项目中的所有Activity继承BaseActivity即可。如让MainActivity继承BaseActivity,这样,在LocalMusicFragment中,获取本地音乐列表的代码如下:

```
    activity = (MainActivity) getActivity();
    resolver=activity.getContentResolver();
    //加载本地音乐需要授予访问SD卡权限,动态申请
    activity.checkPermissions(new String[]{Manifest.permission.READ_
EXTERNAL_STORAGE,
                Manifest.permission.WRITE_EXTERNAL_STORAGE},
        new BaseActivity.OnPermissionResultListener() {
            @Override
            public void onAllow() {
                mDatas = LocalMusicApi.getMusic(resolver);
                adapter.setmDatas(mDatas);
                adapter.notifyDataSetChanged();
            }

            @Override
            public void onReject() {
                activity.showMsg("没有授予相应的权限,无法完成音乐文件遍历!");
            }
        });
```

18.3.3 显示歌曲

在第6章介绍了 ListView 的使用，主要讲述了如何使用 ArrayAdapter 和 SimpleAdapter，它们都是 BaseAdapter 的子类。BaseAdapter 即实用又常用，原因就在于它的全能性，不会像 ArrayAdapter 等封装好的类有那么多局限性，但是使用起来自然会更加麻烦一点。下面通过实例介绍 BaseAdapter 的使用及优化。

（1）创建 Item 的布局：music_item.xml

由图 18-1 和图 18-2 可以看出，每首歌曲需要展示专辑图片、歌曲名称、歌手名字和歌曲时长，所以布局中只需要包含一个 ImageView 和三个 TextView 组件即可。布局代码如下：

```xml
<?xml version="1.0" encoding="utf-8"?>
<LinearLayout xmlns:android="http://schemas.android.com/apk/res/android"
    android:orientation="horizontal" android:layout_width="match_parent"
    android:layout_height="60dp">
    <ImageView
        android:id="@+id/iv_music"
        android:layout_width="50dp"
        android:layout_height="50dp"
        android:src="@mipmap/ic_launcher"
        android:layout_margin="5dp"/>
    <LinearLayout
        android:layout_width="wrap_content"
        android:layout_height="wrap_content"
        android:layout_weight="1"
        android:layout_margin="10dp"
        android:orientation="vertical">
        <TextView
            android:id="@+id/tv_name"
            android:layout_width="wrap_content"
            android:layout_height="wrap_content"
            android:text="歌名"/>

        <TextView
            android:id="@+id/tv_artist"
            android:layout_width="wrap_content"
            android:layout_height="wrap_content"
            android:paddingTop="5dp"
            android:text="歌星" />
    </LinearLayout>
    <TextView
```

```
            android:id="@+id/tv_duration"
            android:layout_width="wrap_content"
            android:layout_height="wrap_content"
            android:layout_gravity="center_vertical"
            android:layout_marginEnd="10dp"
            android:text="3:20"/>
    </LinearLayout>
```

(2)创建MusicAdapter

创建MusicAdapter，继承BaseAdapter，需要重写四个方法：getCount、getItem、getItemId和getView。其中，重要的是getCount和getView方法，getCount确定显示Item的数量，然后为每个Item调用getView方法获取View对象，并进行显示。

```java
    public class MusicAdapter extends BaseAdapter {
        private Context context;//Context对象，主要用于加载布局
        private List<Music> musics;//需要显示的音乐列表
        public MusicAdapter(Context context,List<Music> musics){
            this.context=context;
            this.musics=musics;
        }
        @Override
        public int getCount() {//需要显示的Item的数量
            return musics.size();
        }


        @Override
        public Object getItem(int position) {//获取指定position的对象
            return musics.get(position);
        }


        @Override
        public long getItemId(int position) {
            return position;
        }


        // 为每个Item生成视图。在此可以加载一个布局，并设置组件的相应属性
        @Override
        public View getView(int position, View convertView, ViewGroup parent) {
            //加载布局
            convertView= LayoutInflater.from(context).inflate(R.layout.music_item,parent,false);
            //获取该布局的组件
```

```
        ImageView iv_music=convertView.findViewById(R.id.iv_music);
        TextView tv_name=convertView.findViewById(R.id.tv_name);
        TextView tv_artist=convertView.findViewById(R.id.tv_artist);
        TextView tv_duration=convertView.findViewById(R.id.tv_duration);
        //为每个组件设置相应的属性
        Music music=musics.get(position);
        iv_music.setImageBitmap(LocalMusicApi.getAlbumArt(
    context.getContentResolver(),music.getAlbum_id()));
        tv_name.setText(music.getTitle());
        tv_artist.setText(music.getArtist_name());
        tv_duration.setText(FormatUtil.timeFormat(music.getFile_duration()));
        return convertView;
    }
}
```

其中，LocalMusicApi.getAlbumArt 用于根据专辑ID加载专辑图片，有的歌曲可能没有存储专辑图片；FormatUtil.timeFormat 用于将时长的"秒"转化为"00:00"格式。

（3）显示歌曲

获取本机音乐列表后，添加如下代码即可完成歌曲的显示：

```
MusicAdapter adapter = new MusicAdapter(activity ,mDatas);
listView = root.findViewById(android.R.id.list);
listView.setEmptyView(root.findViewById(android.R.id.empty));
listView.setAdapter(adapter);
```

（4）MusicAdapter 优化

在MusicAdapter 的getView方法中，每个Item都需要加载一次布局，这是没有必要的，可以增加一个判断，在convertView 为null 时加载布局，否则复用convertView。

```
    if(convertView==null) {
        //加载布局
        convertView = LayoutInflater.from(context).inflate(R.layout.music_item, parent, false);
    }
```

getView会被调用多次，那么findViewById一样得调用多次，而ListView的Item一般都是一样的布局，可以对此优化。定义一个ViewHolder 类来对这部分进行性能优化。修改后的代码如下：

```
    // 为每个Item生成视图。在此可以加载一个布局，并设置组件的相应属性
    @Override
    public View getView(int position, View convertView, ViewGroup parent) {
        ViewHolder holder = null;
```

```java
        if(convertView==null) {//第一次使用时加载布局
            //加载布局
            convertView = LayoutInflater.from(context).inflate(
R.layout.music_item, parent, false);
            holder=new ViewHolder();
            //获取该布局的组件
            holder.iv_music=convertView.findViewById(R.id.iv_music);
            holder.tv_name=convertView.findViewById(R.id.tv_name);
            holder.tv_artist=convertView.findViewById(R.id.tv_artist);
            holder.tv_duration=convertView.findViewById(R.id.tv_duration);
            convertView.setTag(holder);//存储Holder对象
        }else {
            holder= (ViewHolder) convertView.getTag();//直接获取Holder对象
        }

        //为每个组件设置相应的属性
        Music music=musics.get(position);
        holder.iv_music.setImageBitmap(LocalMusicApi.getAlbumArt(
context.getContentResolver(),music.getAlbum_id()));
        holder.tv_name.setText(music.getTitle());
        holder.tv_artist.setText(music.getArtist_name());
        holder.tv_duration.setText(FormatUtil.timeFormat(
music.getFile_duration()));
        return convertView;
    }
    static class ViewHolder{
        ImageView iv_music;
        TextView tv_name;
        TextView tv_artist;
        TextView tv_duration;
    }
```

为了便于读者进行对比，优化后的MusicAdapter存储在MusicAdapter1中。

（5）定义通用适配器

由于在项目中会多次使用到ListView，如果对每个都定义一个Adapter，最后项目中将会有很多适配器，增加编码工作量，同时不便于管理。通过对MusicAdapter1进行观察可以发现，适配器中除了getView方法，其他都是固定的写法，可以对MusicAdapter1进行抽象，定义为通用适配器，代码如下：

```java
public abstract class CommonAdapter<T> extends BaseAdapter {
    protected LayoutInflater mInflater;//用来加载布局
    protected Context mContext;//Context对象
    protected List<T> mDatas;//要显示的数据列表
```

```java
    protected final int mItemLayoutId;//Item布局的ID

    public CommonAdapter(Context context, List<T> mDatas, int itemLayoutId) {
        this.mContext = context;
        this.mInflater = LayoutInflater.from(mContext);
        this.mDatas = mDatas;
        this.mItemLayoutId = itemLayoutId;
    }

    @Override
    public int getCount() {
        return mDatas.size();
    }

    @Override
    public T getItem(int position) {
        return mDatas.get(position);
    }

    @Override
    public long getItemId(int position) {
        return position;
    }

    @Override
    public View getView(int position, View convertView, ViewGroup parent) {
        final ViewHolder viewHolder = getViewHolder(position, convertView, parent);
        convert(viewHolder, getItem(position));
        return viewHolder.getConvertView();
    }
    //设置组件的属性
    public abstract void convert(ViewHolder helper, T item);
    //获取ViewHolder对象
    private ViewHolder getViewHolder(int position, View convertView, ViewGroup parent) {
        return ViewHolder.get(mContext, convertView, parent, mItemLayoutId, position);
    }
    //用于异步加载数据，在数据加载成功后，将数据设置到适配器，然后刷新
    public void setmDatas(List<T> mDatas) {
```

```
            this.mDatas = mDatas;
        }
        //用于分页加载，加载完成后，将新数据添加到现有数据的后面
        public void addmDatas(List<T> musics) {
            this.mDatas.addAll(musics);
        }
    }
```

同理，对ViewHolder进行抽象，定义通用ViewHolder类。ViewHolder类的作用就是存储布局中的各种View对象，由于布局中的View对象是不确定的，使用Map来存储对象，具体请查看getView方法。在此使用SparseArray，和HashMap的功能相似，但是性能要优于后者，存储的键值对的键（key）必须为int。

```
    public class ViewHolder {
        //SparseArray相当于HashMap,性能更优。特点：key必须为int
        private final SparseArray<View> mViews;//存储（ID:View）
        private int mPosition;
        private View mConvertView;
        private Context context;
        private ViewHolder(Context context, ViewGroup parent, int layoutId, int position) {
            this.mPosition = position;
            this.mViews = new SparseArray<View>();
            this.context=context;
            mConvertView = LayoutInflater.from(context).inflate(layoutId, parent, false);
            // 存储ViewHolder对象
            mConvertView.setTag(this);
        }

        // 拿到一个ViewHolder对象
        public static ViewHolder get(Context context, View convertView,
                        ViewGroup parent, int layoutId, int position) {
            if (convertView == null) {
                return new ViewHolder(context, parent, layoutId, position);
            }
            return (ViewHolder) convertView.getTag();
        }

        public View getConvertView() {
            return mConvertView;
        }
```

```java
// 通过控件的ID获取对应的控件，如果没有则加入view
public <T extends View> T getView(int viewId) {
    View view = mViews.get(viewId);
    if (view == null) {
        view = mConvertView.findViewById(viewId);
        mViews.put(viewId, view);
    }
    return (T) view;
}

// 为TextView设置字符串
public ViewHolder setText(int viewId, String text) {
    TextView view = getView(viewId);
    view.setText(text);
    return this;
}

// 为ImageView设置图片
public ViewHolder setImageBitmap(int viewId, Bitmap bm) {
    ImageView view = getView(viewId);
    view.setImageBitmap(bm);
    return this;
}
}
```

使用通用适配器后，在LocalMusicFragment中获取数据和显示数据的代码如下：

```java
//加载本地音乐需要授予访问SD卡权限，动态申请
activity.checkPermissions(new String[]{Manifest.permission.READ_EXTERNAL_STORAGE,
        Manifest.permission.WRITE_EXTERNAL_STORAGE},
    new BaseActivity.OnPermissionResultListener() {
        @Override
        public void onAllow() {
            mDatas = LocalMusicApi.getMusic(resolver);
            adapter.setmDatas(mDatas);
            adapter.notifyDataSetChanged();
        }

        @Override
        public void onReject() {
            activity.showMsg("没有授予相应的权限，无法完成音乐文件遍历！");
        }
```

```
            });

    adapter = new CommonAdapter<Music>(getContext(), mDatas, R.layout.music_item) {
        @Override
        public void convert(ViewHolder helper, Music item) {
            helper.setText(R.id.tv_name, item.getTitle());
            helper.setText(R.id.tv_artist, item.getArtist_name());
            helper.setText(R.id.tv_duration, FormatUtil.timeFormat(item.getFile_duration()));
            Bitmap bitmap = LocalMusicApi.getAlbumArt(resolver,item.getAlbum_id());
            if (bitmap != null)
                helper.setImageBitmap(R.id.iv_music, bitmap);
        }
    };
    listView = root.findViewById(android.R.id.list);
    listView.setEmptyView(root.findViewById(android.R.id.empty));
    listView.setAdapter(adapter);
```

18.3.4 刷新歌曲

Android系统在开机时，会对系统内的文件进行扫描，把扫描到的音乐文件添加到MediaStore中。开机后添加到手机的音乐文件，无法从MediaStore获取到，如果要求用户重启手机，非常不方便，可以编写代码扫描指定的文件夹（因为添加到手机的文件存储在特定的文件夹），然后将音乐文件添加到MediaStore中。

第一，在LocalMusicApi类中添加mediaScan方法，用来扫描指定的文件夹，并将音乐添加到MediaStore中。代码如下：

```
/**
 * 将歌曲添加到模拟器时，重修扫描文件，将歌曲添加到MediaStore中
 *
 * @param context 上下文
 * @param file    要扫描的文件夹
 */
public static void mediaScan(final Context context, File file) {
    System.out.println(file.getAbsolutePath());
    ArrayList<String> fileList = new ArrayList<>();
    //遍历文件夹的所有文件
    File[] files = file.listFiles();
    for (File ff : files) {
        if (ff.getName().endsWith("mp3")) {
            fileList.add(ff.getAbsolutePath());
        }
```

```java
        }
        int size = fileList.size();
        //将List转为String[]
        String[] array = (String[]) fileList.toArray(new String[size]);
        //启动扫描
        MediaScannerConnection.scanFile(context,
                array, null,
                new MediaScannerConnection.OnScanCompletedListener() {
                    @Override
                    public void onScanCompleted(String path, Uri uri) {
                        //将扫描的音乐文件（uri对应）添加到MediaStore
                        Intent scanIntent = new Intent(Intent.ACTION_MEDIA_SCANNER_SCAN_FILE);
                        scanIntent.setData(uri);
                        context.sendBroadcast(scanIntent);
                    }
                });
    }
```

第二，在MainActivity的onCreate方法中，为侧滑菜单进行设置，并为NavigationView添加事件处理。代码如下：

```java
    //抽屉布局设置（左侧的侧滑）
    final DrawerLayout drawer = findViewById(R.id.drawer_layout);
    NavigationView navigationView = findViewById(R.id.nav_view);
    navigationView.setNavigationItemSelectedListener(
    new NavigationView.OnNavigationItemSelectedListener() {
        @Override
        public boolean onNavigationItemSelected(@NonNull MenuItem menuItem) {
            switch (menuItem.getItemId()) {
                case R.id.nav_localMusic:
                    getSupportFragmentManager().beginTransaction()
                            .replace(R.id.container, LocalMusicFragment.newInstance())
                            .commitNow();
                    break;
                case R.id.nav_netMusic:
                    getSupportFragmentManager().beginTransaction()
                            .replace(R.id.container, NetMusicFragment.newInstance())
                            .commitNow();
                    break;
                case R.id.nav_musicList:
```

```
                getSupportFragmentManager().beginTransaction()
                        .replace(R.id.container, MusicListFragment.newInstance())
                        .commitNow();
                break;
            case R.id.nav_musicScan:
                File file = Environment.getExternalStoragePublicDirectory(
   Environment.DIRECTORY_DOWNLOADS);
                LocalMusicApi.mediaScan(MainActivity.this,file);
                break;
        }
        drawer.closeDrawers();
        return false;
    }
});
mDrawerTogger = new ActionBarDrawerToggle(this, drawer, toolbar,
        R.string.navigation_drawer_open, R.string.navigation_drawer_close);
mDrawerTogger.syncState();
drawer.addDrawerListener(mDrawerTogger);
```

第三，使ListView重新加载数据。方法有多种，在此使用SwipeRefreshLayout + ListView实现。首先，修改fragment_local_music.xml布局文件，将ListView放到SwipeRefreshLayout中，其他不变。

```
<android.support.v4.widget.SwipeRefreshLayout
    android:id="@+id/swipe"
    android:layout_width="match_parent"
    android:layout_height="0dp"
    android:layout_weight="1">
    <ListView
        android:id="@id/android:list"
        android:layout_width="match_parent"
        android:layout_height="match_parent"
        />
    <TextView android:id="@id/android:empty"
        android:layout_width="wrap_content"
        android:layout_height="match_parent"
        android:gravity="center_vertical"
        android:layout_gravity="center"
        android:text="对不起，没有数据显示，请添加音乐文件！ "/>
</android.support.v4.widget.SwipeRefreshLayout>
```

在LocalMusicFragment的onCreate方法中，对SwipeRefreshLayout进行设置，并添加事

件处理。代码如下：

```java
swipe = root.findViewById(R.id.swipe);
swipe.setSize(SwipeRefreshLayout.DEFAULT);//设置加载默认图标
if(Build.VERSION.SDK_INT >= 23) {
    swipe.setProgressBackgroundColorSchemeColor(
            ContextCompat.getColor(activity,android.R.color.holo_orange_light));
    swipe.setColorSchemeColors(//刷新控件动画中的颜色
            ContextCompat.getColor(activity,android.R.color.holo_blue_dark),
            ContextCompat.getColor(activity,android.R.color.holo_red_dark),
            ContextCompat.getColor(activity,android.R.color.holo_green_dark)
    );
}
//事件处理，重新加载数据
swipe.setOnRefreshListener(new SwipeRefreshLayout.OnRefreshListener() {
    @Override
    public void onRefresh() {
        mDatas = LocalMusicApi.getMusic(resolver);//加载数据
        adapter.setmDatas(mDatas);//将数据设置到适配器
        adapter.notifyDataSetChanged();//通知刷新
        swipe.setRefreshing(false);//刷新完毕，图标消失
    }
});
```

同理，可以实现艺术家列表（图18-8）和专辑列表（图18-9）对应的Fragment。

图18-8　艺术家列表

图18-9　专辑列表

18.4 网络歌曲列表

播放网络歌曲是卓卓音乐盒不可或缺的功能，然而自己搭建一个音乐服务网站，任务量大，也超出了本书介绍的范围。为了方便大家练习网络编程，采用网络上提供的免费API接口。

注意：仅供练习，请勿商业应用。

18.4.1 音乐接口介绍

在网上搜索“音乐API接口”，可以找到很多免费的音乐API接口，以图18-10为例介绍如何调用音乐API接口。

百度音乐api接口

百度音乐全接口 会利用使用接口找歌简单又快捷

http://tingapi.ting.baidu.com/v1/restserver/ting

获取方式：GET

参数：format=json或xml&calback=&from=webapp_music&method=以下不同的参数获得不同的数据

PS：format根据开发需要可选择json或xmml，其他参数对应填入，calback是等于空的。

一、获取列表

> 例：method=baidu.ting.billboard.billList&type=1&size=10&offset=0

> 参数： type = 1-新歌榜,2-热歌榜,11-摇滚榜,12-爵士,16-流行,21-欧美金曲榜,22-经典老歌榜,23-情歌对唱榜,24-影视金曲榜,25-网络歌曲榜

> size = 10 //返回条目数量

> offset = 0 //获取偏移

图18-10 音乐API接口

根据图18-10的说明，可以构造出访问新歌榜的URL：

将该URL复制到浏览器中，就可以查看到返回的数据。

注意：如果需要获取其他的榜单，修改type的值即可；size代表每次返回的数据数量，offset代表偏移量，二者配合可以实现歌曲的分页加载。如size=10、offset=0，加载的是第一页的数据，offset=10将加载第二页的数据。

18.4.2 JSON解析

JSON（JavaScript Object Notation，JS对象简谱）是一种轻量级的数据交换格式。它基于 ECMAScript（欧洲计算机协会制定的JS规范）的一个子集，采用完全独立于编程语言的文本格式来存储和表示数据。简洁和清晰的层次结构使得JSON成为理想的数据交换语言，

易于人阅读和编写，同时也易于机器解析和生成，并有效地提升网络传输效率。

JSON以"key:value"的形式保存数据，多个数据以逗号分隔，花括号保存对象，如：

```
{ "firstName":"John" , "lastName":"Doe" }
```

方括号保存数组，如：

```
{"employees": [
    { "firstName":"John" , "lastName":"Doe" },
    { "firstName":"Anna" , "lastName":"Smith" },
    { "firstName":"Peter" , "lastName":"Jones" }
    ]
}
```

在例子中，对象"employees"是包含三个对象的数组，每个对象代表一条关于某人（有姓和名）的记录。

在Android中提供了解析JSON对象的API：JSONObject。

- JSONObject(String json)：将json字符串解析成JSON对象。
- getXxx(String name)/optXxx(String name)：根据name在JSON对象中得到相应的value值。

解析JSON数组的API：JSONArray。

- JSONArray(String json)：将json字符串解析成JSON数组。
- int length()：得到JSON数组中元素的个数。
- getXxx(String name)/optXxx(String name)：根据name得到JSON数组中对应的元素数据。

getXxx(String name)和optXxx(String name)二者的区别：getXxx取值不正确或者类型不正确时会抛出异常，必须用try catch或者throw捕获；optXxx取值不正确时会试图进行转化或者返回默认值，不会抛出异常。

请根据需要选择合适的方法，建议使用optXxx方法。

以JSON数组为例，熟悉JSON解析。代码如下：

```java
private void parseJson(String string) {
    try {
        JSONObject obj=new JSONObject(string);
        JSONArray array = obj.optJSONArray("employees");
        for (int i = 0; i < array.length(); i++) {
            JSONObject object = array.optJSONObject(i);
            String fname = object.getString("firstName");
            String lname = object.getString("lastName");
            Log.e("1", "fname:" + fname + "  lname:" + lname);
        }
    } catch (JSONException e) {
        e.printStackTrace();
    }
}
```

18.4.3 封装工具类

在util包中创建工具类MusicApi,用来封装获取网络音乐的常用方法。部分代码如下:

```java
public class MusicApi {
    private static final String TAG = "MusicApi";
    //获取歌曲列表
    private static final String BASE_URL = "http://tingapi.ting.baidu.com/v1/restserver/ting?format=json&calback=&from=webapp_music&method=baidu.ting.billboard.billList&type=1&size=10&offset=";
    //歌曲播放,添加歌曲ID
    private static String playpath = "http://tingapi.ting.baidu.com/v1/restserver/ting?format=json&calback=&from=webapp_music&method=baidu.ting.song.play&songid=";
    //获取歌曲id,本地歌曲获取歌词,需要先获取歌曲ID
    private static String playsearch = "http://tingapi.ting.baidu.com/v1/restserver/ting?format=json&calback=&from=webapp_music&method=baidu.ting.search.catalogSug&query=";
    //获取歌词URL
    private static String playlrc = "http://tingapi.ting.baidu.com/v1/restserver/ting?format=json&calback=&from=webapp_music&method=baidu.ting.song.lry&songid=";

    /**
     * 查询歌曲信息
     */
    public static ArrayList<Music> getBaiduMusic(int count, int page) {
        String fetchUrl = BASE_URL + page * count;
        ArrayList<Music> musics = new ArrayList<>();
        try {
            URL url = new URL(fetchUrl);
            HttpURLConnection conn = (HttpURLConnection) url.openConnection();
            conn.setConnectTimeout(5000);
            conn.setRequestMethod("GET");
            int code = conn.getResponseCode();
            if (code == 200) {
                InputStream in = conn.getInputStream();
                byte[] data = readFromStream(in);
                String result = new String(data, "UTF-8");
                musics = parseMusic(result);
```

```java
            } else {
                Log.e(TAG, "请求失败:" + code);
            }
        } catch (Exception e) {
            e.printStackTrace();
        }
        return musics;
    }
    /**
     * 解析返回JSON数据的方法
     */
    public static ArrayList<Music> parseMusic(String content)
throws Exception {
        ArrayList<Music> musics = new ArrayList<>();
        JSONObject object = new JSONObject(content);
        JSONArray array = object.getJSONArray("song_list");
        for (int i = 0; i < array.length(); i++) {
            JSONObject results = (JSONObject) array.get(i);
            Music music = new Music();
            music.setSong_id(results.getString("song_id"));
            music.setTitle(results.getString("title"));
            music.setAlbum_id(results.getString("album_id"));
            music.setAlbum_title(results.getString("album_title"));
            music.setFile_duration(results.getInt("file_duration"));
            music.setArtist_name(results.getString("artist_name"));
            music.setLrclink(results.getString("lrclink"));
            music.setPic_big(results.getString("pic_big"));
            music.setPic_small(results.getString("pic_small"));
            music.setPic_huge(results.getString("pic_huge"));
            music.setUrl(getMusicUrl(music.getSong_id()));
            music.setLocal(false);
            musics.add(music);
        }
        return musics;
    }

    /**
     * 读取流中数据的方法
     */
    public static byte[] readFromStream(InputStream inputStream) throws
Exception {
        ByteArrayOutputStream outputStream = new ByteArrayOutputStr
```

第18章 Android应用项目的设计与开发

```java
eam();
        byte[] buffer = new byte[1024];
        int len;
        while ((len = inputStream.read(buffer)) != -1) {
            outputStream.write(buffer, 0, len);
        }
        inputStream.close();
        return outputStream.toByteArray();
    }
    //获取指定strUrl的图片
    public static Bitmap getImageByUrl(String strUrl) {
        Bitmap bitmap = null;
        try {
            URL url = new URL(strUrl);
            HttpURLConnection conn = (HttpURLConnection) url.openConnection();
            conn.setConnectTimeout(5000);
            conn.setRequestMethod("GET");
            int code = conn.getResponseCode();
            if (code == 200) {
                InputStream in = conn.getInputStream();
                bitmap = BitmapFactory.decodeStream(in);
            } else {
                Log.e(TAG, "请求失败:" + code);
            }
        } catch (Exception e) {
            e.printStackTrace();
        }
        return bitmap;
    }

    /**
     * 根据歌曲ID，获取播放该歌曲的URL
     */
    public static String getMusicUrl(String song_id) {
        String fetchUrl = playpath + song_id;
        String ret = null;
        try {
        URL url = new URL(fetchUrl);
        HttpURLConnection conn = (HttpURLConnection) url.openConnection();
            conn.setConnectTimeout(5000);
```

```
                conn.setRequestMethod("GET");
                int code = conn.getResponseCode();
                if (code == 200) {
                    InputStream in = conn.getInputStream();
                    byte[] data = readFromStream(in);
                    String result = new String(data, "UTF-8");
                    JSONObject object = new JSONObject(result);
                    ret = object.optJSONObject("bitrate").optString
("file_link");
                } else {
                    Log.e(TAG, "请求失败：" + code);
                }
            } catch (Exception e) {
                e.printStackTrace();
            }
            return ret;
        }
    }
```

18.4.4 获取音乐数据及显示

首先，在ui包中创建NetMusicFragment，由于和本地歌曲列表大体相同，在此就不再说明。主要的区别在两个方面。

① 网络歌曲列表的获取属于耗时的操作，需要新开线程进行操作；操作完成后，需要通知主线程刷新数据。

② 为了节省手机流量，每次只获取10首歌曲，当列表滑动到底部时，需要加载下一页。加载下一页只需要修改loadMusic的第二个参数即可；判断列表是否滑动到底部，需要添加OnScrollListener事件处理。在onScroll方法中有三个参数。

- firstVisibleItem：当前第一个可见的Item。
- visibleItemCount：当前总共有多少个可见的Item。
- totalItemCount：总的Item。

很明显，当"firstVisibleItem+visibleItemCount==totalItemCount"时，说明已经到达底部，就可以开始加载下一页。

NetMusicFragment 的主要代码如下：

```
public class NetMusicFragment extends Fragment {
    private MainActivity activity;
    private ListView listView;
    private CommonAdapter<Music> adapter;
    private List<Music> musics = new ArrayList<>();

    private int lastVisibleItem;//最后一个可见的Item
    private int itemCount;//总的Item
```

第18章 Android应用项目的设计与开发

```java
    private boolean isLoading = false;//是否正在加载数据
    private int curPage=0;//记录已经加载的页码
    public static NetMusicFragment newInstance() {
        return new NetMusicFragment();
    }

    @Nullable
    @Override
    public View onCreateView(@NonNull LayoutInflater inflater, @Nullable ViewGroup container, @Nullable Bundle savedInstanceState) {
        View view = inflater.inflate(R.layout.fragment_net_music, container, false);
        activity = (MainActivity) getActivity();
        listView = view.findViewById(R.id.list);
        adapter = new CommonAdapter<Music>(getContext(), musics, R.layout.music_item) {
            @Override
            public void convert(ViewHolder helper, Music item) {
                helper.setText(R.id.tv_name, item.getTitle());
                helper.setText(R.id.tv_artist, item.getArtist_name());
                helper.setText(R.id.tv_duration, FormatUtil.timeFormat(item.getFile_duration()));
                helper.setImageByUrl(R.id.iv_music, item.getPic_small());
            }
        };
        listView.setAdapter(adapter);
        //当滑动到底部时，加载下一页
        listView.setOnScrollListener(new AbsListView.OnScrollListener() {

            //参数scrollState表示滑动的状态
            public void onScrollStateChanged(AbsListView view, int scrollState) {
                //如果最后一个可见Item等于总的Item，且当前滚动状态为滚动停止，就应该开始加载数据了
                if (lastVisibleItem == itemCount && scrollState == SCROLL_STATE_IDLE) {
                    if (!isLoading) {
                        isLoading = true;
                        //加载数据
                        loadMusic(10,++curPage);
                    }
```

```java
                }
            }
            /***
             * 该方法用来监听实时滚动中的Item
             * firstVisibleItem:当前第一个可见的Item
             * visibleItemCount:当前总共有多少个可见的Item
             * totalItemCount:总的Item
             */
            public void onScroll(AbsListView view, int firstVisibleItem,
                            int visibleItemCount, int totalItemCount) {
                lastVisibleItem = firstVisibleItem + visibleItemCount;
                itemCount = totalItemCount;
            }
        });
        loadMusic(10,0);
        return view;
    }
    //加载音乐列表
    private void loadMusic(final int count, final int page) {
        //由于网络访问是一个耗时的操作,需要新开一个线程进行
        new Thread() {
            @Override
            public void run() {
                musics = MusicApi.getBaiduMusic(count, page);
                if(page==0) {//第一次获取
                    adapter.setmDatas(musics);
                }else {//将数据添加到列表末尾
                    adapter.addmDatas(musics);
                }
                //回到主线程,进行刷新数据
                getActivity().runOnUiThread(new Runnable() {
                    @Override
                    public void run() {
                        adapter.notifyDataSetChanged();
                        isLoading=false;//表示当前加载结束
                    }
                });
            }
        }.start();
    }
}
```

第18章　Android应用项目的设计与开发

注意：如果手机的Android版本为28（9.0）及以后版本，使用HTTP请求报错"Not Permitted By Network Security Policy"。这是因为Google公司认为使用HTTP请求是不安全的，建议使用HTTPS，区别是后者在发送请求时进行了数据加密。

解决方法：

① 在res下新建一个xml目录，然后创建一个名为network_security_config.xml的文件，该文件内容如下：

```xml
<?xml version="1.0" encoding="utf-8"?>
<network-security-config>
  <base-config cleartextTrafficPermitted="true" />
</network-security-config>
```

在AndroidManifest.xml文件的application节点内增加配置android:networkSecurityConfig="@xml/network_security_config"即可。

② 服务器和本地应用都改用HTTPS(推荐)。

③ targetSdkVersion降级回到27。

由于我们使用的接口是HTTP的，使用第一和第三种方法均可。

18.5　音乐播放

音乐播放是卓卓音乐盒的核心功能，需要和多个组件进行交互。为了在Activity退出时可以继续播放音乐，需要在Service中进行音乐播放。大体的流程如下。

① 在本地歌曲列表（图18-1）或者网络歌曲列表（图18-2）中点击一首歌曲，使用服务播放指定歌曲。歌曲开始播放或者切换歌曲时，需要发送广播，更改歌曲列表下方的音乐信息（图18-11），启动定时器，发送歌曲播放的进度，更新ProgrossBar的进度（图18-11上方）。发送通知，在Activity退出时，可以使用通知控制音乐播放（图18-12）。

图18-11　音乐播放控制

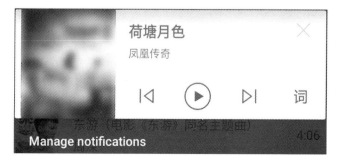

图18-12　通知

② 用户在图18-11或者图18-12中点击图标，实现音乐的播放、暂停、上一首和下一首控制。

③ 当来电或者去电时，暂停音乐播放；挂断电话，继续播放音乐。

18.5.1 使用Service播放音乐

在service包中创建PlayService类，继承Service类。由于该类代码太多，完整代码请查看光盘中本章源代码。

（1）添加音乐播放控制

在PlayService类中，添加代码实现音乐的播放、暂停、下一首、上一首和继续播放功能，代码如下：

```java
//播放
public void play(int position) {
    if (position >= 0 && position < musics.size()) {
        Music music = musics.get(position);
        //进行播放，播放前判断
        try {
            mPlayer.reset();//复位
            mPlayer.setDataSource(this, Uri.parse(music.getUrl()));
            //资源解析,Mp3地址
            mPlayer.prepare();//准备
            mPlayer.setOnPreparedListener(MediaPlayer::start);
            //保存当前位置到currentPosition,比如第一首,currentPosition = 0
            currentPosition = position;
            mDuration = music.getFile_duration();
            //获取歌曲的专辑图片，然后发送通知和广播
            getPicByUrl(music);
            //如果定时器没有启动，则启动定时器，开始播放进度广播发送
            if (!timerStart) {
                timer.schedule(task, 0, 1000);
                timerStart = true;
            }
        } catch (IOException e) {
            e.printStackTrace();
        }
    }
}

//暂停
public void pause() {
    if (mPlayer.isPlaying()) {
```

```java
            mPlayer.pause();
            sendNotification(false);
        }
    }

    //下一首
    public void next() {
        if (currentPosition >= musics.size() - 1) {
            //如果超出最大值,(因为第一首是0),说明已经是最后一首
            currentPosition = 0;
            //回到第一首
        } else {
            currentPosition++;//下一首
        }
        play(currentPosition);
    }

    //上一首 previous
    public void prev() {
        if (currentPosition - 1 < 0) {
            //如果上一首小于0,说明已经是第一首
            currentPosition = musics.size() - 1;
            //回到最后一首
        } else {
            currentPosition--;//上一首
        }
        play(currentPosition);
    }

    //继续当前歌曲播放
    public void start() {
        if (mPlayer != null && !mPlayer.isPlaying()) {
            //判断当前歌曲不等于空,并且没有在播放的状态
            mPlayer.start();
            sendNotification(true);
        }
    }
```

（2）startService方式交互

通过前面的学习，我们知道使用startService启动服务，只在第一次启动时会执行onCreate方法，然后调用onStartCommand方法，再次调用startService，只调用onStartCommand方法。可以通过intent携带参数，来控制音乐播放。在PlayService类中添加或者修改onStartCommand方法，代码如下：

```java
@Override
public int onStartCommand(Intent intent, int flags, int startId) {
    String action = intent.getStringExtra("action");
    if (action != null) {
      if ("next".equals(action)) {
          next();
      }else if ("progress".equals(action)) {
          int pos = intent.getIntExtra("pos",0);
          mPlayer.seekTo(pos*1000);
      }
    }
    return super.onStartCommand(intent, flags, startId);
}
```

当需要切换到下一首歌曲时，再次启动服务即可，代码如下：

```java
Intent intentNext = new Intent(context, PlayService.class);
intentNext.putExtra("action", "next");
startService(intentNext);
```

（3）bindService方式交互

使用bindService方式启动服务，可以通过onBind方法获取PlayService对象，然后就可以使用该对象直接调用歌曲的播放、暂停等功能。在PlayService中添加内部类PlayBinder，用于返回PlayService对象。修改后的代码如下：

```java
//内部类PlayBinder实现Binder
public class PlayBinder extends Binder {
    public PlayService getPlayService() {
        return PlayService.this;
    }
}


@Override
public IBinder onBind(Intent intent) {
    return new PlayBinder();
    //通过PlayBinder拿到PlayService,给Activity调用
}
```

在MainActivity中，使用bindService方式绑定服务，MainActivity的部分代码如下：

```java
protected PlayService playService;
private boolean playing = false;
private Intent service;
//是否已经绑定
private boolean isBound = false;
```

```java
//绑定Service
private ServiceConnection conn = new ServiceConnection() {
    @Override
    public void onServiceConnected(ComponentName name, IBinder service) {
        //转换
        PlayService.PlayBinder playBinder = (PlayService.PlayBinder) service;
        playService = playBinder.getPlayService();
    }

    @Override
    public void onServiceDisconnected(ComponentName name) {
        playService = null;
        isBound = false;
    }
};

//绑定服务
public void bindPlayService() {
    if (!isBound) {
        bindService(service, conn, Context.BIND_AUTO_CREATE);
        isBound = true;
    }
}

//解除绑定服务
public void unbindPlayService() {
    if (isBound) {
        unbindService(conn);
        isBound = false;
    }
}
@Override
protected void onCreate(Bundle savedInstanceState) {
    ......
    //启动服务,用于音乐播放
    service = new Intent(this, PlayService.class);
    startService(service);
    bindPlayService();
    ......
}
```

```java
@Override
protected void onDestroy() {
    super.onDestroy();
    unbindPlayService();
    unregisterReceiver(receiver);
}

public void playMusic(List<Music> musicList, int position) {
    playService.setMusics(musicList);
    playService.play(position);
}
```

18.5.2 发送通知

在play方法中，调用了getPicByUrl方法，该方法用来发送通知和歌曲变化的广播。单独写成一个方法，主要是因为网络歌曲获取图片需要异步调用。代码如下：

```java
/**
 * 开始播放歌曲，发送通知时，需要获取专辑图片
 * 根据本地歌曲和网络歌曲不同来获取图片，然后发送通知
 * @param bean
 */
private void getPicByUrl(Music bean) {
    if (bean.isLocal()){
        //本地歌曲
        Bitmap bitmap= LocalMusicApi.getAlbumArt(getContentResolver(),bean.getAlbum_id());
        NotificationContentWrapper wrapper = new NotificationContentWrapper(
                bitmap, bean.getTitle(),
                bean.getArtist_name(),
                bean.getLrclink());
         sendCustomViewNotification(getApplicationContext(),
notificationManager, wrapper
                , true);

        Intent intent = new Intent("Change_Music");
        intent.putExtra("duration",bean.getFile_duration());
        ((MusicApp) getApplication()).setWrapper(wrapper);
        sendBroadcast(intent);
    }else {
        //网络歌曲，启动异步任务获取歌曲专辑图片
        new AsyncTask<String, Void, Bitmap>() {
```

第18章 Android应用项目的设计与开发

```java
            @Override
            protected Bitmap doInBackground(String... strings) {
                return new MusicApi().getImageByUrl(strings[0]);
            }

            @Override
            protected void onPostExecute(Bitmap bitmap) {
                NotificationContentWrapper wrapper = new NotificationContentWrapper(
                        bitmap, bean.getTitle(),
                        bean.getArtist_name(),
                        bean.getLrclink());
                sendCustomViewNotification(getApplicationContext(), notificationManager, wrapper
                        , true);

                Intent intent = new Intent("Change_Music");
                intent.putExtra("duration",bean.getFile_duration());
                ((MusicApp) getApplication()).setWrapper(wrapper);
                sendBroadcast(intent);
            }
        }.execute(bean.getPic_small());
    }
}
    public void sendCustomViewNotification(Context context, NotificationManager nm, NotificationContentWrapper content, Boolean isPlaying) {
        //创建点击通知时发送的广播
        Intent intent = new Intent(context, MainActivity.class);
        PendingIntent pi = PendingIntent.getActivity(context, 0, intent, 0);
        //创建各个按钮的点击响应广播
        Intent intentPre = new Intent();
        intentPre.setAction(ACTION_PRE);
        PendingIntent piPre = PendingIntent.getBroadcast(context, 0, intentPre, PendingIntent.FLAG_UPDATE_CURRENT);

        Intent intentPlayOrPause = new Intent();
        intentPlayOrPause.setAction(ACTION_PLAY_OR_PAUSE);
        PendingIntent piPlayOrPause = PendingIntent.getBroadcast(context,0,intentPlayOrPause,PendingIntent.FLAG_UPDATE_CURRENT);
        Intent intentNext = new Intent(context, PlayService.class);
```

```java
        intentNext.putExtra("action", "next");
        PendingIntent piNext = PendingIntent.getService(context, 0,
intentNext, PendingIntent.FLAG_UPDATE_CURRENT);

        Intent intentLyrics = new Intent(context, MusicPlayActivity.
class);
        PendingIntent piLyrics = PendingIntent.getActivity(context, 0,
intentLyrics, 0);

        Intent intentCancel = new Intent();
        intentCancel.setAction(ACTION_CANCEL);
        PendingIntent piCancel = PendingIntent.getBroadcast(context,
0, intentCancel, PendingIntent.FLAG_UPDATE_CURRENT);
        //创建自定义小视图
        RemoteViews customView = new RemoteViews(context.
getPackageName(), R.layout.custom_view_layout);
        customView.setImageViewBitmap(R.id.iv_content, content.
bitmap);
        customView.setTextViewText(R.id.tv_title, content.title);
        customView.setTextViewText(R.id.tv_summery, content.summery);
        customView.setImageViewBitmap(R.id.iv_play_or_pause,
BitmapFactory.decodeResource(context.getResources(),
                isPlaying ? R.mipmap.ic_pause : R.mipmap.ic_play));
        customView.setOnClickPendingIntent(R.id.iv_play_or_pause,
piPlayOrPause);
        customView.setOnClickPendingIntent(R.id.iv_next, piNext);
        customView.setOnClickPendingIntent(R.id.iv_lyrics, piLyrics);
        customView.setOnClickPendingIntent(R.id.iv_cancel, piCancel);
        //创建自定义大视图
        RemoteViews customBigView = new RemoteViews(context.
getPackageName(), R.layout.custom_big_view_layout);
        customBigView.setImageViewBitmap(R.id.iv_content_big, content.
bitmap);
        customBigView.setTextViewText(R.id.tv_title_big, content.
title);
        customBigView.setTextViewText(R.id.tv_summery_big, content.
summery);
        customBigView.setImageViewBitmap(R.id.iv_play_or_pause_big,
BitmapFactory.decodeResource(context.getResources(),
                isPlaying ? R.mipmap.ic_pause : R.mipmap.ic_play));
        customBigView.setOnClickPendingIntent(R.id.iv_pre_big, piPre);
        customBigView.setOnClickPendingIntent(R.id.iv_play_or_pause_
```

```
big, piPlayOrPause);
    customBigView.setOnClickPendingIntent(R.id.iv_next_big, piNext);
    customBigView.setOnClickPendingIntent(R.id.iv_lyrics_big, piLyrics);
    customBigView.setOnClickPendingIntent(R.id.iv_cancel_big, piCancel);
    //创建通知
    //设置通知左侧的小图标
    nb.setSmallIcon(R.mipmap.ic_notification)
            //设置通知标题
            .setContentTitle("卓卓音乐盒")
            //设置通知内容
            .setContentText("卓卓音乐盒! ")
            //设置通知不可删除
            .setOngoing(true)
            //设置显示通知时间
            .setShowWhen(true)
            //设置点击通知时的响应事件
            .setContentIntent(pi)
            //设置自定义小视图
            .setCustomContentView(customView)
            //设置自定义大视图
            .setCustomBigContentView(customBigView);
    //发送通知
    nm.notify(NOTIFICATION_CUSTOM, nb.build());
}
```

通知使用自定义布局，两个布局如图18-13和图18-14所示。系统会自动选择合适的视图进行显示。

图18-13　custom_view_layout布局

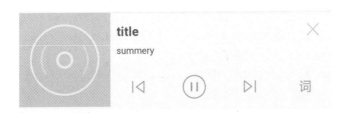

图18-14　custom_big_view_layout布局

在通知中，按钮的事件处理需要使用PendingIntent。PendingIntent是对Intent的封装，但它不是立刻执行某个行为，而是满足某些条件或触发某些事件后才执行指定的行为。

```
    Intent intentNext = new Intent(context, PlayService.class);
        intentNext.putExtra("action", "next");
        PendingIntent piNext = PendingIntent.getService(context, 0,
 intentNext, PendingIntent.FLAG_UPDATE_CURRENT);
```

相当于调用了

```
    startService(intentNext);
    Intent intentPlayOrPause = new Intent();
    intentPlayOrPause.setAction(ACTION_PLAY_OR_PAUSE);
    PendingIntent piPlayOrPause = PendingIntent.getBroadcast
(context,0,
    intentPlayOrPause,PendingIntent.FLAG_UPDATE_CURRENT);
```

相当于

```
    sendBroadcast(intentPlayOrPause );
```

18.5.3 广播接收者

18.5.2节中，歌曲的播放控制需要通过广播进行控制，需要在PlayService中定义广播接收者PlayReceiver。在手机来电或去电时，需要暂停音乐播放，定义广播接收者PhoneStateReceiver用来实现对电话状态进行监听。部分代码如下：

```
    //用于使用广播控制音乐的播放、暂停、下一首和关闭
    public class PlayReceiver extends BroadcastReceiver {
      @Override
      public void onReceive(Context context, Intent intent) {
        if (ACTION_PLAY_OR_PAUSE.equals(intent.getAction())) {
            if (mPlayer.isPlaying()) {//如果正在播放，则暂停
                pause();
            } else {//如果是暂停，则继续播放
                start();
            }
        } else if (ACTION_NEXT.equals(intent.getAction())) {
            next();//下一首
        }else if (ACTION_PRE.equals(intent.getAction())) {
            prev();//上一首
        }else if (ACTION_MODE.equals(intent.getAction())) {
            //用户修改播放模式
            int mode = intent.getIntExtra("playMode",1);
            setPlay_mode(mode);
        }else if (ACTION_PROGRESS.equals(intent.getAction())) {
            //当用户拖动播放进度时，设置新的进度，继续播放
```

```java
            int progress = intent.getIntExtra("progress",1);
            pause();
            mPlayer.seekTo(progress);
            mPlayer.start();
        } else if (ACTION_CANCEL.equals(intent.getAction())) {
            //关闭音乐播放
            pause();
            notificationManager.cancel(NOTIFICATION_CUSTOM);
        }
    }
}
//定义广播接收者，对来电和去电进行监听。来电和去电时，暂停音乐播放，挂断电话
继续播放音乐
public class PhoneStateReceiver extends BroadcastReceiver {
    @Override
    public void onReceive(Context context, Intent intent) {
        // 如果是打电话
        if (Intent.ACTION_NEW_OUTGOING_CALL.equals(intent.getAction())) {
            pause();
        } else {// 如果是来电
            TelephonyManager tm = (TelephonyManager) context
                    .getSystemService(Service.TELEPHONY_SERVICE);
            switch (tm.getCallState()) {
                case TelephonyManager.CALL_STATE_RINGING:// 响铃
                case TelephonyManager.CALL_STATE_OFFHOOK:// 摘机
                    pause();
                    break;
                case TelephonyManager.CALL_STATE_IDLE:// 空闲
                    start();
                    break;
            }
        }
    }
}
```

在onCreate方法中进行动态注册即可。代码如下：

```java
//注册广播接收者，用于响应音乐播放控制
playReceiver = new PlayReceiver();
IntentFilter filter = new IntentFilter();
filter.addAction(ACTION_PLAY_OR_PAUSE);//播放、暂停
```

```
filter.addAction(ACTION_NEXT);//下一个
filter.addAction(ACTION_PRE);//上一首
filter.addAction(ACTION_MODE);//播放模式设置
filter.addAction(ACTION_PROGRESS);//修改播放进度
filter.addAction(ACTION_CANCEL);//关闭
registerReceiver(playReceiver,filter);

mReceiver = new PhoneStateReceiver();
//注册广播接收者，用于来电话时，暂停音乐播放
IntentFilter intentFilter = new IntentFilter();
intentFilter.addAction(Intent.ACTION_NEW_OUTGOING_CALL);//去电
intentFilter.addAction("android.intent.action.PHONE_STATE");
registerReceiver(mReceiver, intentFilter);
```

本章小结

本章通过一个简单音乐播放器的分析设计与实现，让读者来了解一个系统的开发流程。为了便于初学者学习，本章逐步实现本地音乐列表显示、网络音乐列表显示、音乐播放功能，还应该具有自定义播放列表、歌词显示等功能。由于篇幅的限制，文中只介绍了核心模块的设计与实现。不过，只要读者理解了这部分内容，完全有能力设计出其他功能模块和更复杂的系统。本实例提供的源代码已经实现了播放列表的创建和歌词显示等功能，可供大家参考。为了加强大家对 Android 基础的掌握，在项目中没有使用第三方 Jar 包进行开发，实际开发中，可以使用第三方 Jar 包来简化代码。

本章应用软件工程的设计思想，带领读者走完了一个系统的开发流程，相信读者通过本例的学习，会在 Android 应用开发技术上提升一个新的台阶。

附录　配套学习资源

强化练习参考答案

全书实例程序源文件